BROADBAND ACCESS, WDM METRO AND NETWORK MANAGEMENT

Proceedings of the European Conference on Networks and Optical Communications 2000 (NOC 2000)

Published in two parts:

WDM and Photonic Networks

Broadband Access, WDM Metro and Network Management

Editors: D.W. Faulkner and A.L. Harmer

Broadband Access, WDM Metro and Network Management

NOC 2000

Edited by

D.W. Faulkner

BT Adastral Park, Ipswich, United Kingdom

and

A.L. Harmer

I.E.C., Geneva, Switzerland

IOS Press

Ohmsha

Amsterdam • Berlin • Oxford • Tokyo • Washington, DC

ISBN 1 58603 067 1 (IOS Press)
ISBN 1 58603 068 X (IOS Press, complete set)
ISBN 4 274 90368 0 C3055 (Ohmsha)
ISBN 3 905084 63 5 (AKM AG, Basel)
Library of Congress Catalog Card Number: 00-103404

Publisher
IOS Press
Nieuwe Hemweg 6B
1013 BG Amsterdam
The Netherlands
fax: +31 20 620 3419
e-mail: order@iospress.nl

Distributor in the UK and Ireland
IOS Press/Lavis Marketing
73 Lime Walk
Headington
Oxford OX3 7AD
England
fax: +44 1865 75 0079

Distributor in the USA and Canada
IOS Press, Inc.
5795-G Burke Centre Parkway
Burke, VA 22015
USA
fax: +1 703 323 3668
e-mail: iosbooks@iospress.com

Distributor in Germany
IOS Press
Spandauer Strasse 2
D-10178 Berlin
Germany
fax: +49 30 242 3113

Distributor in Japan
Ohmsha, Ltd.
3-1 Kanda Nishiki-cho
Chiyoda-ku, Tokyo 101
Japan
fax: +81 3 3233 2426

PRINTED IN THE NETHERLANDS

Contents

Broadband Access, WDM Metro and Network Management
D.W. Faulkner and A.L. Harmer (Eds.)
IOS Press, 2000 © AKM

High Speed Communications Systems for Cars - Dream or Reality?

Dung Tu
Texas Instruments Deutschland, Haggertystr.1, 85356 Freising, Germany
dung-tu@ti.com

1. Introduction

In the last few years the speed of communications systems has been improved significantly. New broadband digital subscriber line (DSL) modems boost the speed of Internet connection by factor of 100 or more compared with the 56kbps of analog modems. In the digital wireless communications world General Packet Radio Services (GPRS) or Enhanced Data Rates for Global Evolution (EDGE) enable data rates up to 384kbps, a real improvement compared with the 9.6kbps of the current GSM data rate. Future third generation Universal Mobile Telephone System (UMTS) will allow data rates up to 2Mbps.

In the same time we see also a surging market for automotive electronics. According to Forward Concept global demand for vehicle electronics systems is expected to climb from $46 billion in 1998 to $53.5 billion by the year 2000 and nearly $75 billion in 2005. Semiconductor content in automotive systems will grow from $7.9 billion in 1998 to nearly $18.9 billion in 2005. A significant portion of this market will be in the area of communications and multimedia systems.

The cabin of a high-end modern car is full with electronics. Entertainment, mobile phone, navigation, board computer, telematics systems are either available as optional accessories from carmakers or can be bought as after-market products. The question here is whether automotive communications and multimedia systems will keep pace with those in the office or at home. Are high-speed communications systems for cars a dream or reality?

2. Car Electronics Systems

Fig.1 shows a generic architecture of electronics systems in a car. The mission-critical electronics include control systems for chassis, safety, power train, and body. The Control Area Network (CAN) bus that is capable of up to 1Mbps data transfer rate is often used in this area. For emerging safety and body electronics that require higher speed there are new bus protocols that are capable of up to 10Mbps.

The multimedia world of a car includes a car radio as basic equipment that can be complemented with additional components for CD, DVD, TV, navigation, camera, radar, mobile phone, telematics, MP3, digital audio broadcast (DAB), and digital multimedia broadcast (DMB).

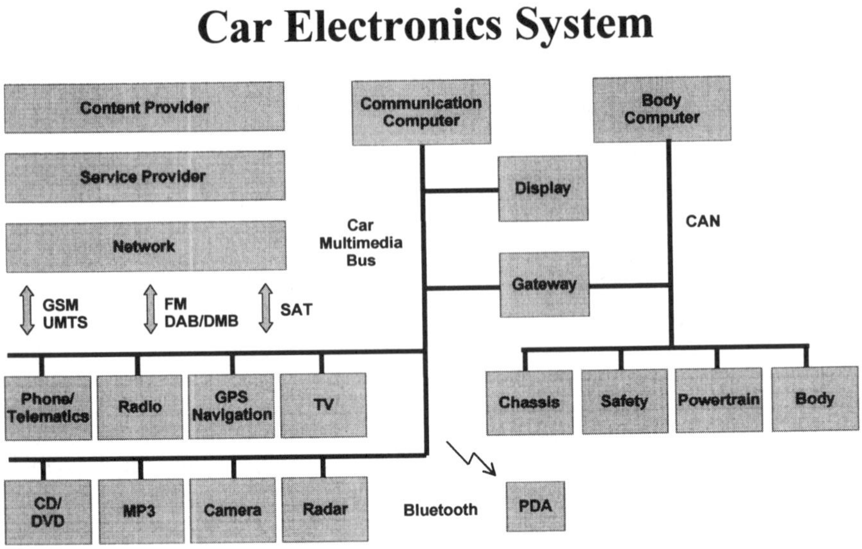

Fig.1: Car electronics systems

Looking at these systems we can classify them into several categories: unidirectional broadcast systems like radio, bidirectional communications systems like mobile phone or telematics units, entertainment systems using compressed video or audio data like CD, DVD or MP3, and complementary safety systems using uncompressed real-time data like camera or radar.

Whether a system can establish itself as standard equipment in every car is not a question of technological feasibility but a question of the value end-users get for the money they pay.

3. Value Chain

Figure 2 shows the value chain how the content of a service is transported through an infrastructure to reach an on-board terminal to probably satisfy the needs of an end-user. Although at the end of the chain, the starting point should be the needs of a driver and his willingness to pay for useful, affordable services and equipment. The needs determine the content of services.

Information and entertainment are basic needs of drivers. Up-to-date traffic information that helps to avoid congestion is desirable. Fast help in case of breakdown and especially in emergency case improves safety and can save life. With the widespread use of Internet in daily life people will want to have in a car the same capabilities e.g. to get emails, look for points of interest, plan a trip or buy things. For carmakers these new communications systems can be useful to offer new services.

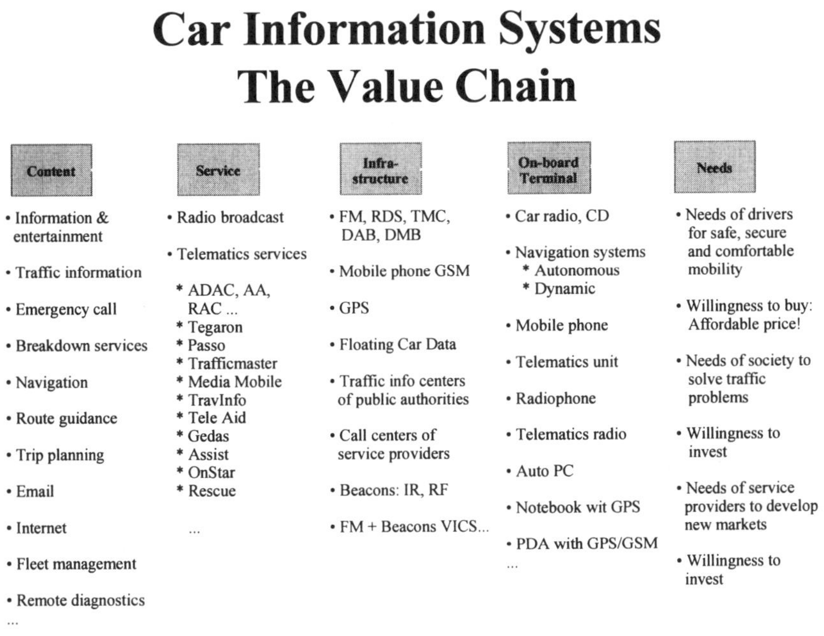

Fig.2: The value chain

Conventional radio broadcast is the most popular medium used by drivers to get information and entertainment. Emerging telematics services include automated emergency call in case of accidents, roadside assistance in case of breakdown, real-time traffic information, route guidance, and normal mobile phone function. Telematics services are offered not only by automobile clubs like ADAC and AA or service providers like Tegaron, Passo, Media Mobile, or Trafficmaster but also by carmakers. DaimlerChrysler works together with Tegaron, BMW offers Assist, VW has Gedas, GM/Opel has OnStar, and Ford offers Rescue as telematics services. Renault, PSA, Fiat have also their telematics services.

4. On-board Terminals

There is a large variety of products as on-board terminals starting from a simple car radio to radiophone, autonomous and dynamic navigation systems, mobile phone with telematics box, integrated telematics unit, AutoPC, notebook or PDA with GPS and GSM, etc.

The car radio is now almost standard equipment in every car. Autonomous navigation systems include usually a high-end radio with a CD changer and a board computer; they are therefore relatively expensive. Autonomous navigation systems can guide a driver on an unknown route very well how to get from A to B. However without the link to a communications system an autonomous navigation system is not aware whether the route is free or congested. Dynamic navigation systems use Traffic Message Channel or the link to a mobile phone to update drivers with "real-time" traffic information and to give dynamic route guidance. Navigation systems are usually sold as optional equipment for a car although there are now high-end models with navigation as standard equipment.

On the other side telematics systems are based on a mobile phone and a GPS unit that can be integrated into a car radio to deliver telematics services. Telematics systems are less expensive and the business model is to sell services rather than to sell equipment.

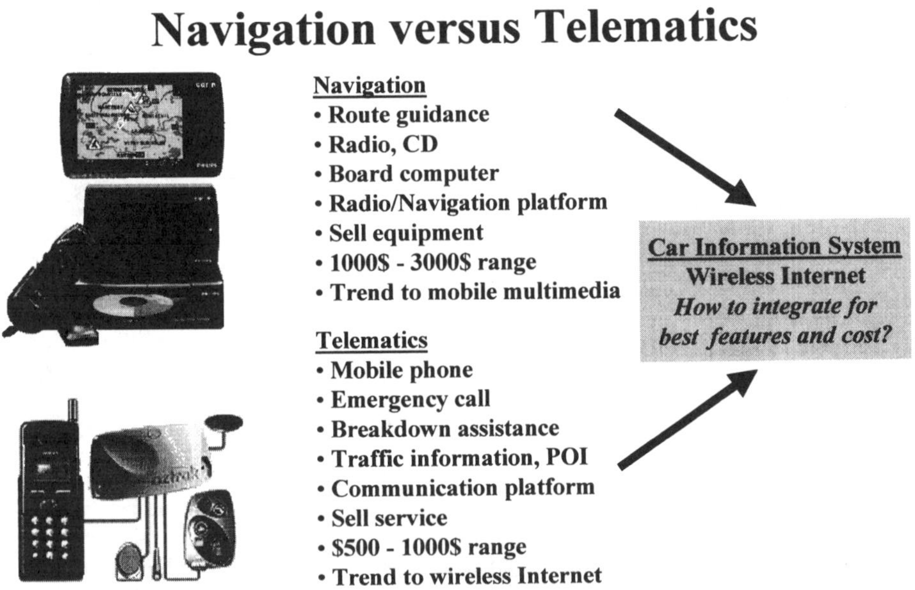

Fig.3 Navigation versus Telematics

5. Telematics

Currently wireless mobile communication means voice communication. Short Message Service is the first well-known wireless data service. Despite its limited capability of only 160 characters, SMS has become a cult communication media especially among young people. The Internet is without doubt the megatrend of our age. Wireless Application Protocol (WAP) makes possible the first step towards mobile Internet. WAP-enabled phones and services are starting to trigger a new market for wireless data.

Telematics services in Europe use GSM data as a bearer. Fig. 4 shows GSM data services today and in the near future, and how these fit to telematics services. Some telematics services today are using SMS that has proven to be quite useful. With higher bandwidth, GPRS will be much more suitable for delivering this kind of services.

Telematics services or in general wireless communication services require seamless operability between terminals and networks. It is crucial that the specific interfaces and functionality are standardised. Currently, there are two proposals for standardisation: Global Automotive Telematic Standard (GATS) and Intelligent Transport Systems over Wireless Application Protocol (ITSWAP).

	GSM data services today				Future GSM data services				
Service	voice	data	SMS		USSD	HSCSD	GPRS		
			PTP	CB			PTP	PTM	Anonymous
Emergency Breakdown	+/-	+/-	+	N/A	++	-	++	-	N/A
Traffic Information	-	-	+	++	+	+	+	++	N/A
Floating Car Data	N/A	N/A	+	+	+	-	++	+	++
Route Guidance	-	+	+	-	+	++	++	-	-
Fleet Management	-	+	+	-	+	+	++	++	N/A
Information Services	+	+	+	+	+	++	++	-	N/A
Theft Detection and Recovery	N/A	+/-	+	N/A	+	+/-	++	-	N/A
Remote Diagnostics	N/A	+/-	+	N/A	+	+/-	++	-	N/A

++ very appropriate   + appropriate   +/- satisfactory   - not appropriate

SMS	Short Message Services
USSD	Unstructured Supplementary Data Services
HSCSD	High Speed Circuit Switched Data Services
GPRS	General Packet Radio Services
PTP	Point to Point
PTM	Point to Multipoint

Fig. 4 GSM data services for telematics

6. Mobile Internet and Digital Media Broadcast

It's a fact that automotive equipment has its own rules due to the limited available space in the dashboard and the harsh temperature and environment. The human machine interface in a car has different requirement than at home or in the office. However a car radio must use in principle the same technology as a radio at home, otherwise the cost would be prohibitively high.

Fig.5 shows the evolution of wireless communications from a mobile voice terminal to mobile multimedia and wireless Web. With increasing bandwidth wireless communications will be able to offer new features and services that are not available today.

In the US the E-911 mandate of the Federal Communications Commission imposed that until 10/2001 when a mobile phone caller makes an emergency call his position must be located within 125 meters in 67% of all calls. This mandate has triggered the development of new wireless location technologies as well as accelerates the integration of GPS functionality in the wireless handset. The first WAP phone with built-in GPS receiver has now reached the market.

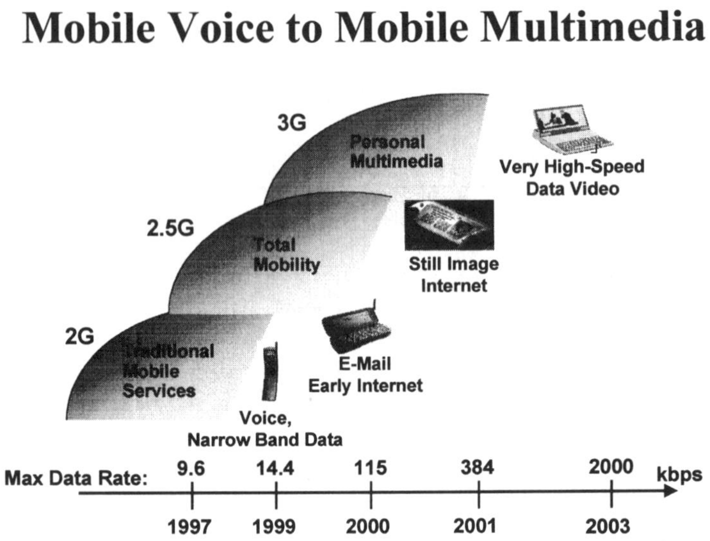

Fig.5: From mobile voice to mobile multimedia

It is expected that positioning will be an optional or built-in feature in future wireless handsets. There is already the first provider of value-added real-time traffic information, maps, and navigation services on mobile Internet phones. With the availability of higher bandwidth future navigation help can be downloaded from the net to an automotive navigation system or to a wireless information device for a pedestrian.

Mobile Internet can be considered as one medium of digital multimedia broadcast. Other digital broadcast media are digital audio broadcast and digital video broadcast. Digital radio technology gives consumers clear CD-like sound quality and can be used to offer telematics services as well. In the US the In Band On Channel (IBOC) digital audio broadcast technology is being developed. In Europe and many other parts of the world the Eureka 147 DAB standard is used. There is an interesting project coordinated by ERTICO - the European consortium for Intelligent Transportation Systems - to develop a technology that uses DAB in combination with GSM for delivering telematics services. GSM will be used for the uplink and DAB for the downlink.

To maximise the value for consumers, telematics, car communications or multimedia systems must take into account these developments and reuse as much as possible technologies developed elsewhere for communications, consumer or computer electronics.

7. Open Multimedia Application Platform

Fig.6 shows the increasing need for real-time computing power in term of million instructions per second for various wireless communications applications. Compared to the first generation GSM, second generation (2G, 2.5G) and especially third generation (3G) require very high computing power.

While general-purpose processors are suitable for running operating systems and graphics user interface, digital signal processors are optimised for real-time voice, image and video signal processing. Smart phones, mobile multimedia, wireless information devices require both. A platform that includes a high performance DSP core combined with a RISC core is well suited for future high-speed wireless communications systems.

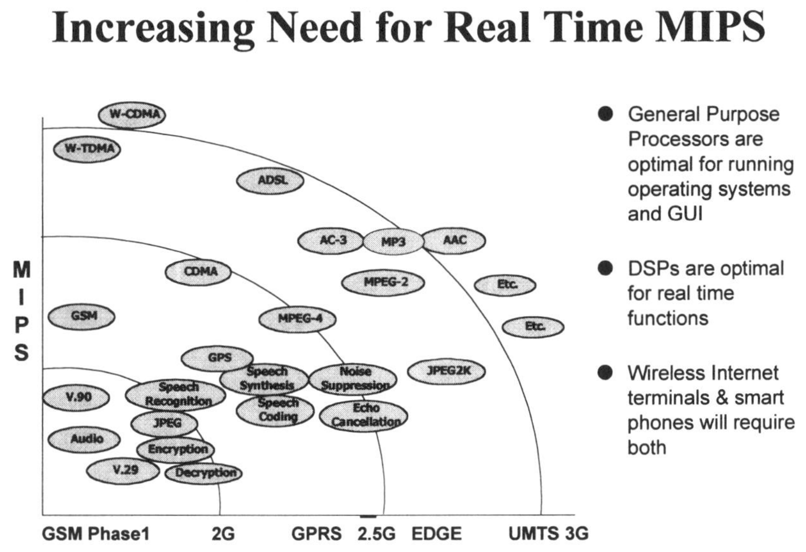

Fig.6: Increasing needs for real time MIPS

Texas Instruments has announced shortly OMAP™, an Open Multimedia Application Platform that includes a low-power DSP core at 320MIPS (C55x core) and a RISC processor core at 130MHz (ARM925 core) plus high-speed dedicated logic-blocks on a single-chip, using 0.15 micron effective CMOS process technology.

TI will support this platform with development chips, a software development kit and clearly defined Application Programming Interfaces (APIs). The APIs will be open and can be implemented by other vendors to ensure an attractive DSP accelerated multimedia environment for independent software vendors. The new offering will provide software developers with a wealth of capabilities until now unavailable. OMAP™ supports also various operating systems for smart phones, PDAs and wireless information devices.

On the hardware side, OMAP™ is a complete set of building blocks consisting of complex megacells and peripherals which are characterised so that they may easily be connected and merged with a customer's own blocks.

System integration is the key to optimise performance and at the same time reduces system cost. On the other side flexibility or in other words programmability is essential to adapt to market changes. With its open hardware and software architecture OMAP™ offers an ideal development and product platform for telematics and auto multimedia systems.

8. Car Multimedia Bus Standardisation

One of the biggest headaches and cost factor for automotive system suppliers is the customisation of their products to each carmaker. Without standardised bus architecture car multimedia electronics need to be developed several years before the vehicles hit the market. This is a big problem, because electronic devices in new cars might use old technology. In such a dynamic market like communications and multimedia this is unacceptable because consumers do not want to buy out-of-date products.

Carmakers have realised this problem. They have founded the Automotive Multimedia Interface Collaboration (AMIC) with the goal to create a set of standards for auto multimedia electronics. The industry group consists of 12 manufacturers: BMW, DaimlerChrysler, Ford, Fiat, General Motors, Honda, Mitsubishi, Nissan, PSA/Peugeot-Citroen, Renault, Toyota and Volkswagen.

Fig.7 shows a summary of several car multimedia bus solutions that are either in production or in development. To avoid problems with EMI and to reduce weight, carmakers believe that a car multimedia bus should use plastic optical fibers (POF).

Car Multimedia Bus Solutions

Bus	Owner	Topology	Bandwidth Mbps	Status
MOST	MOST Cooperation	Ring	22	in preproduction
D2B	C&C Electronics	Ring	5 - (43)	5 Mbps in production
MML	Delphi	Star	100	in development
HiQOS	Matsushita	Ring/Star	54	in development
FireWire IEEE1394	Apple Open standard	Tree	100-400 (1394.a) 100-3200 (1394.b)	computer, consumer in production, POF in development

Fig.7: Car multimedia bus solutions

D2B is the first car multimedia bus in production. It uses a ring topology. The current bandwidth is 5.6Mbps.

MOST (Multimedia Oriented System Transfer) uses a ring as preferred topology. The bandwidth is currently 22 Mbps; a higher speed version has been announced.

MML (Multimedia Mobile Link) uses a passive combiner/splitter in a star configuration with a bandwidth of 100 Mbps.

HIQOS bandwidth is 54 Mbps. The topology is a ring or hybrid of ring and star.

IEEE1394 is an open multimedia bus standard for consumer and computer electronics. An automotive solution for POF is in development. IEEE1394 has the highest bandwidth (100-3200 Mbps), and it uses a tree topology. Currently IEEE1394 is the only real open bus architecture that is compatible with consumers and computer electronics and capable to work with both compressed as well as uncompressed video data.

The IDB Forum is a consortium that is working together with AMIC on the standardization of a car multimedia bus. IDB stands for Intelligent Transportation System Data Bus. The IDB Forum includes 50 companies and institutions from automotive, consumer, and computer industries. Fig.8 shows the simplified IDB bus architecture. IDB-C is a low speed control bus with CAN physical layer. IDB-M is a high-speed multimedia bus. Apparently MOST and IEEE1394 are good candidates for IDB-M.

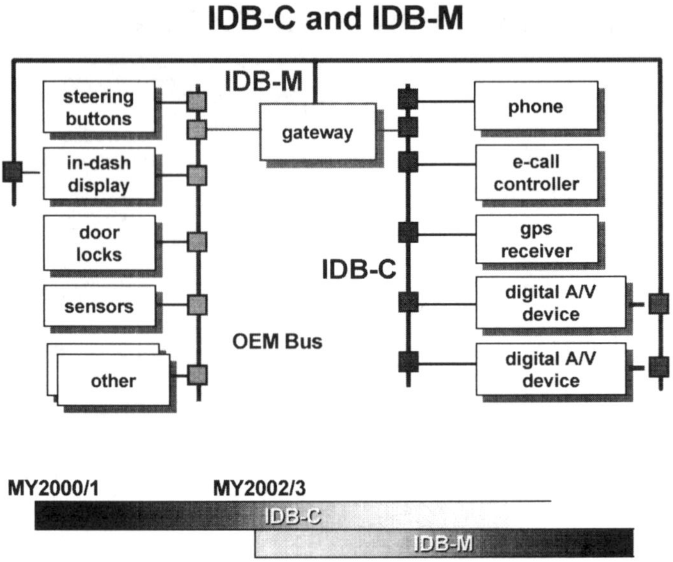

Fig.8: IDB-C and IDB-M

9. Summary

High-speed communications for cars is not a dream. It is anticipated that by 2005 every car will be capable of wireless Internet. This is logical because the car is an important part of the personal network like the home or the office. Useful and affordable services, system integration for optimum performance and cost, well-designed ergonomic human machine interface are the keys to make this market a success.

Network performance modeling of fiber-radio systems with realistic hardware constraints

Graham H. Smith*, Jennifer M. Yates+, Ampalavanapillai Nirmalathas and Dalma Novak
**Bell Laboratories, Lucent Technologies, Holmdel, N.J., 07733, U.S.A.*[1]
Tel: +1 732 332 6934 Fax: +1 732 949 2473 Email: grsm@lucent.com
+AT&T Laboratories – Research, AT&T Shannon Laboratory, Florham Park, N.J. 07932, U.S.A.[2]
Australian Photonics Cooperative Research Centre, Photonics Research Laboratory, Department of Electrical and Electronic Engineering, The University of Melbourne, Parkville VIC 3052, Australia

Abstract: Fiber-radio systems have been proposed for future high-speed radio networks. In such systems, optical fiber links can be used to transport radio signals between remote antenna base stations and a central office where centralized signal routing and processing are performed. This paper extends traditional teletraffic modeling techniques to investigate the performance of a fiber-radio network with realistic hardware constraints. The effect of component non-linearities on network performance is determined for a variety of factors including call admission algorithms, radio cell size, cell capacity and receiver bandwidth.

1. Introduction

Broadband radio access technologies such as PCS, MMDS, LMDS, and LVDS allow for quick and simple installation of networks that provide high bandwidth services, such as interactive multimedia services. In such systems, optical fiber links have been proposed to transport radio signals between antenna base stations (BSs) and a central office (CO) where centralized signal routing and processing can be performed [1-3]. The optical links also provide large bandwidth with low loss and immunity to EMI and enable smaller and simpler base stations to be implemented.

In a fiber-radio system with a large number of active users, significant crosstalk can result from the mixing of signals from different users in non-linear components. Third-order non-linearities are especially significant, since they may cause intermodulation distortion (IMD) terms that fall within the receiver bandwidth. For bi-directional fiber-radio links, the linearity performance is often more stringent for upstream transmission than downstream, due to the varying received signal strengths at the BS. It has been shown that the dominant non-linearity is the optical source at the remote BS [1]. In this paper we examine the effect that IMD has on network performance.

The linearity of a fiber-optic or radio link is commonly specified by the two-tone IMD performance with two equal power input tones. The spurious-free dynamic range (SFDR) is then used as a measure of the link linearity and is defined as the range of input powers over which the IMD powers at the link output remain below the noise floor. The SFDR requirements for a fiber-radio link have commonly been determined via a worst-case system analysis [1]. This analysis ensures that a single user with minimum received signal power has adequate signal quality when all other users are received at maximum signal strength. However, these specifications satisfy system requirements that correspond to rarely occurring worst-case scenarios. To significantly reduce the device non-linearity requirements it is possible to accept a slight degradation in system performance. A number of system analyses have recently been published which have examined this tradeoff between system performance and device non-linearity requirements [1-9].

If components with reduced linearity performance are incorporated into a fiber-radio system, it is possible for some user channels to experience unacceptable signal quality. The signal quality is defined here in terms of the carrier-to-noise-plus-intermodulation distortion ratio (CNDR). If a new call arrives, then all of the available frequencies are searched to find one with an acceptable CNDR. If no such frequency exists, then the new call is rejected, and is deemed to be *blocked*. Deciding if a new call is accepted or rejected is referred to as *call admission control*. If the signal quality of an already established call becomes unacceptable, then the call is said to experience *outage*. Thus, having relaxed linearity requirements can result in an increased probability of blocking and/or outage.

Cutrer *et al.* [4] first showed that reductions in the device non-linearity requirements can be achieved by accepting a slight degradation in outage. More recently Koshy *et al.* [8] used a similar model to Cutrer *et al.* [4] to examine the effect of laser non-linearities on outage in fiber-fed microcellular networks designed for GMSK systems. They also showed that the non-linearity of the laser created a tradeoff between the outage and the GMSK system parameters such as capacity and threshold BER. Sabella [7, 9] also reported on the analysis of directly modulated upstream fiber-radio links for PCS systems and for mm-wave radio

[1] This work was performed whilst at the Photonics Research Laboratory, The University of Melbourne, Australia.
[2] This work was performed whilst at the Software Engineering Research Centre, RMIT, Australia.

systems that incorporate downconversion at the remote BSs. He qualitatively examined the combined laser non-linearities effects, such as clipping, chirp, and resonance frequency, on higher-order modulation schemes such as QPSK and 16QAM. He showed in [7] that when the laser is operated at high optical modulation indexes, the clipping of the signal has a greater effect on performance when compared to IMD effects. In results similar to Cutrer *et al.* [4], Sabella [9] showed that a tradeoff exists between the number of users and the channel bandwidth and furthermore highlighted the significant impact that laser chirp could have on performance.

In all of these analyses [4, 7-9] the system performance is simulated using a fixed numbers of calls. We refer to this as the *fixed load model*. This is a simple model that enables fast simulation of various systems. However, it does not account for the statistical nature of call arrivals and departures. To accurately predict the system performance, an analysis should also take into account the fluctuating number of users in a system.

This paper presents the first analysis of fiber-radio systems that models both call arrivals / departures and realistic hardware constraints. We refer to this as the *teletraffic model*. In Section 2 we describe the system model for which we extend traditional teletraffic modeling techniques to account for device non-linearities. In Section 3 we show that using this more realistic model of user behavior results in device non-linearity requirements that often differ significantly from those predicted using the simpler fixed load model. In Sections 4, 5 and 6 we examine other factors that affect system performance, including call admission algorithms, CNDR requirements, radio propagation environment, cell size, cell capacity and receiver bandwidth.

2. System model

A *fixed access* fiber-radio network is modeled, with large cell frequency reuse factors. As mobility is not an issue in this access environment, each cell is assumed to operate independently. This allows us to approximate network performance by that in a single cell, as shown in Fig.1.

A radio cell is defined to have a maximum number of channels, N, that correspond to N unique frequencies. Each user accepted into the system is assigned a unique frequency for the duration of their call. The arrivals and departures of calls in the fiber-radio system are statistically modeled with new calls assumed to arrive independently at a rate, λ, with interarrival times drawn from a negative exponential distribution. Each call exists in the system for a length of time denoted by the call holding time, drawn from a negative exponential distribution with mean $1/\mu$. With these assumptions, the number of users in a system without any hardware limitations (IMD-free) can be described by a Poisson distribution [10].

Calls offered to the system are assumed to be distributed uniformly and randomly throughout the radio cell with users constrained to being more than 5 m from the remote BS. Calls that are not blocked and are accepted into the system are assigned a channel frequency on which the link from the user to the BS is established. For this modeling work, it is assumed that channel frequencies in the radio spectrum are equally spaced and that a continuous frequency spectrum exists for the radio cell. These assumptions ensure that maximum utilization of the frequency spectrum is achieved. However, they also mean that IMD terms will fall directly on to the channel frequencies, making it impossible to filter them out.

The cross-talk powers on each channel are calculated by examining the in-band two-tone and three-tone IMD terms. The power in the in-band two-tone and three-tone IMD terms generated by the optical transmitter are calculated for each channel using [4]:

$$P_{IMD} = \gamma \left(\sum_{2f_i - f_j} P_i^2 P_j + \sum_{f_i + f_j - f_k} 4\, P_i P_j P_k \right) \tag{1}$$

where γ is a proportionality constant determined from a two-tone characterization of the device (directly related to the SFDR), P_1, P_2, and P_3 are the powers in the three tones, and the summations are over all of the combinations of the two and three-tone frequencies f_1, f_2, and f_3.

For our system model, the smallest received carrier power within the cell is amplified to a minimum level above the noise floor using an automatic gain control (AGC) amplifier, as shown in Fig. 1. The amplifier gain is chosen so that the carrier-to-noise ratio (CNR) exceeds the minimum required CNDR with an additional 0.2 dB to allow for extra IMD terms. The system performance of the fiber-radio network is measured in terms of the percentage blocking, P_B and percentage outage, P_O. P_B represents the percentage of calls that are offered to the system but are not accepted, and is defined as:

$$P_B(\%) = \left(\frac{\textit{total number of new calls blocked}}{\textit{total number of calls offered to system}} \right).100 \tag{2}$$

P_O represents the average amount of time that calls have their signal quality deteriorate below an acceptable value after they have been accepted into the system. P_O is defined as:

$$P_O(\%) = time\ average\left(\frac{total\ number\ of\ calls\ in\ outage}{total\ number\ of\ calls\ in\ system}\right).100 \quad (3)$$

In all of the investigations of linearity requirements described in the literature, the fiber links transported radio frequencies in the low microwave bands (< 2 GHz) and the radio cells were assumed to be either macrocellular (> 1 km) or large microcellular (> 500 m). For such links, it has been shown that the radio propagation environment can be approximated by a d^{-4} relationship [1, 11, 12]. This is because a generic macrocell or large radio microcell can have many objects blocking the direct path, thereby producing a non-line-of-sight (NLOS) link [1]. However, in systems where a direct line-of-sight (LOS) link is always maintained, a more accurate propagation model is given by a d^{-2} relationship [13, 14]. For a fixed radio cell size, there will be a greater difference between the maximum and minimum received powers at the remote BS with the NLOS model. For the simulations in this paper, we assume a d^{-2} LOS model.

The performance of the system described was investigated using a network simulation tool. This tool was used to investigate the impact that device non-linearities have on a realistic fiber-radio network.

3. Comparison of teletraffic model to existing fixed load model

As discussed above, previous analyses have modeled a fixed number of calls in the system, in contrast with the realistic call arrival and departure statistics used here. In order to highlight the significance of this effect, Fig. 2 compares the simulated outage for a 20 channel fiber-radio system modeled using both the teletraffic model, and using the simple fixed load model with 20 active users. In the teletraffic model, calls are accepted into the system whenever there are available channels, ignoring received signal quality. The two curves produced with the teletraffic model correspond to different offered loads.

Fig. 2 shows that for a fixed SFDR, the outages predicted with the teletraffic model are smaller than those predicted using the fixed load model. This observation can be explained in terms of the number of users in the system. In the fixed load model, there are a fixed 20 calls established in the system at all times. However, in the teletraffic model, the number of calls in the system varies over time, with a maximum of 20 established calls. There is thus a lower mean number of calls in the system in which user behavior is modeled, whilst the number of calls never exceeds that with the fixed load model. In general, outage increases with increasing numbers of calls in the system, due to increased numbers of IMD terms. Thus, with fewer calls in the system it is expected that the model with realistic user behavior would experience lower outage for a fixed SFDR than a system with a constant (maximum) number of calls.

The increase in outage with increasing number of calls also explains the observation that outage increases with an increase in mean offered load in the teletraffic model. This can be observed from Fig. 2, in which the outage for an offered load of 12.03 Erlangs is higher than for an offered load of 10.07 Erlangs. However, in all cases, using the more realistic teletraffic model allows SFDR requirements to be reduced compared to those predicted when using the fully loaded fixed load model presented by Cutrer [4].

Calls are accepted independent of signal quality in all of the scenarios modeled for Fig. 2. Blocking is always zero in the fixed load model, as exactly 20 calls are always offered to the 20 channel system. However, in the more realistic teletraffic model, blocking probability is a function of offered load. The blocking probability is given by the Erlang-B loss formula [10], giving a blocking of 0.2 % for the 10.07 Erlangs system, and 1 % for the 12.03 Erlangs system, independent of SFDR.

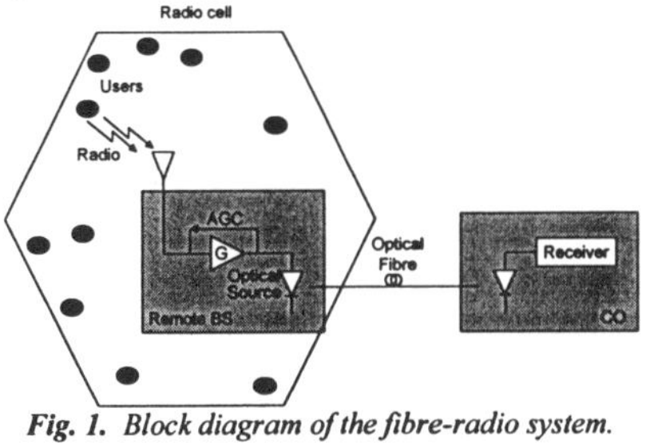

Fig. 1. Block diagram of the fibre-radio system.

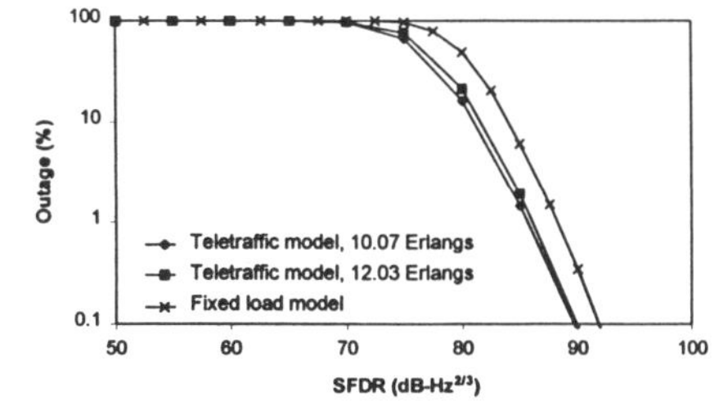

Fig. 2. Simulated outage vs SFDR for a fiber-radio system; 20 channels, 300 m cell size.

4. Call admission algorithms

In the previous section, call admission ignored the signal quality of a new call, so that calls with unacceptable signal quality could enter the system and immediately be in outage. In order to avoid this, we propose a new call admission algorithm in which the channel quality is examined before accepting a new call. If there are no channels with an acceptable signal quality, then the call is blocked.

Channels are searched in order of increasing frequency in the radio spectrum with a new call allocated to the first available channel that is deemed to be acceptable. A channel is defined to be acceptable if the new call, when established, experiences an acceptable CNDR. Alternatively, the call admission algorithm can additionally constrain calls to be accepted only if accepting the new call into the system does not cause any already established call to go into outage.

Figs. 3 and 4 compare the simulation results obtained for a 20 channel with the different call admission algorithms. The offered load to the system is fixed at 10.07 Erlangs, corresponding to a blocking of 0.2 % in an ideal system without hardware limitations.

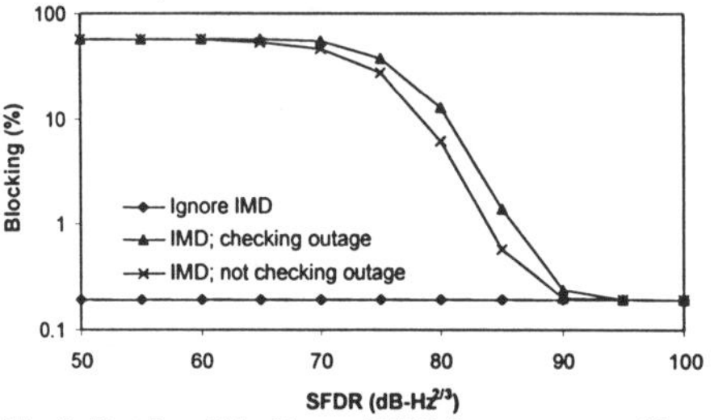

Fig. 3. Simulated blocking vs SFDR for a mm-wave fiber-radio link; 20 channels, 300 m cell size.

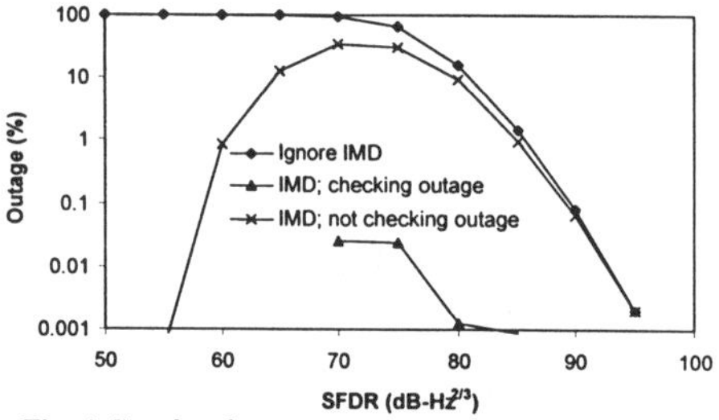

Fig. 4. Simulated outage vs SFDR for a mm-wave fiber-radio link; 20 channels, 300 m cell size.

As discussed above, blocking is constant with varying SFDR for the call admission algorithm that ignores IMD. However, Fig. 3 shows the severe effect on the corresponding outage, with excessive outage experienced for small SFDR. Outage then decreases with increasing SFDR requirements.

Blocking becomes a function of SFDR when call admission is based on received signal quality, as illustrated in Fig. 3. When SFDR is small, the resulting IMD terms will be so large that any channel with an IMD term will be unable to accept calls. This means that calls can only be allocated on a small subset of the total channels, leading to high blocking probabilities. As the SFDR increases, the blocking decreases. In this range, blocking is higher when the call admission algorithm checks the CNDR on all channels before accepting a call, than when existing call signal qualities are ignored. At high SFDRs, IMD effects become negligible, and blocking occurs only when the system is full.

It is also possible for accepted calls to experience outage if their signal is degraded below acceptable levels. Different minimum acceptable CNDR values are defined for blocking and for outage. In these simulation results, it is assumed that a call goes into outage when the CNDR drops more than 0.2 dB below the acceptable channel CNDR defined for call admission. Fig. 4 shows the simulated outage results for the teletraffic analysis showing that the outage peaks for both of the call admission algorithms that use IMD. These peaks occur at around 70 – 75 dB-Hz$^{2/3}$. From Fig. 3, it can be seen that a SFDR of 70 dB-Hz$^{2/3}$ corresponds to the point where the IMD terms reduce enough that some channels with IMD terms have acceptable channel quality and the call blocking therefore begins to reduce. From Figs. 3 and 4, it is also extremely important to note that outage becomes negligible once the blocking falls below approximately 10 % for call admission which verifies signal quality on all channels before accepting a new call. Therefore, using this call admission algorithm, outage caused by IMD can be neglected within the expected 0.5 – 5 % operating blocking requirements of existing and future cellular systems. In contrast, if call admission neglects to examine the signal quality of established calls, significantly higher outage is experienced for a fixed SFDR. At SFDRs where both blocking and outage performances are acceptable, the outage experienced with this call admission algorithm is only marginally lower than that achieved when signal quality is completely ignored in call admission. The added complexity and increased blocking achieved with the call admission algorithm based on IMD, but neglecting the effect of accepting a new call on existing calls, is difficult to justify. This algorithm is thus not examined further within this paper.

In order to select optical transmitters based on the SFDR requirements predicted from the teletraffic model, it is necessary to be able to implement the channel allocation scheme used in the simulations. For the algorithm based on IMD, this requires the separate measurement of both the carrier power and IMD on all of the user channels in the frequency spectrum. However, the powers measured on each channel will include both carrier power and IMD terms. In the fixed load model, it was shown [4] that simply measuring the peak powers of channels does not necessarily equate to acceptable signal quality. However, with the teletraffic model, the measured powers on each channel (signal power + IMD + noise) can be a close approximation to the signal power, because users are only ever accepted into the system when the CNDR is acceptable in the first place. This implies that the signal power is at least a factor of CNDR higher than the IMD + noise terms at the time when the call is accepted. The carrier power can thus be accurately approximated by the measured received power. The approximation becomes less valid, however, as the CNDR for the system is reduced and in a system with large amounts of time-varying fading.

The call admission algorithm described above can be used in a realistic system by approximating the received signal power by the measured power within a channel. The IMD terms on each channel are then calculated using the approximated signal powers. The signal quality for each channel in the system is finally approximated by the ratio of the "measured" received signal power and the calculated IMD on that channel. Simulation results examining the performance of this algorithm show negligible degradation using the approximated signal powers compared with using the real signal powers. We thus conclude that call admission based on measured signal powers is practical within a fibre-radio system.

In the following sections we continue to study the performance of fiber-radio networks with IMD. Unless otherwise stated, the call admission algorithm used is that which examines the signal quality of both the new call *and* already established calls before accepting a new call into the network.

5. *Other factors affecting performance*

There are a number of other factors that influence system performance. Firstly, the minimum specified channel CNDR required for call admission affects the linearity requirements of the optical sources. Figs. 5 and 6 show simulation results corresponding to minimum call admission CNDR specifications of 18 and 7 dB. The plots show the expected result that the higher the required CNDR, the more stringent the SFDR requirement. Considering the 20 channel system, for a fixed blocking, better devices are required (i.e. more stringent SFDR) in order to reduce IMD terms and therefore improve CNDR. For example, at a system blocking of 1 %, specifying a CNDR of 18 dB instead of 7 dB results in a 10 dB increase in the SFDR requirement. The corresponding outage curves shown in Fig. 6 also indicate that the peaks in the outage performance are affected by the CNDR specifications with similar 10 dB differences between the 18 and 7 dB cases. However, as Fig. 6 highlights, these outage values are small anyway, especially for the SFDR requirements corresponding to realistic blocking.

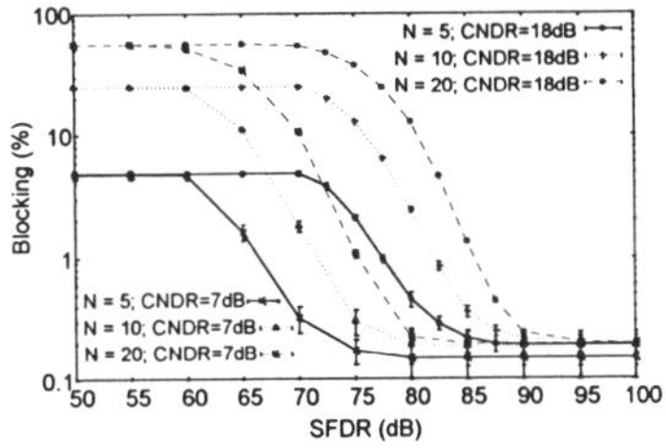

Fig. 5. Blocking vs SFDR for a fiber-radio system with a 300 m cell size, and with varying numbers of channels and minimum CNDR specifications.

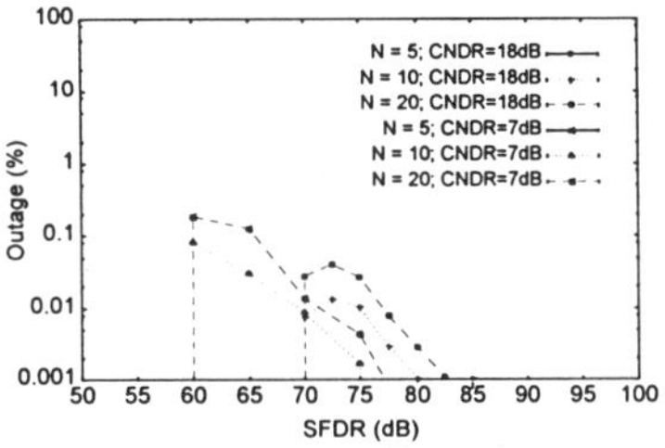

Fig. 6. Outage vs SFDR for a fiber-radio system with a 300 m cell size, and with varying numbers of channels and minimum CNDR specifications.

Figs. 5 and 6 also show the simulation results for the link performance when there are 5, 10, and 20 channels in the system. The load for each system is calculated to give a fixed blocking probability of 0.2 % when ideal devices are used (i.e. infinite SFDR). As expected, when there are more potential channels in the system, there are also even more IMD terms so that the percentage blocking also increases. In the extreme case of small SFDRs (extreme IMD), 3 channels are usable in the 5 channel system, as compared with only 6 channels usable in the 20 channel system. Fig. 5 shows this with an increase in blocking from 5 % to 58 % in going from 5 channels to 20 channels (remembering that the offered loads are chosen to achieve the same blocking at high SFDR). Fig. 5 also shows that, for a fixed blocking of 1 %, the 10 dB relaxation in the SFDR requirements achieved when reducing the minimum required CNDR for call admission from 18 dB to 7 dB is basically independent of the number of users in the system. Finally, there is again negligible outage experienced in the systems with 1 % blocking for all system capacities, as depicted in Fig. 6. From Figs. 3 and 6, it is clear that outage can be neglected and so will be ignored for the rest of the paper.

The outage model results from [4, 7-9] indicate that the receiver bandwidth is a critical parameter in determining SFDR requirements. In this section, the performance of a 20 channel fiber-radio system is shown for various channel bandwidths ranging from 30 kHz to 155 MHz. It is assumed that the traffic load corresponds to a 0.2 % blocking for an IMD-free system. Fig. 7 indicates that as the receiver bandwidth increases, the SFDR requirements for a fixed blocking also increase. For example, at a system blocking of 1%, the requirement on the SFDR increases from 68 dB for a receiver bandwidth of 30 kHz, to 93.5 dB for a receiver bandwidth of 155 MHz. It is important to realize that there is a direct relationship between the receiver bandwidth and the maximum per-channel capacity. The largest data rate that a radio channel can have is determined by the receiver bandwidth. As expected, Fig. 7 shows that there is a strong dependence of the SFDR requirement or the system performance on the receiver bandwidth or equivalently the maximum radio channel data rate. For systems that require large channel rates, tighter constraints on the SFDR of the optical sources are therefore necessary to achieve a specified system performance.

Fig. 8 shows the simulated system performance for various cell sizes. For a fixed blocking it can be seen that the SFDR constraints are weakest for small cell sizes due to the small variations in received signal qualities. The smaller cell sizes mean that the signal power variation is smaller and therefore the rate of percentage blocking increase is faster once IMD terms dominate. The curves in Fig. 8 show marginal increases in SFDR requirements as cell size increases, with the most significant effects being for small cell sizes. Almost no distinguishable tightening of the SFDR requirements is observed for when the cell size is increased from a radius of 300 m to 1 km.

The smaller cell size cases can also be thought of as equivalent to a fixed cell size but with different ranges of power control. In the limit, where the maximum cell size is equal to the minimum cell radius of 5 m, perfect power control is emulated. In this case, for a fixed channel number, a step transition would result where the IMD terms below a particular SFDR specification on the optical transmitter would allow acceptable performance while any transmitter with a SFDR smaller than a specific point would be blocked to the limit.

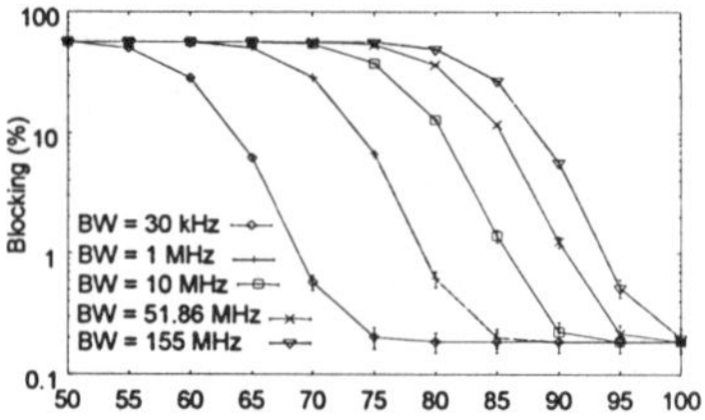

Fig. 7. Blocking vs SFDR for a fiber-radio system with 20 channels and a 300 m radio link, operating with different channel bandwidths.

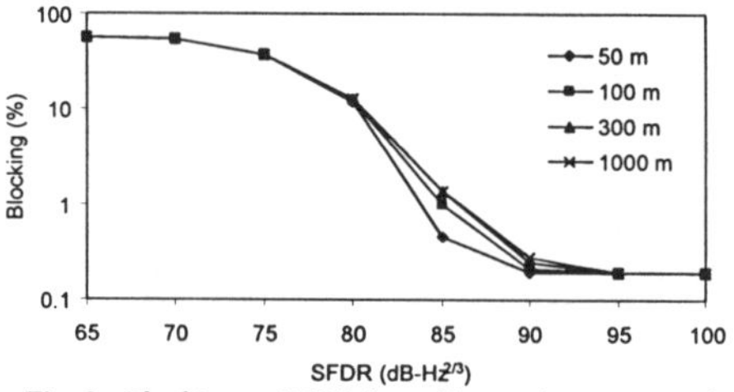

Fig. 8. Blocking vs SFDR for a fiber-radio system with 20 channels and various radio cell sizes.

The previous curves examined blocking as a function of cell size, and demonstrated only marginal increases in blocking with increasing cell size. However, within a single cell, users further from the BS will in general be received with lower signal powers. There is thus expected to be a dependence between blocking and / or outage as a function of user position within the cell.

To examine this effect, the cell radius is divided into 10 evenly spaced bands, achieved by dividing the cell radius into fixed intervals. Note that bands further from the BS have a larger area, and thus have a higher average number of users. Fig. 9 plots blocking and outage as a function of the distance bands, with increasing band index corresponding to increasing distance from the BS. The curves indicate that blocking is a strong function of user position. It can be seen that blocking is particularly high for users close to the BS. These users are generally received with large powers at the BS. Mixing of these high received power signals with other signals can result in significant IMD products falling on other channels, potentially sending already established calls into outage should the call be accepted. The call admission algorithm used here will block such calls, hence the high blocking for calls in the lower distance bands.

Blocking decreases with increasing distance from the BS up to a point, where it starts to increase slightly again. The users in these outer bands are generally received with very low signal powers, and can experience difficulty in obtaining adequate signal qualities in the face of IMD.

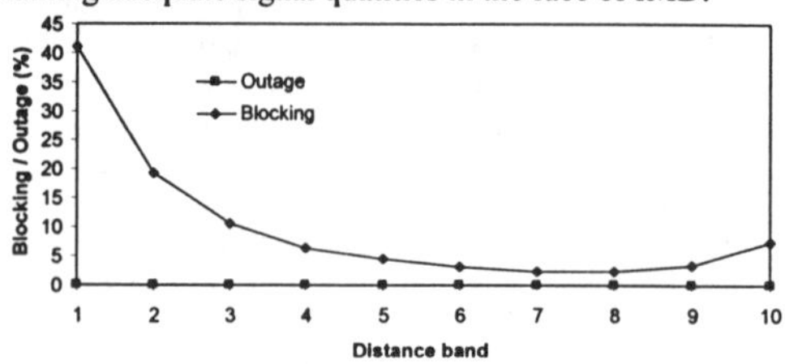

Fig. 9. Blocking and outage vs distance for a fibre-radio system with 20 channels and a 300 m radio system.

The relationship between blocking and/or outage as a function of distance depends on the call assignment algorithm. However, blocking was a strong function of user position in all of the algorithms studied. It is thus important to account for this effect when dimensioning a network. For example, in the above scenario the users closest to the BS received the highest blocking. In a fixed access environment, the same users will always be closest to the BS – and thus their performance cannot be averaged over a range of different positions within the cell as would be the case in a mobile environment. In this case, if a guaranteed quality of service is to be provided for *all* users within the cell, then the cell size, device quality and other such parameters must be chosen so as to ensure that those users closest to the BS receive adequate performance.

6. MM-wave fiber-radio model

For outdoor mm-wave radio links, LOS is essential in order to provide sufficient signal strength for reliable service. Significant attenuation of the mm-wave radio signals can result if there are any obstacles between the remote BS and users, thereby destroying any possible link connection. At mm-wave frequencies, it is also extremely important to consider the effects of oxygen (O_2) absorption and water absorption (for example, from rain or fog). The International Telecommunications Union (ITU) have a draft proposal [14] which describes the relationship for mm-wave radio line-of-sight loss, L_{LOS}, as:

$$L_{LOS}(dB) = 32.4 + 30\alpha + 10\alpha \log_{10}\left(\frac{d}{1000}\right) + \log_{10}(1000f) + 20\log_{10}(f) + \gamma_o d + \gamma_r d \qquad (4)$$

where the first five terms represent the free-space propagation loss over a transmission distance d (in m) and for a radio frequency f (in GHz). The additional decay-rate term, α, in Eqn 4 is used to fit the above equation to various propagation environments. It is indicated in the ITU proposal [14] that measurements have shown that α is typically in the range of 1.4 to 2.5. This proposal also recommends that for a first model independent of scenario type, an α value of 2.2 should be used (a $d^{-2.2}$ relationship). This is very close to the d^{-2} LOS model used earlier. It should also be noted that the influence of α increases dramatically at larger distances since larger values of α contribute to a faster increase in rf power loss. The final two terms in Eqn. 4 account for the O_2 and rain absorption at mm-wave frequencies. These are modeled with the oxygen and rain attenuation coefficients γ_o and γ_r (in dB/m). For mm-wave systems, the ITU proposal [14] recommends O_2 attenuation rates of 0.0001 dB/m and 0.015 dB/m for 40 and 60 GHz links, respectively. It also indicates that the rain attenuation coefficients are highly dependent on the rainfall rate. A CCIR study group reported such values [15]. For low rainfall rates of approximately 5 mm/hr these coefficients were 0.001 dB/m and 0.003 dB/m for 40 GHz and 60 GHz respectively. Equivalent values for moderate rainfall rates of 25 mm/hr were reported as 0.006 dB/m and 0.010 dB/m, respectively. Finally, for high rainfall rates of 150 mm/hr, the rainfall attenuation coefficients were reported as 0.035 dB/m and 0.045 dB/m for the 40 and 60 GHz links, respectively. These values indicate that there is a large variation in the rain attenuation coefficient dependent on the rainfall rate. It is therefore obvious that for mm-wave links, it is extremely important to characterize the rainfall conditions of geographical regions where systems are to be installed.

In order to obtain some guide as to the propagation characteristics of mm-wave links, Eqn 4 was used to generate the loss characteristics for a 60 GHz system. This represents a regulated/preassigned band for mm-wave communication systems. Fig. 10 illustrates the predicted propagation losses showing that the effects of rain become increasingly significant for increasing rain rates. Fig. 10 also shows that the combined effects of O_2 absorption and rain attenuation result in extremely large increases in propagation losses with increasing distance. It therefore indicates that the effects of O_2 and rain cannot be neglected for a cell size of 300 m. The teletraffic model simulation results shown in Fig. 11 confirm the strong effect that the mm-wave radio environment has on the system performance. It shows that for a fixed blocking of 1 %, the SFDR requirements vary by up to 11 dB for the different propagation environments examined. The results in Fig. 11 show that blocking is significantly increased in the presence of rain, with an 8 dB variation in SFDR requirements being observed in going from no rainfall to heavy rainfall. These results show that rain can therefore pose a significant problem for system designers when dimensioning a mm-wave fiber-radio network. Figs. 10 and 11 indicate that for such systems, the geographical rainfall statistics will be required in order to select an appropriate rainfall rate to dimension the network with. We have also simulated 40 GHz mm-wave fiber-radio systems with similar, but less dramatic results as the 60 GHz ones presented here.

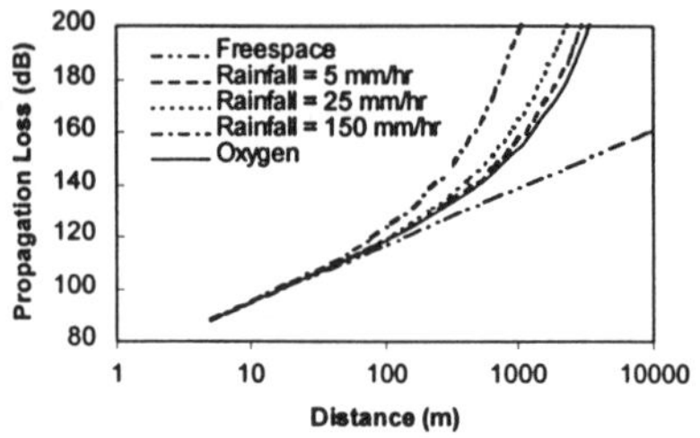

Fig. 10. Propagation loss for a 60 GHz radio link indicating the effects of O_2 *and rain attenuation.*

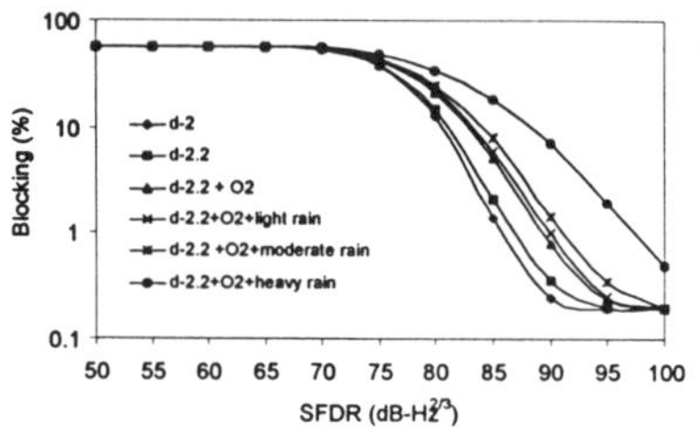

Fig. 11. Blocking vs SFDR for a 60 GHz fiber-radio system; 20 channels, 300 m cell size.

7. Conclusions

In this paper we have predicted the network performance for fiber-radio systems including both the effect of users arriving and departing from the system as well as the linearity of the optical sources at the BS. This teletraffic model was firstly used to show that different call admission algorithms introduce a tradeoff between blocking and outage. When IMD is ignored in call admission, outage was a strong function of the link SFDR, whilst blocking is independent of SFDR. In contrast, if call admission examines the quality of a channel before accepting a new call, then blocking and outage are both a function of SFDR. If the quality of established calls is verified before accepting a new call, then we showed that blocking is the dominant effect. This paper has also shown the extent to which the system performance degrades as the network load, cell size, number and bandwidth of radio channels, and minimum signal quality all increase. Finally the paper investigated the system performance and linearity requirements when the propagation environment is considered. We showed that this is particularly important for mm-wave fiber-radio systems which can be effected by oxygen absorption and rain attenuation.

8. Acknowledgments

The authors would like to thank Lachlan Andrew for providing us with the basic simulation tool that was extended to account for device non-linearities so as to perform this investigation.

9. References

[1] W. I. Way, "Optical fiber-based microcellular systems - an overview," *IEICE Transactions on Communications*, vol. E76B, pp. 1091-1102, 1993.
[2] M. Shibutani, T. Kanai, W. Domom, K. Emura, and J. Namiki, "Optical fiber feeder for microcellular mobile communication systems (H-015)," *IEEE Journal on Selected Areas in Communications*, vol. 11, pp. 1118-1126, 1993.
[3] H. Harada, H. J. Lee, S. Komaki, and N. Morinaga, "Performance analysis of fiber-optic millimeter-wave band radio subscriber loop," *IEICE Trans. on Communications*, vol. E76B, pp. 1128-1135, 1993.
[4] D. M. Cutrer, J. B. Georges, T. H. Le, and K. Y. Lau, "Dynamic range requirements for optical transmitters in fiber-fed microcellular networks," *IEEE Photonics Technol. Lett.*, vol. 7, pp. 564-566, 1995.
[5] H. M. Salgado and J. J. Oreilly, "Accurate performance modeling of subcarrier multiplexed fiber/radio systems - implications of laser nonlinear distortion and wide dynamic range," *IEEE Trans. on Communications*, vol. 44, pp. 988-996, 1996.
[6] J. C. Fan, C. L. Lu, and L. G. Kazovsky, "Dynamic range requirements for microcellular personal communication systems using analog fiber-optic links," *IEEE Trans. on Micro. Theory & Techn.*, vol. 45, pp. 1390-1397, 1997.
[7] R. Sabella, "Performance evaluation of optical fiber link in micocellular mobile communications systems," *Micro. and Optical Technol. Lett.*, vol. 18, pp. 233-239, 1998.
[8] B. J. Koshy and P. M. Shankar, "Effect of non-linearities on outage performance of fiber-fed microcellular networks for GMSK systems," *IEEE Trans. on Vehicular Technol.*, vol. 47, pp. 819-829, 1998.
[9] R. Sabella, "Performance analysis of wireless broadband systems employing optical fiber links," *IEEE Trans. on Commun.*, vol. 47, pp. 715-721, 1999.
[10] D. Bertsekas and R. Gallager, *Data Networks*. Upper Saddle River: Prentice-Hall Inc., 1992.
[11] V. Erceg, "Urban/suburban out-of-sight propagation modeling," *IEEE Commun. Mag.*, vol. 30, pp. 56-61, 1992.
[12] W. T. Webb, "Sizing up the microcell for mobile radio communications," *Electronics & Communication Engineering Journal*, vol. 5, pp. 133-140, 1993.
[13] ITU-R, "Propagation data and prediction models for the planning of indoor radio communication systems and radio local area networks in the frequency range 900 MHz to 100 GHz," *Recommendation ITU-R P.1238*, Geneva, Switzerland, 1997.
[14] ITU-R, "Propagation data and prediction models for the planning of outdoor radio communication systems and radio local area networks in the frequency range 100 MHz to 300 GHz," *Working Party 3K, ITU-R 211/3*, Geneva, Switzerland, 1998.
[15] CCIR, "Attenuation by precipitation and other atmospheric particles," CCIR Study Group Document 721, 1978.
[16] R. L. V. Tuyl, "Opportunities and challenges for 60 GHz wireless communication," *presented at Asia Pacific Micro. Conf.*, Yokohama, Japan, 1998.

Broadband Access, WDM Metro and Network Management
D.W. Faulkner and A.L. Harmer (Eds.)
IOS Press, 2000 © AKM

Bidirectional Single Fibre Links for Base Station Remote Antenna Feeding

Guido STEINER, Werner BAECHTOLD
Swiss Federal Institute of Technology ETH,
Laboratory for Electromagnetic Fields and Microwave Electronics,
Gloriastrasse 35, CH-8092 Zurich, Switzerland; steiner@ifh.ee.ethz.ch

Stephan HUNZIKER
Contraves Space AG, Dept. W-ES,
Schaffhauserstrasse 580, CH-8052 Zurich, Switzerland; czhs@ocag.ch

Abstract. In this contribution the feasibility and performance of low-cost bidirectional single fibre links for analogue RF fibre-optic (FO) antenna feeding are investigated. Wavelength division multiplexing (WDM) allows transmission of uplink (λ_{opt}=1310nm) and downlink (λ_{opt}=1550nm) signals over one single standard fibre. Designs using low cost multiple quantum well Fabry-Pérot laser diodes (MQW FP LDs) in bidirectional transceiver modules or in separate receiver/transmitter modules are compared. Performance, requirements and costs for FO and conventional coax solutions are discussed.

1. Introduction

Low-cost FO links for remote antenna feeding have become an interesting alternative to coax or high dynamic range DFB solutions in the last years [1-4]. It has been shown that low cost MQW FP lasers are well suited as high dynamic range transmitters [1,4,5]. In the case of remote base station antennas the link span is usually in the range of some 100m up to about 2 or 3 km. Therefore fibre dispersion is no issue for single mode standard fibres at 1310nm laser wavelength even if multi-longitudinal mode laser diodes are used.
Up to now all concepts relied on two separate fibres - one for the uplink and one for the downlink. At fibre spans approaching the km range fibre costs become significant. In many cases it is highly desirable to use as few existing fibres as possible and it would often be advantageous to save fibres at the price of slightly increased transmitter and receiver complexity.

2. Bidirectional Modules for Remote Antenna Feeding Links

Usually cellular base station antennas are operated in a receiving and a transmitting mode. Therefore the feeding link has to be bidirectional. The separation between up- and downlink is done by means of an antenna-duplexer. Known FO remote antenna feeding link concepts apply space diversity (two fibres). We propose WDM (spectral diversity). The main advantage of this scheme is that only one instead of two fibres is required. Bidirectional modules already known from SDH/Sonet and hybrid fibre-coax return path applications are used as transceivers. A bidirectional module consists of a LD, a photo diode (PD) and a WDM beam splitter. Bidirectional modules are used as a pair where one module usually emits at 1310nm and the other module emits at 1550nm (there are bidirectional modules that emit at only one wavelength as well). For the separation and isolation of the two wavelengths a WDM beam splitter is used. Figure 1 shows a cross section of a bidirectional module. As shown in table 3 the dynamic range requirements for the downlink are less demanding than for the uplink. Even if a 1550nm link is used in combination with standard fibres dispersion will be low because of the short link span of less than 5 km.

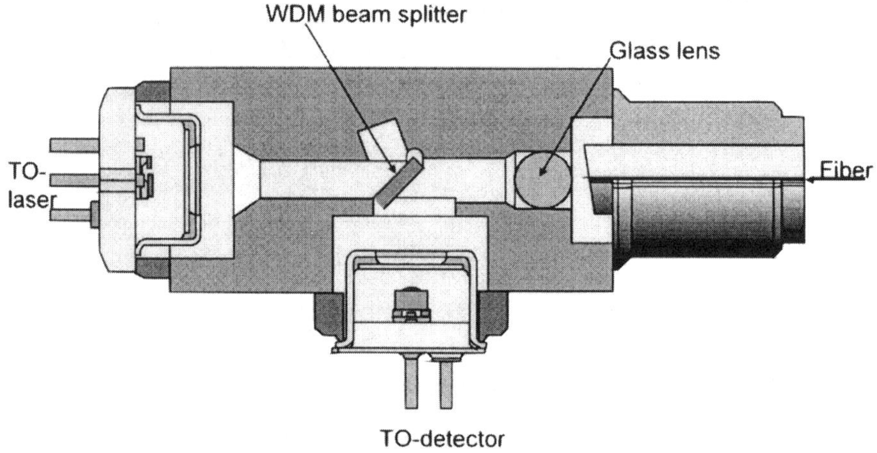

Figure 1: Cross section of a bidirectional module (© Infineon Technologies AG).

3. Measurements

For the measurements the bidirectional modules where mounted on the edge of a Duroid substrate. With this mounting it was possible to keep the highly inductive connecting pins very short. To achieve simple matching to 50Ω a 47Ω resistor was put in front of the LD assuming a LD-resistance of approximately 3Ω. Output matching consists of a 51Ω resistor in parallel to the PD. The following devices have been measured:

- Infineon SBM81314 (1310nm), Infineon SBH51414 (1550nm)
- MRV MRTXLSPS010 (1310nm), MRV MRTXLSPS5010 (1550nm)
- Hitachi HL1326CF (1310nm), Photon PT3543-12-3-1FC (1550nm), Fujitsu FID3Z1KX

The MRV devices are equipped with SC/PC-connectors. All other devices have FC/APC-connectors. The Hitachi and the Photon devices are lasers in a coaxial package. In combination with the Fujitsu PD they form a simple unidirectional optical link. One important parameter for the comparison of the LDs is the optical output power as a function of the applied DC current. This measurement is shown in figure 2. If the optical power at the receiver is in the order of at least 1mW receiver noise is not significant [4] and noise is mainly caused by the LD-RIN (relative intensity noise) and shot noise.

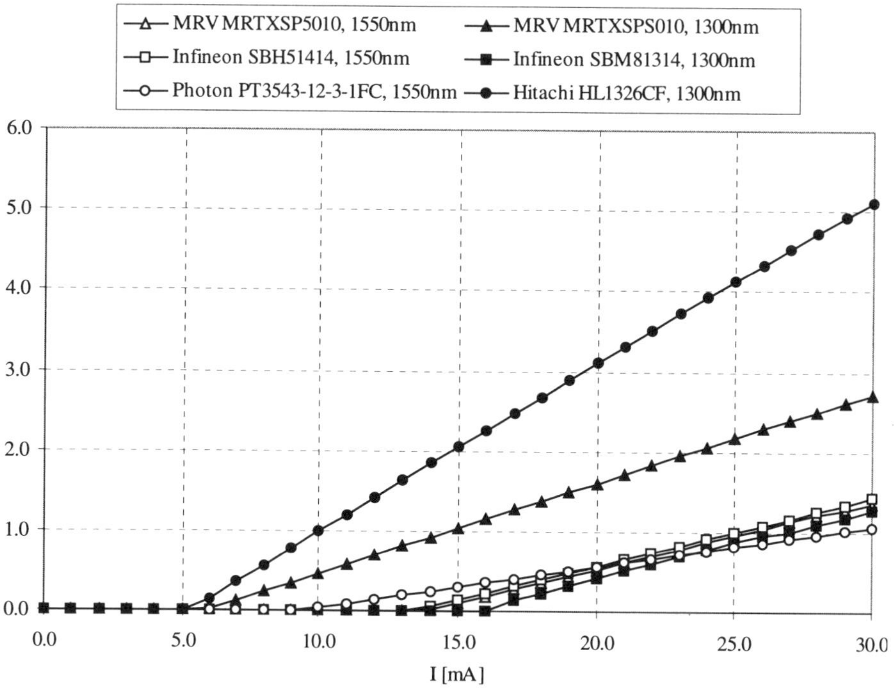

Figure 2: Optical output power of the tested laser diodes as a function of applied DC bias current.

Another important parameter of an optical link is the link gain. To measure the link gain (negative link loss) up to 3GHz a network analyser was used. Figure 3 shows the S-parameter S_{21} of a bidirectional module. In this case the MRV MRTXSPS010 module with the 1310nm LD (biased at 10mA) was used as transmitter. The measurement results of all bidirectional modules are summarised in Table 1.

Transmitter	*Infineon SBH51414 (1550nm)*	*Infineon SBM81314 (1310nm)*	*MRV MRTXLSPS5010 (1550nm)*	*MRV MRTXLSPS010 (1310nm)*
Receiver	*Infineon SBM81314 (1310nm)*	*Infineon SBH51414 (1550nm)*	*MRV MRTXLSPS010 (1310nm)*	*MRV MRTXLSPS5010 (1550nm)*
LD current [mA]	S_{21} 900-1800 MHz [dB]	S_{21} 900-1800 MHz [dB]	S_{21} 900-1800 MHz [dB]	S_{21} 900-1800 MHz [dB]
10	-	-	-	-24.4 ± 0.3
15	-31.8 ± 1.5	-	-	-25.7 ± 0.3
20	-29.0 ± 1.3	-29.5 ± 1.2	-31.1 ± 0.9	-26.3 ± 0.6
25	-30.1 ± 0.7	-30.8 ± 0.6	-32.1 ± 1.1	-26.5 ± 0.7
30	-30.7 ± 0.4	-31.4 ± 0.3	-32.1 ± 1.2	-27.0 ± 0.8

Table 1: S_{21} (link gain) of bidirectional modules between 900 and 1800MHz as a function of LD bias current.

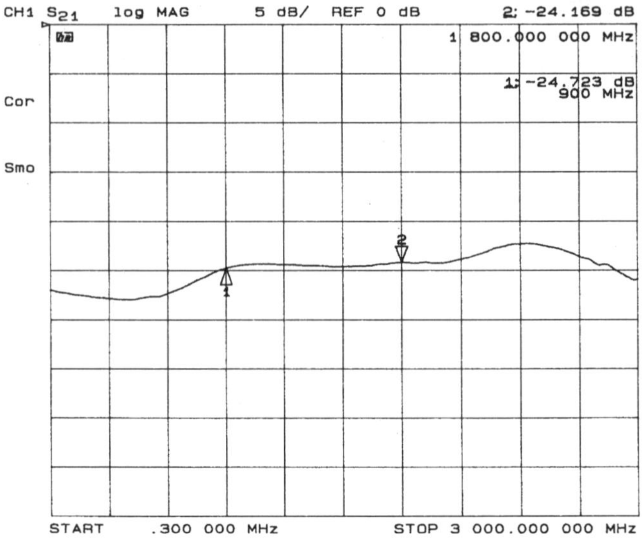

Figure 3: Link gain up to 3GHz with MRV MRTXLSP010 as transmitter (biased at 10mA) and MRV MRTXLSP5010 as receiver.

To characterise the link dynamic range the SFDR (spurious free dynamic range) can be calculated. The SFDR is the difference in dB between the power levels of the third order intermodulation IM_3 (assuming that there is only a third order nonlinearity) and the carrier frequency when the IM_3 power level equals the noise floor at a given noise bandwidth. It can be expressed by the link's third order input intercept point (IIP_3 [dBm]), the equivalent input noise (EIN [dBm/Hz]) and the noise bandwidth (NBW [Hz]) [4]:

$$SFDR = \frac{2}{3}\left[IIP_3 - EIN - 10\log(NBW)\right]$$

To measure the noise floor and the IM_3 a setup according to figure 4 has been chosen. If noise is measured the adjustable external attenuator and the internal attenuator of the spectrum analyser are set to 0dB. Using a setup according to figure 4 the noise generated by the optical link can not be measured directly. In order to quantify the noise floor of the optical link the laser is switched off. Then the noises of the amplifier and the spectrum analyser are measured. Switching the laser on increases the noise floor by the laser noise and therefore the difference between these two noise measurements is the noise generated by the laser.

To get the SFDR the input power is virtually decreased until the IM_3 level equals the noise floor. Decreasing the input power by 1dB decreases the IM_3 level by 3dB. If higher order intermodulation terms are present they will also give a contribution to the $2f_1$- f_2 / $2f_2$- f_1 IM products and therefore in this cases *the SFDR is not calculated correctly.* As can be seen from figure 5 - by checking the $3f_1$- $2f_2$ IM products - there is no IM_5 visible (at 1799.8 or 1800.3 MHz) which means that the IM_5 is at least about 15dB below the IM_3 level.

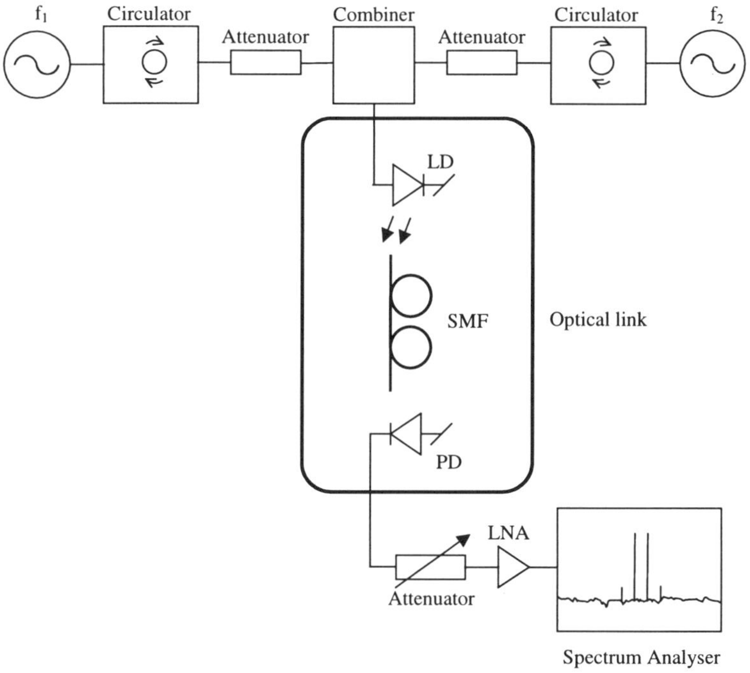

Figure 4: Setup for noise and third order intermodulation measurement.

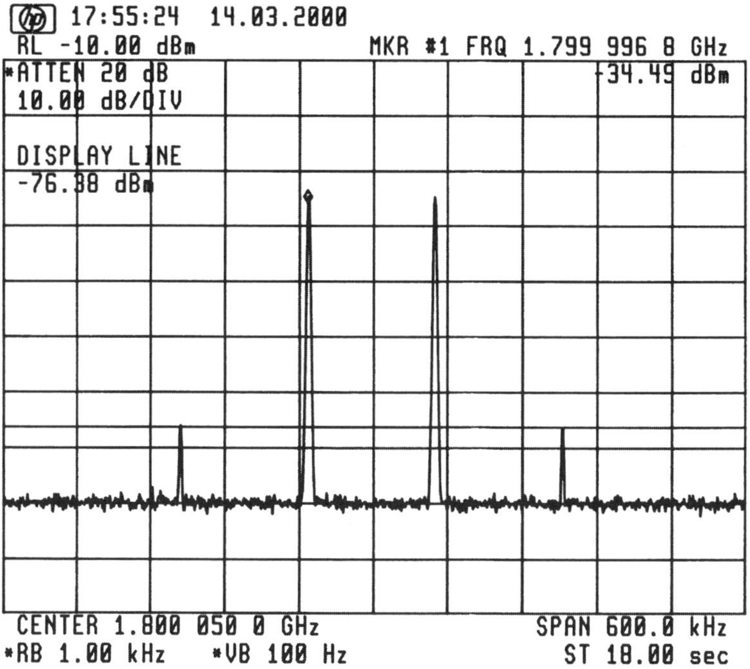

Figure 5: Intermodulation measurement of Infineon SBH51414 biased at 40mA at input frequencies of f_1 = 1800.0 and at f_2 = 1800.1MHz.

It is important to note that switching the laser in the receiving module on does not affect the SFDR of any of the tested transmitting modules. This means that the isolation of the two wavelengths is very good. Measured crosstalk is below 41dB@900MHz / 48dB@1800MHz (MRV) and 55/61dB (Infineon). This crosstalk has been identified as entirely electrical. The SFDR values at a noise bandwidth (NBW) of 200kHz of all the measured links at 900 and 1800MHz are summarised in table 2.

Device, bias current	*SFDR @ 900MHz [dB], NBW 200kHz*	*SFDR @ 1800MHz [dB], NBW 200kHz*
Infineon SBM81314 (1310nm), 40mA	74	69
MRV MRTXSPS010 (1310nm), 31mA	76	73
Hitachi HL1326CF (1310nm), 32mA	76	71
Infineon SBH51414 (1550nm), 40mA	67	64
MRV MRTXLSPS5010 (1550nm), 31mA	76	71
Photon PT3543-12-3-1FC (1550nm), 32mA	78	71

Table 2: SFDR (spurious free dynamic range) of the measured bidirectional/unidirectional modules at 900 and 1800MHz with a noise bandwidth of 200kHz.

The SFDR requirements for cellular remote antenna feeding links are approximately as listed in Table 3 [4].

	SFDR downlink [dB]	SFDR uplink [dB]
indoor	>46	≈ 50-60
outdoor	>46	≈ 65-70

Table 3: SFDR requirements for cellular remote antenna feeding at a NBW of 200kHz [4].

This means that WDM links with bidirectional WDM transceiver modules presented above can be used for indoor as well as outdoor remote antenna feeding in the 900MHz, 1.8GHz and 2GHz ranges. The 1550nm link which shows slightly worse performance than the 1310nm link is used as downlink (lower SFDR requirements) whereas the 1310nm link realises the uplink (higher SFDR requirements). At the upper link span limit of about 5km the downlink SFDR can be degraded by several dB due to dispersion. Comparison of tables 2 and 3 reveals that fairly enough SFDR-margin is left even in this case. It is well known that semiconductor LDs suffer from severe second order distortion. In case of operating a link at 900 and 1800MHz it must be guaranteed that the second harmonics of the 900MHz band do not affect the 1800MHz band. It would also be possible to downconvert the upper band in the range of 100-400MHz before transmission.
It has to be pointed out that all these results have been achieved with Fabry-Pérot multiple quantum well (MQW) laser diodes. No distributed feedback (DFB) lasers have been used for cost reasons. It is very important to mention that good performance with low cost laser diodes can usually only be achieved if angled connectors are used. Then expensive optical isolation is not necessary.

4. Comparison of FO and Coaxial Cable Link

Coaxial and FO links are now compared with respect to hardware costs (transmitters, receivers, repeaters, and cables). Coaxial copper cable links need repeater amplifiers. Moreover they need remote power supply. Particularly if many repeaters are required, this approach will cause reliability and maintenance problems. Figure 6 shows the relative costs for two different coaxial and four single mode FO bidirectional links (up- and downlink). All components are low cost as required for (micro) cellular repeaters. Installation costs are not considered here. For link spans above about 100m FO links are clearly more cost effective than coax links as already known from a previous study [1]. At first sight it seems that single fibre links with bidirectional transceiver modules are only cost effective compared to two fibre links for link lengths above 1km. But the difference between the two is quite small. Therefore it will be advantageous in many cases to use bidirectional transceiver modules to spare one dark fibre for slightly higher cost. If separate transmitters and receivers together with external WDM couplers are used, bidirectional transmission over one single fibre with even higher SFDR is possible. External WDM couplers are rather expensive. So the use of external couplers is only recommended for long outdoor links and exceptionally many carriers (SFDR>75dB) where fibres have to be saved. In FO links it is generally cost effective to use multi-fibre cable as duplex or even 6 or 12 fibre cables, because usually more than only one application is installed in a ducting or tube.

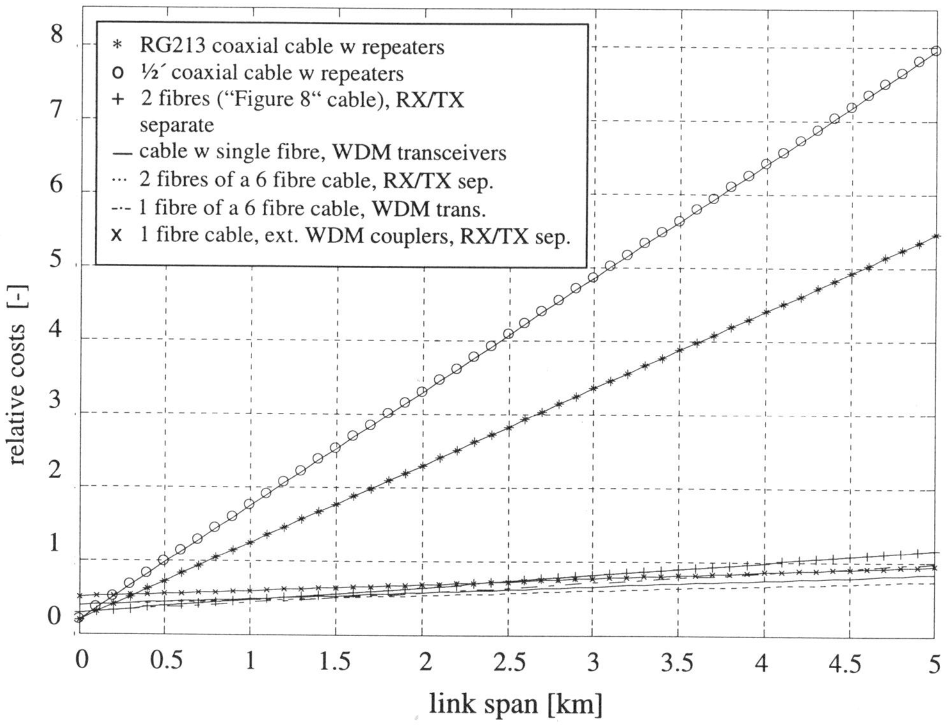

Figure 6: Comparison of relative hardware costs for FO and coaxial in-/outdoor remote antenna feeding links. The following link types are compared: Two coaxial cable links, FO links with one single fibre and WDM transceiver modules, FO links with one single fibre, separate RX and TX modules and external WDM couplers, two fibre links with separate RX and TX modules. Curves for single fibre, duplex and six fibre cables as well as RG213 and ½′ coax cables are shown. Installation costs are not considered.

The main cost driver considering hardware cost is the cable. Installation costs are roughly proportional to the cable cross section area. If they were considered too, optical fibre solutions would be even much more advantageous. Coax cables have higher cross sections and higher weight. Fibres can be installed very easily compared to coax cables (e.g. methods using compressed air as the Blown Fibre technique by British Telecom or Blolite by BICC). In distributed antenna systems it is from a system engineering point of view obviously the best way to combine optical fibre and coax technologies to achieve best performance for lowest cost. Distant (>100m) hubs or single distant antennas can be fed by fibre links perhaps combined with optical power splitters. Low cost coaxial cables can then be used for fine distribution or feeding of low distance antennas.

5. Conclusions

It has been shown that FO WDM links are cost effective solutions for remote antenna feeding in GSM900/1800 networks particularly if fibres have to be saved. They cover indoor as well as outdoor applications. Most applications can be covered by bidirectional transceiver modules. Long links for coverage with a high number of carriers can be realised using external WDM couplers and separate receivers and transmitters. For very short links <100m coaxial cables with diplexers may be used.

Acknowledgements

The authors would like to thank the Swiss KTI, Infineon Technologies AG, MRV and Huber+Suhner AG for their support.

References

1. S. Hunziker and W. Baechtold, "Cellular remote antenna feeding: optical fiber or coax cable?," *Electron. Lett.*, vol. 34, 1038-1040, 1997.

2. C.M. Gee, T.R. Chen, N. Bar-Chaim, and I. Urym, "Designing high dynamic range fiber optic links: A comparison between directly-modulated Fabry-Perot and distributed-feedback laser diodes," *Microw. J.* 36, 213-221, 1993.

3. Y. Gao, "Linear fiber-optical transmission systems for cellular telephone signals with wide dynamic range," *Proc. EFOC&N '95*, Brighton, UK, vol. 3, 128-131, 1995.

4. S. Hunziker, "Low cost fibre optic links for cellular remote antenna feeding," to appear in *Radio Over Fibre, ed. by H. Al-Raweshidy and S. Komaki*, Artech House, 2000.

5. C. Cox III, E. Ackermann, R. Helkey, and G.E. Betts "Techniques and Performance of Intensity-Modulation Direct-Detection Analog Optical Links, "; *Trans. MTT-S '97*, vol 45, 1375-1383, 1997

6. Infineon Optoelectronics Data Book, BIDI™ Transceivers.

Broadband Access, WDM Metro and Network Management
D.W. Faulkner and A.L. Harmer (Eds.)
IOS Press, 2000 © AKM

A Way to Extract Information from CENELEC EN50083-10 Multiple Interference Measurements

V. MOEYAERT*, L. ROBETTE, J.C. FROIDURE, J. HANOTEAU*, M. BLONDEL*

FACULTE POLYTECHNIQUE DE MONS,
Service d'Electromagnétisme et de Télécommunications - MULTITEL Telecom*
Boulevard Dolez, 31, B-7000 Mons, BELGIUM

Abstract. In its document EN50083-10, CENELEC has defined a way to realize multiple interference measurements, based on a regular spectrum analyzer trace acquisition and post-processing. It also defines a way to process the data in order to obtain a practical parameter for an interactive service implementation: the 'channel availability'. Nevertheless, for monitoring or pre-implementation purposes, other information can be extracted from systematic spectrum analyzer measurements. Measurements of multiple interference on a real network were performed following CENELEC EN50083-10. This paper will present the way we handle the collected data in order to highlight noise level variation, periodicity, ... by coloured level, statistical centile, minimum, maximum and average analysis in the frequency domain.

1 Introduction

Due to the high demand of new services emerging from residential subscribers (Internet access, second telephone line, ...) and small industries (LAN to LAN interconnection, ...), classical broadcast tree-and-branch CATV (Community Antenna TeleVision) networks are evolving towards an interactive HFC (Hybrid Fibre Coax) topology, able to support video, voice and data [1].

Since reliable communications must be established on a shared medium, the upstream path quality becomes a very important issue. Indeed, there are many factors that make cable system performance in the upstream direction lower relative to the downstream direction. One factor is the sub-split band's choice (5 to maximum 65 MHz [2]) where man-made noise is prevalent, particularly in the 5 to 15 MHz portion of the spectrum. An other impairment is the noise funnelling phenomenon: it is the summation of all unwanted noises (Gaussian noise, ingress noise and impulse noise) from both the subscribers and the cable plant in the return transmission system. Gaussian noise is identified as filtered white noise coming from active equipments within the network, while ingress and impulse noises refer to signals picked up by the network.

CENELEC [2] defines ingress noise as a "*... noise permanently present but slowly varying in its intensity as a function of time*", e.g. short wave broadcasts, amateur radio, citizen bands, ... Ingress noise occurs when shielding is inadequate or has been broken for some reason. Over-the-air RF signals enter the system and interfere with reverse signals. Likewise, impulse noise is defined as a "*noise caused by electromagnetic interference into cable networks characterized by pulses with a duration typically under 10 μs*". Common sources are generally man-made (vehicle ignition, neon signs, power line switching transients, electric motors, electronic switches, household appliances, ...) but also natural (static from lightning, galactic noise, ...).

The work presented in this paper focuses on multiple interference measurement which is defined as a mix of ingress noise and intermodulation distortion products in the case of a cable TV network's return path.

In [2], CENELEC has defined a way to realize multiple interference measurement, based on regular spectrum analyzer traces acquisition and post-processing. It also defines a way to process the data in order to obtain a practical parameter for an interactive service implementation : the "channel availability".

Nevertheless, for monitoring purpose, information other than 'channel availability' can be extracted from systematic spectrum analyzer measurements. An approach of the data handling process, best know under 'DISP (Discrete Interfering signal Probability) Matrix Method' has been described in [3]. It gives so much information that interpreting all results is time consuming. This paper will focus on some important graphs to generate in order to obtain a reasonable statistical knowledge of a return path's multiple interference behaviour.

2 Description of the field trial site and the measurement setup

Measurements on an interactive field trial were performed in Namur (Belgium) during year 1999 (weeks 15, 30 & 31).

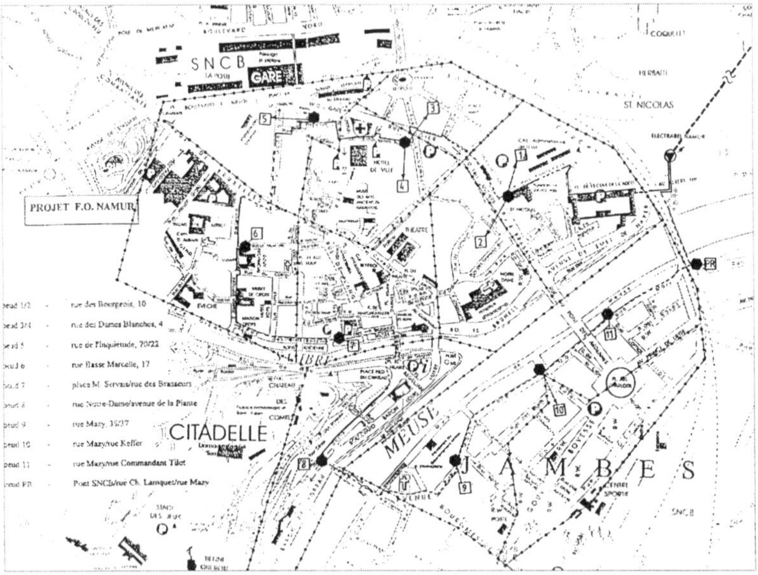

Figure 1: HFC map of Namur town centre

The INATEL CATV network of Namur town centre, operated by ELECTRABEL and its subcontractor CODITEL, has been completely renewed since 1998. It has evolved from a classical coaxial tree-and-branch structure towards a two-way HFC topology. The network was divided in 9 nodes, each having an ONU (Optical Network Unit) being linked to the headend by the way of two optical fibres. Node sizes vary from 500 to 1200 subscribers divided in 4 coaxial links bearing maximum 4 trunk amplifiers and one distribution amplifier in cascade. The biggest node (node 1&2) includes 32 amplifiers. All coaxial cables (except drop cables), amplifiers and taps have been replaced by new equipment for the trial purpose. Available bandwidths are respectively 5 to 25 MHz and 47 to 600 MHz for upstream and downstream transmissions. The return path of Namur's network does not hold any noise filtering equipment, neither at subscribers premises, nor at taps locations.

At the headend location, the output of all optical receivers was gathered to enter the same cable modem headend interface. Measurements on each receiver were performed at a test point (cf. Figure 2) while a range of 75Ω terminated couplers authorized measurements on partial sum of signals (node1&2 + node5; node3&4 + node6; node7 + node8; node9 + node10 + node11) or on the whole sum (representing about 7000 connected households).

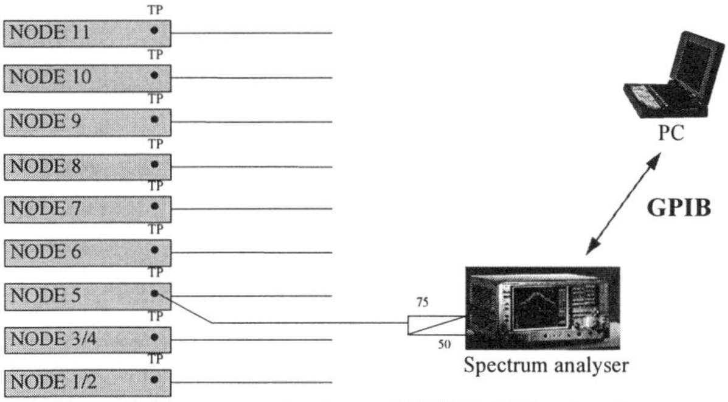

Figure 2: Measurement setup on INATEL HFC network.

The measurement setup is based on a Rohde & Schwarz spectrum analyzer having a GPIB interface connected to a PC. Data were captured following [2] which stipulates that during almost 24 hours, the interference spectrum must be stored with a alternation of 'slow' traces (resolution bandwidth = 3 kHz; video bandwidth = 100 Hz; every hour of the day) and of 'rapid' traces (resolution bandwidth = 30 kHz; video bandwidth = 10 kHz; every 10 seconds). Measurements have to be made in a network which has all the forward channels in operation and no signals on the return path. In the frame of our tests, 'slow' traces were captured as mentioned by the standard but 'rapid' traces were acquired every minute with a 3 kHz video bandwidth. The granularity of the frequency scale between 5 MHz to 35 MHz was 60 kHz (500 points). Following that measurement rhythm, a 1.2 Mbytes text file was generated about every four hours. Figure 3 shows the program interface for data acquisition.

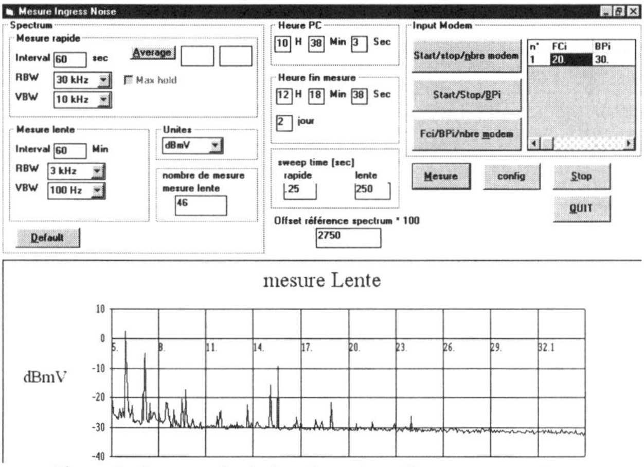

Figure 3: Command window for automatic measurements.

3 Data handling

A Visual Basic program was developed to perform post-processing on each file and to automatically generate graphs in the EXCEL environment using personalized macro. Useful information and presentation of results are, e.g.:

3.1 Spectrograms

This consists of a 2 dimensional graph (x-axis = frequency and y-axis = time) presenting the noise level as a colour code (cf. Figure 4). This way of presenting results may give a straight idea of which frequencies of the return path noise level are the most perturbed and/or when the highest peak occurred. A good choice of colour, like e.g. temperature maps (blue = cold & red = hot), may give this information at a glance!

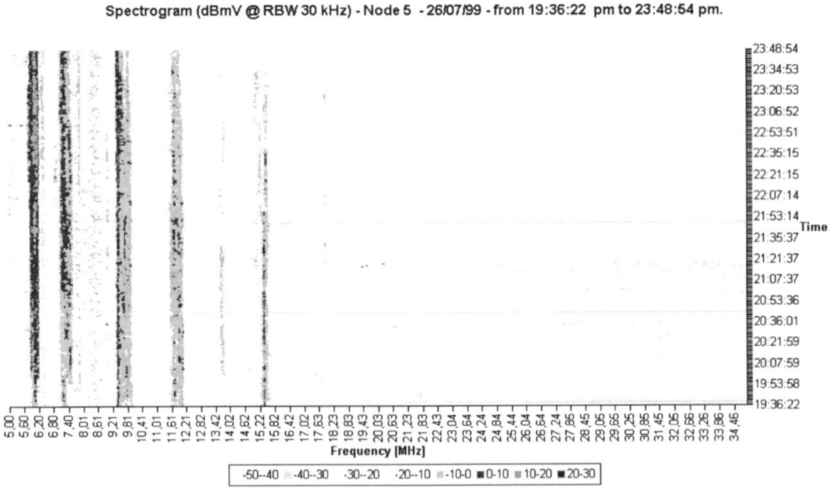

Figure 4: Spectrogram : Lower frequency bands are perturbed, and the 6 MHz band is more impaired. One can see "slow" traces. There are no out of band perturbations under 35 MHz.

3.2 Average, Minimum and Maximum levels in the spectrum analyzer resolution bandwidth vs. Frequency

The knowledge of those 3 parameters gives an idea of the level limits of the multiple interference records during the elapsed time. Figure 5 shows those parameters, as well as the difference between maximum and minimum values. It is noticeable that this last information is about 20 dB and is reproducible whatever the frequency, the node of the HFC network and the day of measurement.

It is also possible to focus the analysis only on the foreseen useful bandwidth (taking into account some necessary spectrum analyzer bandwidth correction factors) in order to be sure of the quality of the upstream frequency portion of interest.

3.3 Centile analysis

The concept of centile analysis is introduced to give a finer analysis of the permanency of measured multiple interference levels. As an example, a 25% centile (= first quartile) analysis of a set of measured spectrum traces gives a calculated level (in log values) that is exceeded during 75% (100 - 25) of the elapsed time. Figure 6 shows some centile analysis: 1% (first centile), 10% (first decile), 25% (first quartile), 75% (third quartile),

90% (last decile), 99% (last centile), as well as the minimum, the maximum and the average.

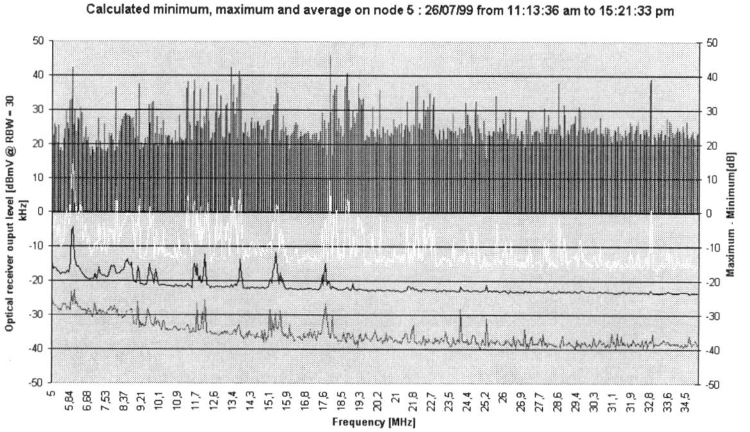

Figure 5: Maximum, Minimum and Average : the minimum value gives an underestimated value of the background noise. The average value gives an idea of the impact (the permanency) of all peaks of the maximum value.

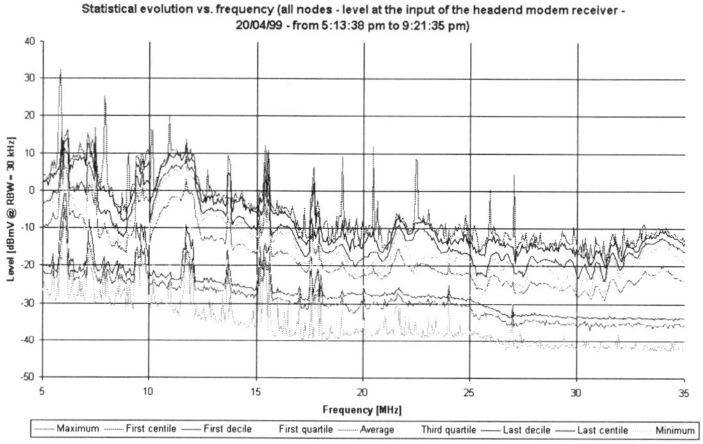

Figure 6: First centile levels give a better idea of the background level than the minimum. The end of the return path filter bandwidth is obvious with low centile analysis. It is clear that the majority of peaks in the lower band are permanently present with a varying level. The comparison between the last centile and the maximum gives and idea on the brief appearance of some peaks (e.g. at 21 MHz).

3.4 Temporal occurrence and frequency occurrence of thresholds crossing

With the setup we used, each stored trace has 500 points (frequencies). It is important to know the temporal evolution of the number of frequencies showing a level higher than a given threshold. This gives an idea of which periods of the day are the most perturbed. It is also important to know the most perturbed frequencies by counting the number of time the value of a threshold was exceeded during a given measurement period. From that basis, correlation with temperature, power consumption, ... could be made. An analysis following different thresholds could also be realized. Figure 7 and Figure 8 are two expressions of the same measurement. Figure 7 shows a temporal occurrence of threshold crossing: one can see an obvious noise activity increasing at the beginning of the evening, and a noise decreasing during the night (at the maximum, 40% of frequencies are crossing the threshold).

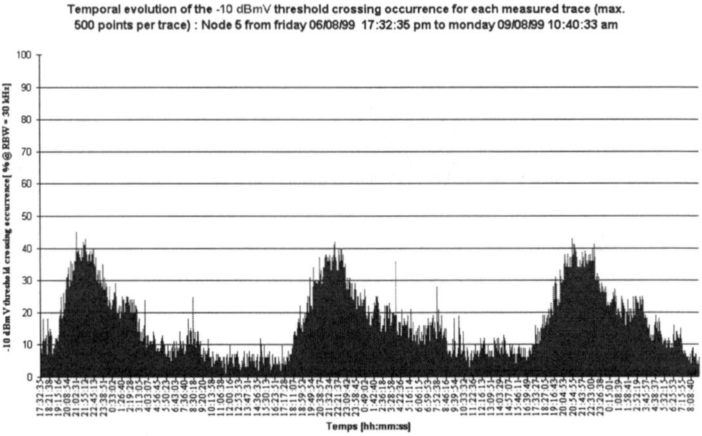

Figure 7: Temporal evolution of the –10 dBmV threshold crossing occurrence.

Figure 8 shows a frequency occurrence of threshold crossing : it is clear that the 6 MHz frequency was the most perturbed (≈ 65% of the time for that given threshold level), which is one of the AM radio band attributed in Belgium. Other bands are also perturbed. They are also part of the AM radio band.

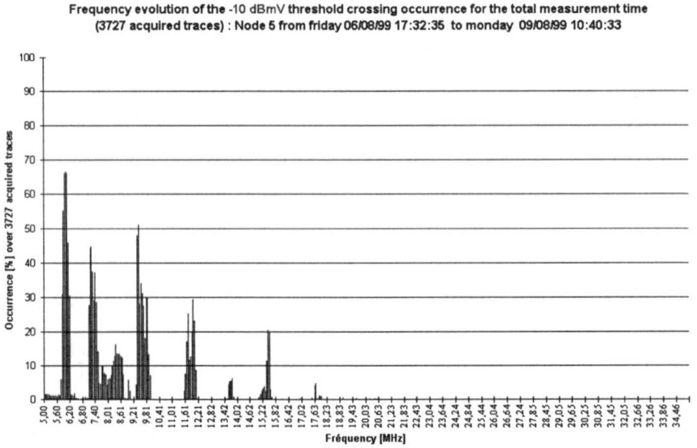

Figure 8: Frequency evolution of the – 10 dBmV threshold crossing occurrence.

4 A case study as an example: impact of the introduction of a directive attenuator on the noise level

Upstream information transmission is generally implemented by the way of burst modems. This implies that, whatever the time slot, each subscriber's emitted signal must reach the modem's receiver located at the headend with a same level. So, a regulation loop is installed in actual modem technology in order to stabilize the reception level of each subscriber at the headend independently of path loss variations. Nevertheless, that loop is unfair to subscribers located on low taps attenuation values in the sense that it reduces their signal level while their noise level remains statistically as high as other subscribers. Therefore, their signal to noise ratio deteriorates.

The introduction of a directive attenuator in the return path implies a signal level reduction at the headend side for all subscribers located before this attenuator. Therefore, due to the regulation loop activity, their emission level will be higher and so becomes their signal to noise ratio. There should also exist a reduction of all noise contributions coming from concerned subscribers.

In Namur's HFC, node 5 is located near the station and is very perturbed. Therefore, we used the measurement setup described by Figure 2 to evaluate the impact of the introduction of a 15 dB directive attenuator at the average distance between two return path amplifiers. Table 1 describes the principal test parameters we used. Figure 9 and Figure 10 show some results.

	July	August	Slow traces	Rapid traces
Beginning	26/07/99 11:13:38	06/08/99 17:32:35		
End	28/07/99 10:36:33	09/08/99 10:40:33		
Note	Without directive attenuator	With 15 dB directive attenuator		
Temperature	> 25°C	>25°C		
Resolution bandwidth			3 kHz	30 kHz
Video bandwidth			10 kHz	100 Hz
Sweep time			250 seconds	0.25 seconds
Rhythm			1 each hour	1 each minutes

Table 1 : Test parameters.

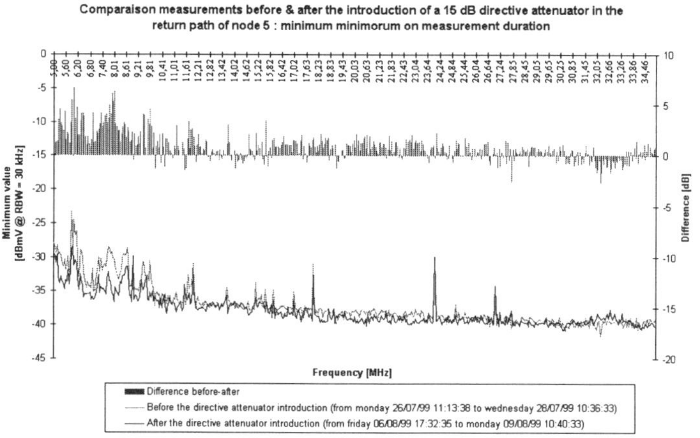

Figure 9: Comparison of minimum values between two network configurations : with and without a directive attenuator.

Figure 9 shows the comparison between the minimum values (be aware that all points may not belong to the same stored trace) during the measurement time. One can see a general improvement of the noise level but not so high that we could model in the case of Gaussian noise (about 6 dB in the present configuration). Figure 10 shows the same tendency : a lowering of the -10 dBmV threshold level crossing occurrence with an installed directive attenuator.

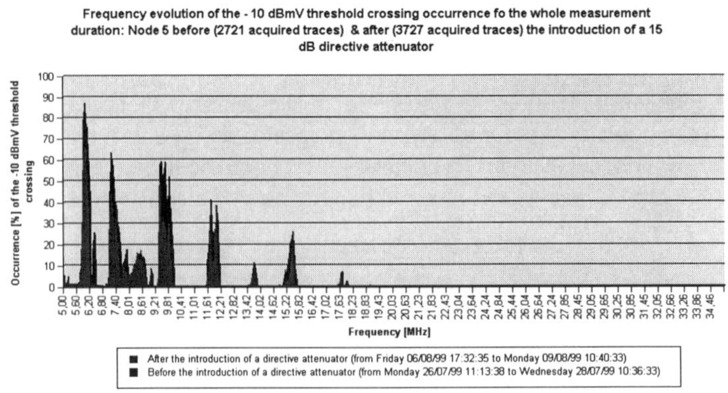

Figure 10: Comparison of the –10 dBmV frequency level crossing between two network configurations : with and without a directive attenuator.

5 Conclusion

The statistical level behaviour analysis of multiple interference in a HFC return path is a very difficult problem to solve. It has to be measured on unused frequency portions of the return path and during long periods of time for confidence in results.

We propose in this paper some tracks to graphically interpret CENELEC EN50083-10 multiple interference measurements results pointing out the most important evolutions to be analyzed.

We used them to evaluate the impact of a 15 dB directive attenuator in the return path of the fifth node on Namur's HFC field trial.

6 Acknowledgement

The authors would like to thank M. Roger VERMEREN from CODITEL (B) and M. Michel LEFORT from ELECTRABEL (B) for their support and help during the measurement process.

Multitel Telecom is a research unit co-funded by the Ministère de la Région Wallonne (B) and by the European Commission (EC).

7 References

[1] V. MOEYAERT et al., 'Physical layer characterization of hybrid fibre coax (HFC) networks', Annals of telecommunications, vol. 54, n° 5-6, May-June 1999

[2] CENELEC EN50083-10 : Cable networks for television signals, sound signals and interactive services – Part 10: System performance of return path.

[3] NCTA Recommended Practices for Measurements on Cable Television Systems: Supplement on Upstream Transport issues, 1997

Broadband Access, WDM Metro and Network Management
D.W. Faulkner and A.L. Harmer (Eds.)
IOS Press, 2000 © AKM

Approaches for the Digitized Fiber Feeder Transport in Hybrid Access Networks

H. Bürkle, H. Haisch, H. Krimmel, J. Schaepperle, M. Tomsu, J. Wolde
Alcatel Corporate Research Center
Stuttgart
Holderäckerstrasse 35, D-70499 Stuttgart
e-mail: hkrimmel@rcs.sel.de

1 Abstract

Approaches to transform the fiber optic transmission in hybrid access networks from the widely used analog format to the more robust digital binary format are discussed. The backgrounds here are the increasing interactive traffic, like telephony and internet access, and the need for enhanced network reliability and service availability. The introduction of binary signaling and the related measures to be taken to form a transparent high-capacity hybrid access network are described.

2 Introduction

Hybrid access networks use the combination of optical fiber and a 'last mile medium' like coax or wireless to provide broadband access for small businesses and residential users. In cable access, the individual traffic of the communication services (cable telephony, cable modem) asks for dedicated fiber optic feeder links to connect the elements of the segmented drop network (cells) to centralized installations at the interface of access and core telecommunication network.
The presented approach for future network upgrades towards a centralized architecture with a high grade of network and service availability capitalizes on the technological progress in the areas of high speed circuit integration, economic high-speed optical fiber transmission and mixed signal analog/digital ICs. The advantages gained through binary signaling on the feeders with respect to robustness, multiplexing, feeder span, network management and service availability is discussed.

3 Narrowcast feeders for interactive services

The provision of interactive services - especially high-quality communicative services such as Voice over IP (VoIP) telephony - involves upgrading the existing distribution networks to transform them into high quality communication networks. Cable operators offering these services are aiming to provide a service quality and network availability comparable to that of the public telecom network operators.

The extensive use of dedicated fiber links close to subscribers' homes helps to achieve this goal by fiber optic bypassing the electrical amplifier chains. The centralization of network equipment is an important measure to improve service availability.

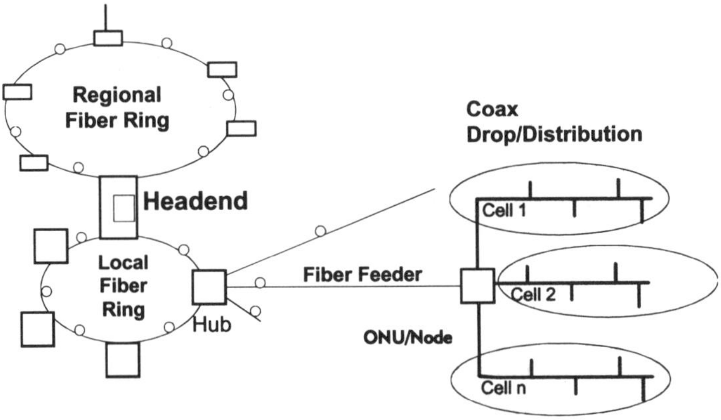

Figure 1 – Hybrid access network with segmented service area and dedicated feeder links

The typical structure of a modern ring protected hybrid fiber access network realising the centralisation paradigm is depicted in Figure 1 with a primary headend and several network hubs. The central headend is connected to the telecommunication network via the regional fiber ring. A local ring interconnects the headend and network hubs. The Optical Network Units (ONU) are the transition points from fiber to coax. To avoid extensive processing at the media interfaces between fiber and the "last mile" medium (coax or wireless), the analog subcarrier format has also been used on the fiber links. However, the analog format, which is adequate for transporting broadcast services, does not meet the diverse needs of communicative narrowcast traffic.

4 Rationale of digitization

The advantages of binary coded transmission are well known from the long-haul voice and data networks and from switching systems. Digitally and especially binary-coded signals are more robust than analog signals, resulting in higher transmission quality and longer link spans. Network monitoring and performance sensing are also significantly enhanced. These features are a prerequisite for a high level of network/service availability. This is achieved by centralizing complex network equipment and by deploying extensive network monitoring and protection on the digital transmission links to ensure continuous operation in the event of a singular failure.

In the past, those advantages could not be exploited in the access network layer, mainly for cost reasons. Now, however, the availability of very high speed integration technologies at lower prices has made digital signal processing and transport in the access network a very attractive proposition.

5 Hybrid access network digitization concepts

Since the last mile analog coax cable will stay in place, at least one digital-to-analog conversion (or vice versa) has to take place when a binary format is used at the headends and on the fiber links.
In a bidirectional multipoint network, it is necessary to differentiate between the downstream and upstream directions.

A point-to-multipoint scheme is used in the downstream direction and a fully integrated digital solution for the quadrature amplitude modulator can be expected to fit into the remote units.
Hence, there are two major candidates for the digital binary transmission of data streams between the headend and the ONUs:

- Waveform-sampled transmission of modulated carriers.
- Transport Stream (TS) transmission and remote modulation.

The basic principle of Nyquist sampling transport digitization with pulse code modulation is shown in Figure 2.

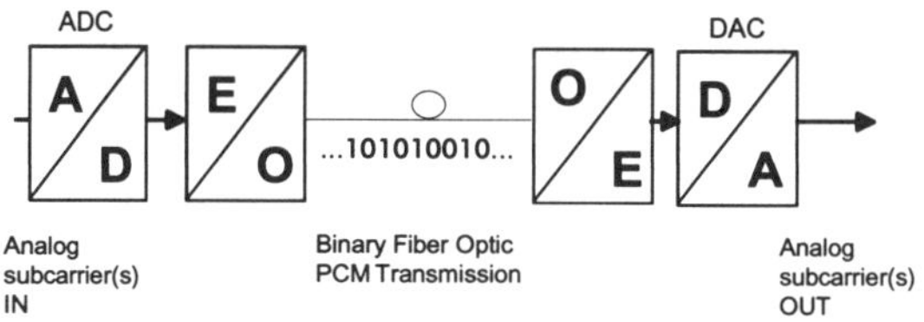

Figure 2 – Broadband signal-transparent digitized feeder with Nyquist-type analog/digital conversion

Complex multiple access schemes (Frequency Division Multiple Access, Time Division Multiple Access, Code Division Multiple Access) are used in the (multipoint-to-point) upstream direction. The power and space needs of the equipment required to demodulate those schemes prevent their termination and data recovery in the remote units. So, in the upstream direction, the best choice is clearly the waveform sampling technique in view of its signal type and multiple access independence.

High speed analog/digital converters (100 to 300 Msamples per second; 8 to 10 bit) became feasible with advanced Integrated Circuit (IC) technologies. The bitrates generated from the upstream band's conversion (5 ... 42 or 5 ... 65 MHz) can be transported on cost-effective gigabit per second fiber optic links. Integrated optical transceiver solutions are available on the datacom (Gigabit Ethernet) and telecom (STM-16) markets.
The downstream choice depends on the availability of integrated subcarrier modulators and frequency upconverters, which will use digital signal processing to allow full integration for output frequencies up to 450 MHz, mainly limited by the digital-to-analog converter. Further, the availability of dark fiber capacity will influence the choice of the downstream transport method.
The wave sampling solution is feasible today for narrowcast sub-bands up to about 100 MHz, which is quite a broad spectrum, since it has to be excluded from

the broadcast part. The 100 MHz narrowcast bandwidth, which is equivalent to a data rate of 500 Mbit/s, can be reused in each coax cell. The size of the coax cell is the scaling parameter then to adapt cell transmission capacity to demand.

6 Building networks

6.1 Streams and pipes

Narrowcast signal transport can be achieved in several ways, depending on whether TS or wave sampling signal transport is used. TS-type data streams are best conveyed by Synchronous Optical Network / Synchronous Digital Hierarchy (SONET/SDH) links, for which building blocks are widely available on the market.
The bitrates generated per single coax cell in the wave sampling case can be approximated from the required Nyquist-bound sampling rate and the quantization requirements determined by signal quality. Typical bitrates for narrowcast downstream transport to a coax cell are 155 Msamples/s $\times$ 8 bit = 1.25 Gbit/s for a 40 MHz sub-band.
Upstream transport consumes 100 Msamples/s $\times$ 8 bit = 800 Mbit/s for a 42 MHz upper band edge or 155 Msamples/s $\times$ 8 bit = 1.2 Gbit/s for a 65 MHz upstream band. A higher resolution may be allocated to allow for cell noise combining carrier-to-noise ratio loss.
Waveform sample streams have a substantially higher bitrate than TS streams and can benefit from a fully synchronous network (isochronous). In contrast to core networks, the access network is a clock island and can be organized hierarchically. This opens the possibility of building fully synchronous networks for the bitrate-hungry wave sample transport and to omit SONET/SDH framing in favor of simple network elements.

6.2 Fiber multiplexing schemes

A mix of Electrical Time Division Multiplexing (ETDM) and Wavelength Division Multiplexing (WDM) is proposed for the fiber multiplexing scheme. The synchronous ETDM approach (see Figures 3 and 4) requires considerably less circuitry for framing and multiplexing than do SONET/SDH networks.

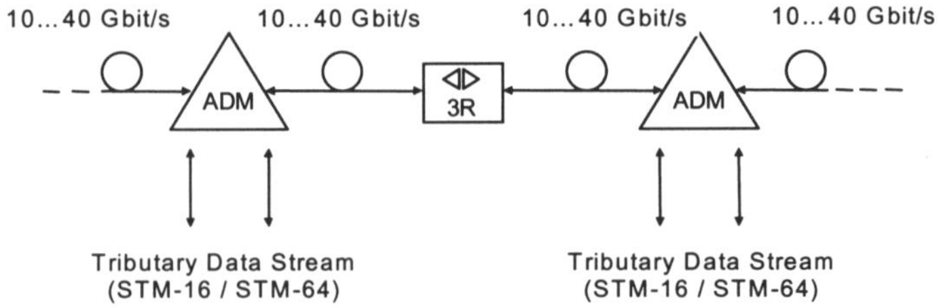

Figure 3 – Simple 10 to 40 Gbit/s ETDM system

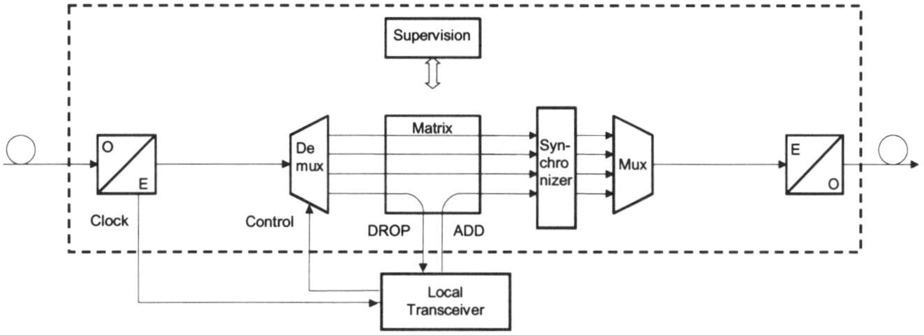

Figure 4 – Simple add-drop multiplexer and local connection

An overlaid WDM scheme with a moderate number of wavelengths (i.e. 4 to 10) on the fiber ring trades off an increase in fiber capacity against the low cost target by eliminating costly Dense WDM (DWDM) optical (de)multiplexers. On the stub fiber links between network hubs and ONUs (Figure 1), a coarse WDM scheme using only two or three wavelengths is foreseen for broadcast services and the narrowcast overlay.

6.3 Clock wander compensation

A fully synchronous fiber network has to cope with temperature-induced clock wander effects. A temperature interval of 100 K (—40°C ... +60°C) on a 100 km link at 2.5 Gbit/s gives rise to a loop bit shift of about 2·1250 bits. Analog to digital conversion tolerates slow phase shifts, but the clock wander effect comes into play when a clock loop is closed by downstream and upstream clock coupling. Data loss can be prevented by inserting special synchronizer modules (see Figure 4) at the multiplexing stations. The amount of memory required is negligible, and this simple device can be arranged as a low speed parallel structure. By this key element synchronism is maintained for the access network case without using SDH/SONET pointers.

7 System overview

An initial digitized narrowcast concept, which is the synthesis of the above elements, is depicted in Figure 5. It is based on analog/digital and digital/analog conversion at the headend and the ONUs, and the use of pulse code modulation on gigabit/s fiber feeders. Hence the modulators and demodulators, which are the most complex and are partly analog elements, can be centralized at the headend. Combining ETDM and WDM at SONET/SDH bitrates on the local fiber ring and simple synchronizer clock wander compensation modules at the add-drop multiplexers results in a cost-effective access network solution. Current building blocks are employed to form a centralized high-capacity, highly reliable solution for broadband cable-based connectivity.

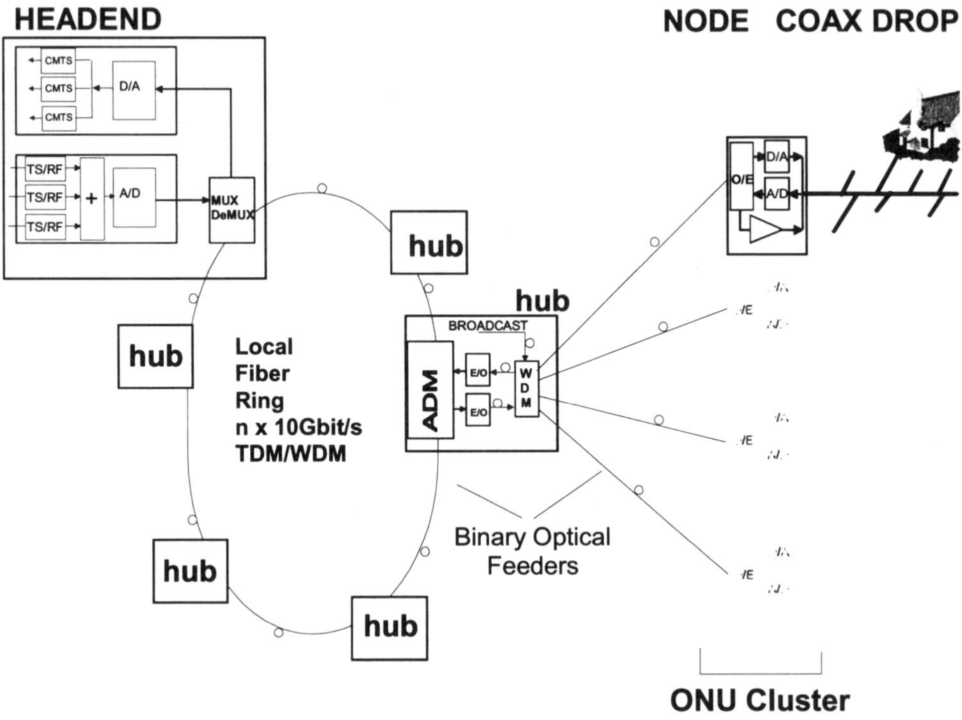

Figure 5 – Digitized hybrid access network system concept

8 Digital domain traffic concentrator

As a first example of the capabilities of broadband digital signal processing, a signal-transparent traffic concentrator has been realized which combines upstream signals from several coax cells. The adder-based building block operates at 155 Msamples/s entirely in the digital domain and can be easily integrated. It exhibits ideal signal transfer characteristics. Two non-overlapping input spectra from two coax cells and the digitally combined output spectrum are depicted in Figure 6.

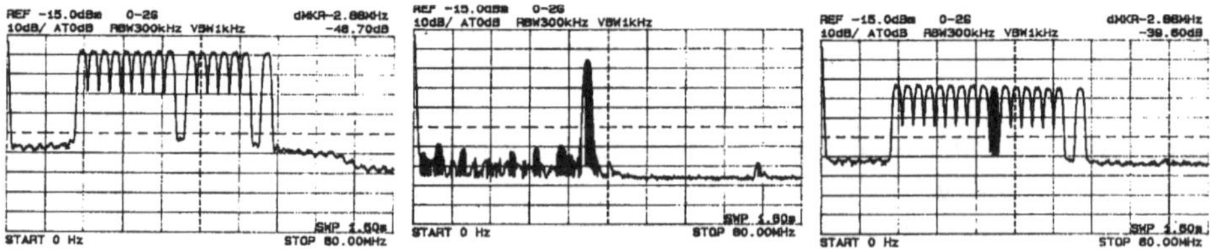

Figure 6 – Upstream frequency spectra at the digital combiner

9 Evolution aspects

The evolutionary path will take account of increased traffic and progress in integration technologies, especially high speed fiber optic transmission and mixed signal circuits. Higher volumes of interactive traffic will require the subcarrier domain processing functions to be fully integrated to achieve a competitive cost. The way forward is through digitization and the application of digital signal processing. The entirely digital pathes between the headend and the analog medium will be completed by purely digital headends. The emerging access networks will be increasingly centralized and will be enhanced by binary-operating and protected ring structures to ensure a high network availability.

10 Conclusion

The design of digitized fiber transport for the narrowcast signals in HFC access networks has been outlined, starting from Nyquist-type digitization and transport of the service carrying sub-bands. Using a simpler networking approach than in core networks, a way has been found to exploit novel high speed microelectronic devices in the high-end access network. Digitization is a catalyst for improving network quality, performance and reliability. These features are required to provide high-quality narrowband and broadband communication services in the shared network. The door has been opened to a fully digital signal processing chain, with purely digital headend processing in the future.

11 Acknowledgements

The work described in this paper was partially funded by the German Government (BMBF) (KomNet project "Giga-Zugang", MEDEA project ICFAN) and the local government of Baden-Wuerttemberg (Forschungs- und Medienverbund Suedwest, project FMS 2 III.4).

12 References

[1] H. Krimmel et al: "Transport meets access: a strategy for the merger of transport and access networks", *Proceedings of ISS '97*, Toronto, 1997

[2] H. Krimmel et al: "Integration of HFC access and the SDH/SONET core telecommunication network for multimedia services", *Proceedings of ISSLS '98*, Venice, 1998.

Broadband Access, WDM Metro and Network Management
D.W. Faulkner and A.L. Harmer (Eds.)
IOS Press, 2000 © AKM

A Detailed Study of the Suitability of Different Access Technologies for VoIP Delivery

S. Vanhastel, B. Duysburgh, F. De Turck,
C. Petrisor, P. Vandaele, D. Dewinne, P. Demeester
Department of Information Technology (INTEC - IMEC)
Ghent University
St.-Pietersnieuwstraat 41, B-9000 Gent, Belgium
{Stefaan.Vanhastel,Piet.Demeester}@intec.rug.ac.be

Abstract

To verify the suitability of different access technologies for Voice over IP transport, an extensive series of measurements were conducted at the Ghent University's Atlantis test network. Access network technologies were evaluated on criteria of delay, delay jitter and throughput. Furthermore, voice fragments were transmitted over these access networks using different codecs and the resulting quality was measured. Tested access technologies include traditional LAN (10Base-T and 100Base-T Ethernet), Wireless LAN (Orinoco), ADSL, ISDN and analogue modems. This paper presents the significant results that were obtained from these measurements.

1 Introduction

By definition, all users are connected to the Internet via an access network, be it a traditional analogue modem, ISDN or xDSL modem, or a Fast Ethernet LAN (the scope of this paper does not encompass PSTN/IP gateways). The quality of a VoIP connection over the Internet will thus be influenced by the limitations imposed by the access network. Therefore, it is necessary to be aware of the performance restrictions associated with the different access network technologies. In a first series of results, the basic performance characteristics of the different access networks are discussed. The results were obtained with an advanced distributed network monitoring tool [1], which uses trace correlation to calculate packet loss, delay and jitter.

Furthermore, users will typically engage in other activities while making a VoIP call, for example looking up facts on the web or sending E-mail. Indeed, many of todays VoIP applications ship complete with Whiteboard or Chat applications. So in a typical VoIP session, the voice traffic will have to share the access link with data traffic. So it is also necessary to look at the performance of access network links in the presence of background traffic. In a second part of the paper, the voice-specific measurement results are presented. For these experiments, actual VoIP traffic generated by different codecs was used. The quality of the voice signal (after transmission through the access networks) is expressed as a MOS (Mean Opinion Score) value.

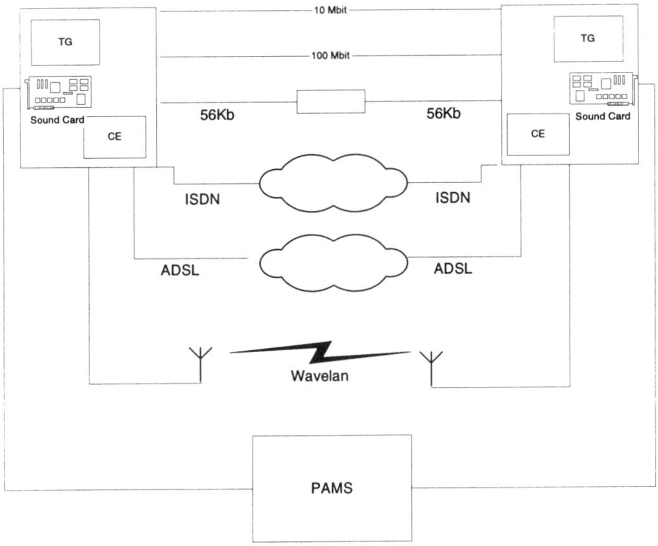

Figure 1: *Test Network Setup.*

This paper is structured as follows: Section 2 explains the setup of the test network used to perform the measurements, Section 3 discusses the access network performance measurement results and Section 4 presents the voice codec specific measurement results. As a conclusion, access technology / codec pairs that perform particularly good or bad will be examined and an explanation for their behavior will be presented.

2 Setup

For the measurements, the following generic setup (displayed in Figure 1) is used: a reference voice fragment, generated by PAMS equipment, is fed into a standard soundcard (Creative Labs Soundblaster 128 PCI). This card is mounted in a standard Pentium III 500MHz PC, running Linux (2.2.14 kernel). The fragment is encoded by the Codec Engine (CE), which offers a choice of codecs. After encoding, the data is packetized and sent over an access network. At the receivers end, the process is repeated in reverse order. A Traffic Generator (TG) is used to generate background traffic.

PAMS Equipment The PAMS equipment (Perceptual Analysis and Measurement System, [2][3]) generates reference voice fragments and is able to compare two audio signals (for example the original fragment and the same fragment after encoding/transmission/decoding). PAMS gives an estimation for the MOS-value (Mean Opinion Score), based on techniques described in [4].

Codec Engine The Codec Engine encodes or decodes the audio signal with a specific codec. A complete list of all tested codecs is presented in Table 1. The Codec Engine reads a codec-specific number of samples (equivalent to one frame) from the sound card and runs it through the codec. When enough encoded frames are available to fill a UDP datagram (the length of which can be specified), the packets are sent over the appropriate access network.

Codec	Bitrate	Framesize	Samples/Frame	Bytes/Frame	Ref.
16bit PCM	128 kbit/s	0.125 ms	1	16	
g711 μ-Law	64 kbit/s	0.125 ms	1	8	[5]
g711 A-Law	64 kbit/s	0.125 ms	1	8	[5]
g726-7 ADPCM 5bit	40 kbit/s	0.125 ms	1	5	[6][7]
g726-7 ADPCM 4bit	32 kbit/s	0.125 ms	1	4	[6][7]
g726-7 ADPCM 3bit	24 kbit/s	0.125 ms	1	3	[6][7]
g729	8 kbit/s	10 ms	80	80	[10]
g723.1	6.3 kbit/s	30 ms	240	192	[8]
g723.1	5.3 kbit/s	30 ms	240	160	[8]
g728	16 kbit/s	0.625 ms	5	10	[9]
gsm-FR	13 kbit/s	20 ms	160	264	[11]

Table 1: *Codec overview*

Access Networks The PC running the Codec Engine also functions as a gateway to different access networks. Table 2 lists the access networks that were tested at the Ghent University's Atlantis testlab. Both the Ethernet and Fast Ethernet NICs were connected back-to-back using a crossed 12ft UTP Cat 5 cable. The two Orinoco adapters (previously known as the Wavelan line of products) were 1ft away from the basestation, the 56k6 modems were connected to a Belgacom Forum R106 PABX and the ISDN modems were both connected to Belgacom's PSTN. The ADSL modems were connected to an ASAM, which in turn was connected to a DANA and a Cisco 7507.

Access Technology	Equipment	Interface
Ethernet (10base-T)	D-Link DFE-530TX	PCI
Fast Ethernet (100base-T)	D-Link DFE-530TX	PCI
ADSL	1000 ADSL	Via Eth. NIC
Wireless (IEEE802.11)	Lucent Orinoco	ISA
Analogue Modem 56k6	Acer Modem 56000	Serial
ISDN	ZyXEL Omni TA128	Serial

Table 2: *Access Technology overview*

Traffic Generator The Traffic Generator is capable of generating various types of streams, including TCP/IP, UDP/IP and RAW IP traffic. For each stream, a number of characteristics can be specified, defining the traffic profile. The parameters burstlength, packetlength and inter arrival time can be characterised by defining their probability distribution function. In this way, a broad range of traffic streams can be defined, corresponding to data, voice or multimedia streams. The traffic generator is embedded in the gateway PCs running the codec engines, in order to get as close as possible to the prototype of a home user surfing or E-mailing while making a VoIP call (i.e. data and voice source on the same machine).

3 Access Network Performance

In a first series of tests, the average delay in function of the packet length is measured per access network technology. From this, the average modem delay is deduced. Next, the influence of background traffic on delay and jitter is measured.

3.1 Delay

An indication of the typical RTT experienced on different access technologies is given in Figure 2. Although network behavior is considered not to be symmetrical [13, 14], division of the RTT by two should give a good idea of the one-way delay.

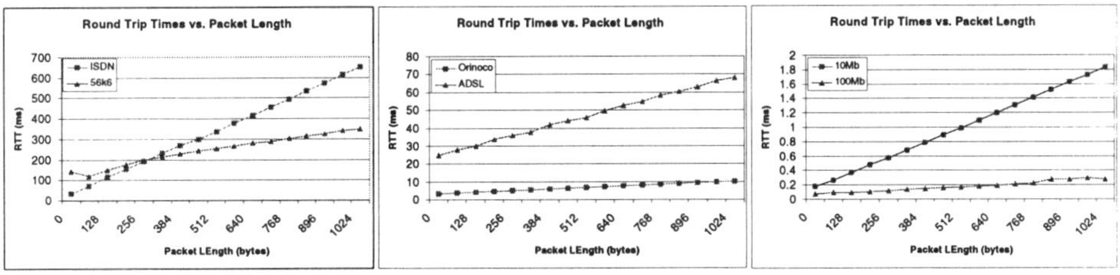

Figure 2: *Average delay in function of packet length.*

Figure 3 shows the evolution of the round trip time with the background load on a link. The background traffic consisted of a constant stream of TCP datagrams (size 500 bytes). The rate was varied to generate 10% to 90% load (relative to the linkspeed of the access network). The RTT was measured with a reference stream, shaped to emulate a 2-minute 8Kbit/s voice connection (one 140-byte packet, headers inclusive, every 100ms). The RTT values are the averages of 1000 measurements. Obviously, especially low-bitrate network technologies such as ISDN and 56k6 suffer from increased delay.

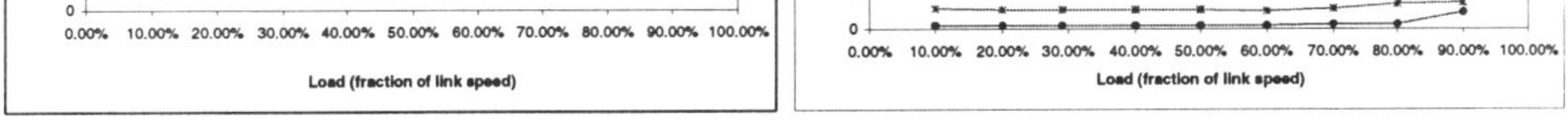

Figure 3: *Round trip time in function of background traffic*

3.2 Modem Delay

Figure 2 shows the average delays for different packet lengths. The delay is measured as the Round Trip Time (RTT) on a specific link. Values for each packet length are the averages of 50 measurements.

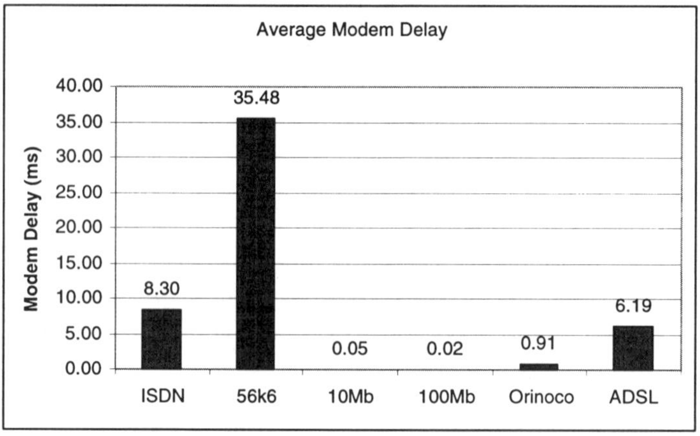

Figure 4: *Average modem melays.*

Figure 4 shows the average modem delays which were calculated by extrapollating the average delays per packet length to a packet length of 0 bytes. The subsequent value was devided by four, since measurements are made on a symmetrical network with identical modems in source and destination, implying that a packet passes 4 times through the modem. Modem delay values are the average of 10 delay vs. packet length measurements per access technology.

The results indicate that ISDN and 56k6 analogue modems are comparable as far as average delay is concerned, though surprisingly the 56k6 modem seems to have a higher throughput. Modem delay of the 56k6 modem however is far worse than that of the ISDN modem. Orinoco and ADSL are situated between the slower modems and LAN equipment, with far higher bitrates than the ISDN and 56k6 modems (closer even to 10Mbit Ethernet), but with modem delays closer to ISDN. Finally, Ethernet and Fast Ethernet have negligable modem delays, certainly in view of the high round trip times (hundreds of ms) typically associated with international connections [12].

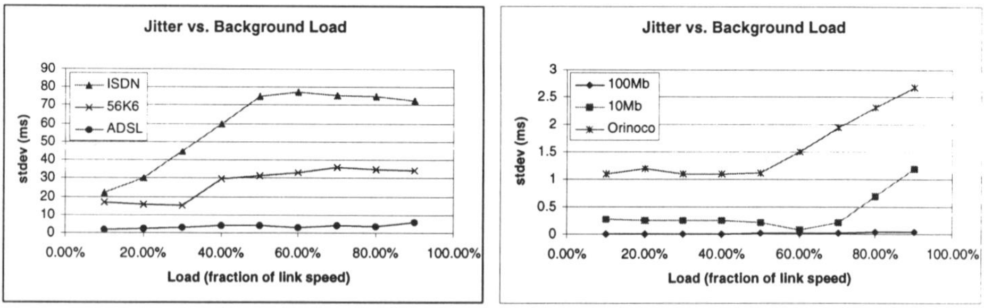

Figure 5: *Jitter in function of background traffic*

3.3 Delay Jitter

The variation of the jitter in function of the background load is illustrated in Figure 5. The load was generated in the same way as for Figure 3. Jitter is expressed as the standard deviation on the RTT. Results indicate that most access technologies have a threshold, above which jitter increases dramatically.

4 Codec Performance

In a first series of test, the influence of the length of IP packets on the MOS value was measured. For the remainder of the measurements, a single packet length was used. The second series of tests shows the influence of packet loss on the MOS value.

4.1 MOS in function of packet length

Figure 6 shows the MOS-value in function of the packet length. Measurements were performed for different codecs, on a 10Mbit Ethernet link. Packet lengths ranged from the equivalent of 10ms of speech to 200ms of speech (packet lengths around 50ms are most often used in voice applications). For each packet length, the average of 15 measurements was taken: 5 each with a male, female and mixed reference voice fragment. To clarify the graph, codecs with similar results (ADPCM 3-, 4-, and 5-bit, and g723 (5.3 and 6.3 kbit/s) were ommitted.

The results show that the MOS value remains fairly constant for packet lengths smaller than 100ms, but drops sharply once past it. The MOS value ranges from 3.0 to 3.9 for the different codecs, about the same as a good telephone conversation (MOS value 3.0-3.5). Although some codecs clearly have higher MOS values than others (higher bitrate codecs perform consistently better than lower bitrate codecs), the difference in quality is hardly noticable to human users.

For the remainder of the measurements, a standard packet length equivalent to 50ms of speech was used.

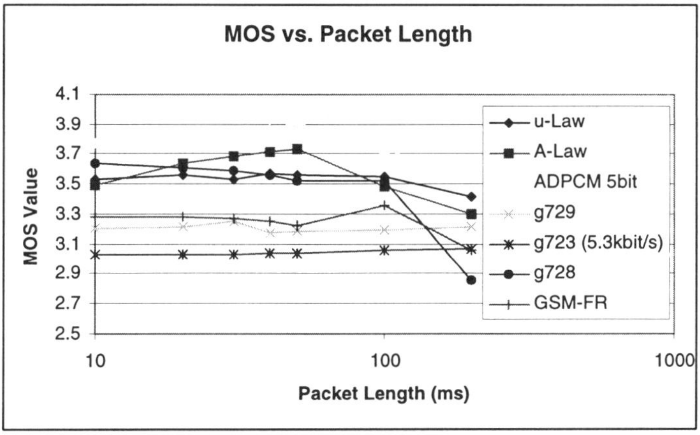

Figure 6: *MOS-value in function of packet length.*

4.2 MOS in function of packet loss

The influence of packet loss on the MOS value was determined using a Network Impairment Node, which is basically a modified router deliberately inducing packet loss in a flow at a specified rate. The PCs were connected to the impairment node through 10Mbit Ethernet links. Packet loss was varied from 0% to 10%. Figure 7 shows the results for these measurements.

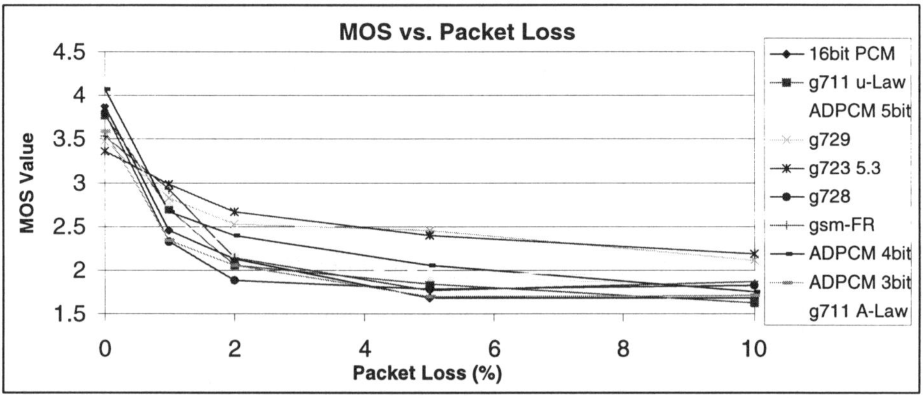

Figure 7: *MOS-value in function of packet loss.*

5 Conclusion

A number of performance parameters of both access network technologies and codecs were measured. Their influence on the quality of a VoIP call can be summarized as follows.

Delay & Modem Delay The total delay experienced on a network consists of a fixed offset (the modem delay) and a part increasing with the length of the packet (the transmission delay). Given that the available delay budget for an interactive VoIP session is about 400ms, analogue modems have too large a modem delay to be really useful. Both ISDN and 56k6 modems suffer from too much transmission delay, certainly at higher link loads.

Delay Jitter Delay jitter is one of the main obstacles for the deployment of VoIP over the Internet. A delay variance of 75ms to 300ms is considered to result in rather poor quality, which excludes ISDN and analogue modems.

Packet Loss A third performance parameter to consider is packet loss. All codecs show a similar evolution of quality with increasing packet loss. 2% is the maximum tolerable packet loss for all codecs.

6 Acknowledgements

Part of this work was funded by the IWT projects Limson and ITA-GBO.

References

[1] S. Vanhastel *et al*, "A Distributed Platform for Monitoring and Shaping Data Flows", International Conference for Telecommunication Systems, Proceedings, 2000

[2] Hollier M P, Hawksford M O, Guard D R, "Characterisation of Communications Systems Using a Speech-Like Test Stimulus", Journal of the Audio Engineering Society., Vol.41, No.12, 1993

[3] Hollier M P, Hawksford M O, Guard D R, "Objective Perceptual Analysis: Comparing the audible performance of data reduction schemes", Presented to the 96th AES Convention in Amsterdam, Preprint No.3879, 1994

[4] ITU-T, "P.56: Telephone transmission quality objective measuring apparatus", 1993

[5] ITU-T "G.711: Pulse Code Modulation (PCM) of Voice Frequencies", 1972

[6] ITU-T "G.726: 40, 32, 24, 16 kbit/s Adaptive Differential Pulse Code Modulation (ADPCM)", 1990

[7] ITU-T "G.727: 5-, 4-, 3-, and 2-bits Sample Embedded Adaptive Differential Pulse Code Modulation", 1990

[8] ITU-T "G.723.1: Dual Rate Speech Coder for Multimedia Communications transmitting at 5.3 and 6.4 kbit/s", 1996

[9] ITU-T "G.728: Coding of Speech at 16kbit/s Using Low-Delay Code Excited Linear Prediction", 1992

[10] ITU-T "G.729: Coding of Speech at 8kbit/s Using Conjugate Structure Algebraic-Code-Excited Linear-Prediction (CS-ACELP)", 1996

[11] ETSI ETS 300 580-1: "Digital Cellular Telecommunications system (Phase 2); Full Rate Speech; Part 1: Processing Functions (GSM 06.01)", 1997

[12] S. Vanhastel, B. Duysburgh, P. Demeester, "Performance Measurements on the current Internet", IFIP99, 1999

[13] K. Claffy, G. Polyzos and H.-W. Braun, "Measurement Considerations for Assessing Unidirectional Latencies", Internetworking: Research and Experimence, vol. 4, no. 3, Sept. 1993, pp. 121-32.

[14] V. Paxson, "End-to-End Routing Behavior in the Internet", IEEE/ACM Trans. Networking, vol. 5, no. 5, Oct. 1997, pp. 601-15

Broadband Access, WDM Metro and Network Management
D.W. Faulkner and A.L. Harmer (Eds.)
IOS Press, 2000 © AKM

Evaluation of ATM Services over Asymmetric Digital Subscriber Line Technology

N. Giouroukos, E. Vasilakopoulou, L. Mandalos, S. Koubias, G. Papadopoulos

Applied Electronics Laboratory,
Department of Electrical Engineering & Computer Technology,
University of Patras,
Patras 26500, Greece

Abstract. Digital Subscriber Line (xDSL) is a new rich set of technologies, rapidly developed to deliver broadband services to domestic and business environments using the existing twisted-pair telephone infrastructure. A wide variety of high bandwidth applications are supported, such as high-speed Internet access, telecommuting and virtual private networking. Deployed systems by many vendors are currently on the focus of technical and market trials worldwide. The authors present an overview of current and new xDSL access technologies continuing with a study, concerning the developed and installed systems word wide as well as the conducted trials. Finally, the employment of ATM technology for interfacing to the broadband access networks is presented and a platform to evaluate the application of ATM services over an ADSL network and to explore the capabilities of the modems provided from different vendors, is discussed thoroughly.

1. Introduction

The accelerated growth of applications that demand high bandwidth utilization has changed the nature of information networks. High-speed communications is an ordinary requirement through business, academic and domestic environments. Internet access, telecommuting and remote LAN access are three of the services that network access providers offer at the moment. These rapidly growing applications are placing a new level of demand on the existing telephone infrastructure that was initially designed to handle short duration voice calls. Moreover, the connection from the subscriber's premise to the central office has been provisioned with copper cabling that cannot be easily used for employing broadband applications. Therefore, the existing environment is being stressed by the demand of higher bandwidth capacities.

In the past decade, several technologies (fiber to the home, hybrid fiber coax) offered increased bandwidth but failed prone to the lack of applications. Moreover, the cost to replace the existing loops with fiber cables would be unaffordable for today's business models and more importantly the time to accomplish such a transition would be unacceptable. In light of these drawbacks, xDSL technologies are an efficient, possible solution [1].

xDSL technology encompasses a wide range of access methods over the existing single-pair copper cables, allowing the extension of bandwidth capacity from the service provider to the customer premise without the need for massive infrastructure replacements and investments. Employing advanced modulation techniques, such as Discrete Multitone (DMT) and Carrierless Amplitude and Phase (CAP), the data transmission rates are extended into the broadband range [2,3]. Both methods use line coding to reduce the transmitted symbol rate ensuring lower levels of signal loss due to spectrum shifting at lower frequencies.

In a typical application xDSL modems are deployed in two pairs, utilizing the spectrum resources beyond the 3.3KHz voice channel. One modem is placed at the telephone's company central office and the other at the customer premise location, receiving digitized data streams, overlaid into a high-speed analog signal. Symmetric as well asymmetric bandwidth configurations can be achieved to support both one and two way high bandwidth requirements.

High bit rate Digital Subscriber Line (HDSL) is a reliable and cost effective variant of xDSL that provides repeaterless T1 or E1 services over two-twisted copper loops. On the other hand, symmetrical digital subscriber line (SDSL) is a single-pair deployment of the HDSL technology, allowing easy implementation of applications requiring symmetric data rates over a single loop. Rate adaptive DSL (RADSL) supports a wide range of data rates optimizing the line quality through wire condition sensing. This is an important feature because the service provider is not obliged to measure a line and manually adjust the modem, while the management system can efficient fix the data rate to match a particular service. Very high bit rate DSL (VDSL) delivers multimegabit data rates asymmetrically over short cable loops. Although several variations of VDSL have been proposed and trailed, standardization is still in early stages [4].

Asymmetric DSL (ADSL) is the most commonly known member of the xDSL group that delivers a full DS1 downstream to the subscriber over a single 18Kft long, unloaded wire pair and higher rates over shorter loops. It offers fast data rates, from 1.5 to 8 Mbits/sec at the downstream direction while it provides slower data rates from 32 to 768 Kbits/sec upstream. In advance, ADSL was developed to support voice calls at the same wire pair with the use of an external device, called splitter that separates the data spectrum from the dedicated voice band [5]. Recently a lower speed, splitter less version of ADSL, referred to as G.Lite, has emerged that is intended to support speeds of 96 Kbits/sec to possibly 256 Kbits/sec downstream with 32 to 128 Kbit/sec upstream. This technology is primarily targeted for residential customers for delivering combined data streams and voice calls, facilitating the modem's installation with the exclusion of the external splitter.

2. ADSL Deployment

ADSL was originally designed to provide video on demand services. The proposed network model is shown in figure 1. The DSLAM is configured as a typical central office and has the ability to support plenty of lines. It is designed to be connected to an ATM switch or an ATM cross-connect via the 155 Mb/s OC-3 interface. On the other hand, the POTS splitter is installed to the central office as well as to the customer premise, providing data and voice messages concurrently.

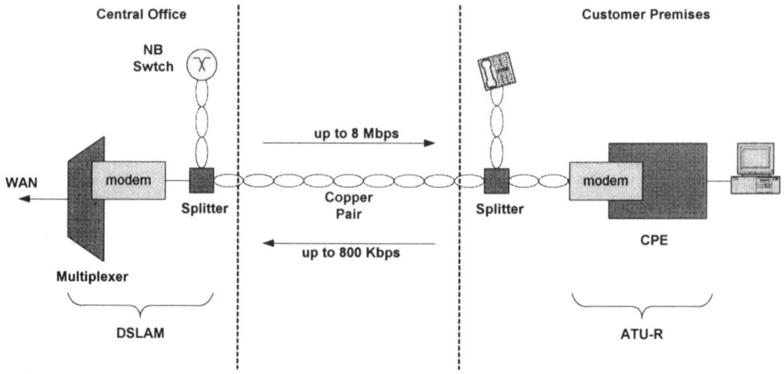

Figure 1. The ADSL connection model

Comparing xDSL technologies, ADSL is believed to be the most efficient and convenient solution for SOHO users due to the acquired effort for the equipment installation and management, providing multimedia and fast Internet access services. With closer examination of the others technologies, G. Lite ADSL solution seems to be the key to universal and successful deployment of ADSL, prone to its cost effectiveness as well as the lack of additional equipment inside a building.

Although ADSL's standardization has been completed some years ago, end-to-end products interoperability must be taken in mind during system development. With interoperability, a multi-vendor environment can be employed leading to the most affordable solution for the user and the service provider. Significant progress has been made to accomplish ADSL physical layer interoperability. In contrast, many new ADSL models have been proposed lately, due to trials adaptation to market and technical requirements.

Packet-based data can be transferred through an ADSL network, employing PPP and FR protocols that support IP and IPX layer 3 protocols. PPP packets are transferred between the ATU-R and the ATU-C using HDLC frames as determined at the RFC 1662 proposal. PPP encapsulation on the other hand, allows other protocols to operate over point-to-point connections. However, frame user network interface (FUNI) would be a more efficient method to implement packets transfer. A simple FUNI connection supports multiplexing of multiple network protocols using LLC encapsulation or multiplexed virtual paths [6].

Concluding the ADSL network management is accomplished using Simple Network Management Protocol (SNMP). This protocol is extensively employed for the management of routers, Ethernet switches, ATM switches, workstations and network adapters.

3. ATM services over ADSL

ATM should be used as the standard interface technology because most backbone networks are based on ATM technology. Some of the features that make ATM the most suitable access technology are pointed out. Firstly, ATM integrates a wide range of applications with different QoS regardless of the underlying physical media. Secondly, ATM enables efficient management of bandwidth, carrying different traffic types. Finally, ATM enhances access to Internet services, in terms of performance, load sharing, and redundancy [7]. Using an ATM over ADSL architecture, SOHO environments and residential users can gain access to broadband Internet environments and corporate Intranets. The transportation of ATM cells over ADSL links ensures the desired QoS, in terms of stable delay and guaranteed bandwidth.

3.1. The employed infrastructure

ATM application and performance is tested over an end-to-end ADSL network constructed in a laboratory environment. The testing and evaluation were carried out using a testbed especially developed for this purpose, based on an infrastructure consisting of systems provided from different vendors. More specifically complete ADSL access equipment from each vendor was used, consisting of the DSLAM multiplexer, the customer premise modems (all types of modems with different interfaces, like NIC, External modem, and USB) and the Management System (used to perform the overall control and monitoring of the DSLAM). The necessary splitters that are used to separate the lower band frequencies (voice channel) from the higher frequencies (data channel) were employed as well. Ethernet and ATMF-25 interface were supported on the modems used that have two operational modes, POTS and ISDN. Moreover, to establish a connection, the presence of an ATM switch was mandatory, due to the failure of DSLAMs to support switching functionalities. For simple loop testing of data rates, noise margins, and rate adaptation cooper wire was used. At last, an external line simulator was used to conduct extensive tests, evolving the emulation of standard noise models described at ANSI T1.413 Issue 2 and at ITU G.992.1 Full-Rate ADSL standard. A Network Analyzer, a Traffic Generator and advanced monitoring software, provided by modem vendors, completed the equipment used.

To evaluate the performance of an ATM over an ADSL network, interoperability, priority of services, multiple connections per user support were taken in mind. The end-to-end Network architecture that was employed during the conducted experiments is shown in Figure 2.

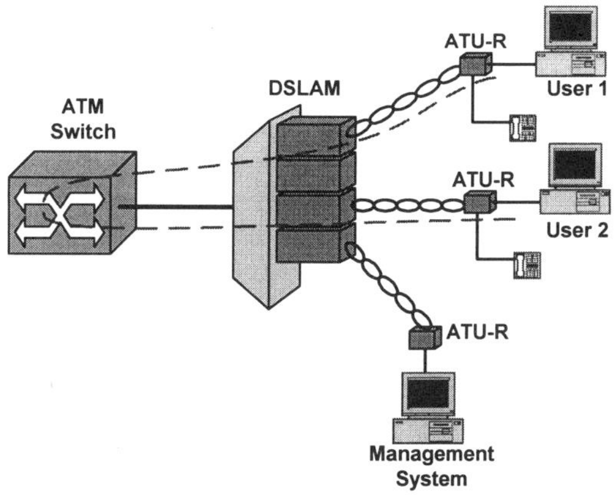

Figure 2. The Network architecture in the Lab environment

One of the PCs acts as a server and another one acts as a client. The connection between the end-to-end devices is established through the DSLAM and the ATM switch. Each connection has a VPI/VCI identifier on the Network side and one at the User side.

The management system is connected to the DSLAM using an ADSL modem employing a defined VPI/VCI value. To establish an end-to-end connection the type of service, the desired data rate, the noise margin, the VPI/VCI identifiers, and the interface employed must be defined. With completion of initialization process, the system tries to establish the connection. During the line-up procedure the system tries to achieve the specified rate. It is also possible for a system to adapt it's transmission rate to the line conditions (during the line-up phase) providing the maximum achievable rate.

3.2. Evaluation of basic ATM services

For massive deployment of ADSL technology, the access systems should support the following features [8,9]:

- ATM mapping
- UNI 3.1 signaling (UNI 4.0 optional)
- QoS support, CBR, UBR, nrt-VBR, rt-VBR, ABR traffic support
- VC cross-connect (support of PVC)
- VP cross-connect (support of PVC and SVC)
- Multiple connections per user
- Multiple QoS per user
- Traffic policing per VC
- Traffic shaping
- EPD, PPD (Traf. Man. 4.0)

The end-point of the ADSL network should support:

- UNI 3.1 signaling
- ATM or Ethernet interfaces
- QoS support, CBR, UBR, nrt-VBR, rt-VBR, ABR traffic support
- Traffic shaping

To realize the above requirements it is assumed that higher layers are transported transparently by the ADSL based network. Testing and evaluation of the explored ADSL systems was conducted assuming that the above-mentioned requirements were met. Table 1 depicts briefly the evaluation results, referring to all systems evaluated.

Considering the experimental results, the provision of DSLAMs multiplexing functionalities can be mentioned firstly. This means that at the Network side the interface is an STM-1 and at the User side the four supported ADSL lines are concentrated at a line board. Furthermore, the management of more than one connection per line (usually around 10 connections) can be achieved although the modems don't support that amount of connections (usually 4 are supported). In advance, the multiplexing and demultiplexing of VPI/VCI is supported. Developed DSLAMs support only two types of QoS, the CBR and the UBR. Some of them support multiple QoS at one ADSL line as well. This means that at the user premise two types of services can be supported, one with CBR and one with UBR QoS. Taking in account that the public ATM networks do not fully support SVCs, the existing DSLAM's do not support SVCs. However, they provide simple VP cross-connect functions and each PVC must be configured manually before any communication can take place over the ATM network. User data are transferred to the ATM backbone network and then to the ISP's using the PVC and the Access Multiplexer. The access system support EPD and PPD to control the user traffic. Extensive policing tests were carried out exploring all possible service type combinations, with the objective of bandwidth preservation during operation.

Function	Result	Comment
Multipexing/Demultiplexing of ADSL signals	Yes using a 155 Mbps STM-1 interface	4 ADSL Lines / LT board
Multipexing/Demultiplexing of VPI/VCI	YES	
Managing and control of ADSL links	YES	Through an Workstation connected to the DSLAM or through an web based
ATM mapping	YES	VPI/VCI User side ↔ VPI/VCI Network side
Supported UNI interface	UNI 3.1	UNI 4.0 is until now not supported
Supported QoS	CBR and UBR	rt-VBR, nrt-VBR, ABR are not supported from any system
Multiple QoS per user	SOME	Some systems do support multiple QoS per ADSL line
Multiple connections per user	YES	The number of connections varies from vendor to vendor
VC cross-connect (support of PVC)	YES	
VP cross-connect (support of PVC & SVC)	Only PVC, SVC is not supported	
Traffic policing per connection	YES	CBR-CBR / CBR-UBR / UBR-UBR
Rate adaptation mode	YES	Operational controlled and Automative Rate Adaptation
Operational mode	POTS / ISDN	Both are supported from all vendors
Support of Traffic management 4.0	YES	
Channel type	Fast και Interleaved	
Control of frequencies (tones)	SOME	Some systems have the ability to control (activate & deactivate) the frequencies
EPD/PPD (TM 4.0)	ALL	

Table 1. Basic ATM features

Provisioning of constant bit rate services under noisy environments, is realized at the ADSL systems, gradually lowering the noise margins to adapt to the new line conditions and preserving the negotiated bandwidth. The connection is released at the case that the system is unable to provide negotiated data rate due to line impairments. With UBR service the rate is reduced. To be more specific, we noticed that, once the transmission rate was lowered the systems were unable to increase their rate even when the line conditions were improved. This means that after initialization real time adaptation of the transmission rate is not supported. Thus to increase the data rate the connection must be reconfigured. With multiple QoS per ADSL line support, the DSLAM must ensure the priority of the CBR class traffic. Various ADSL events (e.g., Fast Retrain) may have impact on the traffic management and QoS.

All configurations are conducted using the management system that is directly connected to the DSLAM, employing an Ethernet interface or an ADSL connection. In advance, some vendors provide WEB based management systems with fewer capabilities as well. Using a WEB based management system the connections can be configured, released or reconfigured.

4. Conclusion

xDSL is a new platform of copper broadband access technologies available on the market today. On the other hand, the use of ATM over an ADSL network seems to be the technological solution for integrated services at both the service node and end-user interfaces, due to traffic multiplexing capabilities, multicast support and end-to-end QoS management ability.

This paper presented an overview of current and new xDSL access technologies discussing over the deployed products, the installed systems worldwide as well as the conducted trials. Several issues concerning technology standardization, products interoperability and management issues were addressed as well.

The authors have also described the network architecture of a typical ADSL network and proposed a platform for evaluation of ATM services. Using the suggested test bench and modems provided from different vendors, we measured the network performance based on the supported QoS as well as traffic management and multiplexing. Finally, a discussion concerning the experimental results followed, revealing the status of ADSL products deployment.

References

[1] Maurice Gagnaire, "An Overview of Broad-Band Access Technologies", *Proceedings of the IEE*E, Vol. 85, No. 12, December 1997.

[2] "Network and Customer Installation Interfaces, Asymmetric Digital Subscriber Line (ADSL), Metallic Interface" ANSI, T1.413-1998, 1998.

[3] J. Bingham, "Multicarrier Modulation for data Transmission: An idea whose time has come", *IEEE Communications Magazine*, Vol.28, No 5, May 1990, pp 5-14.

[4] V. Oksman Ed., "VDSL Draft Specification", *ANSI cont. T1E1.1/98-045R1*, June 1-5, 1998.

[5] Walter Y. Chen, "The Development and Standardization of Asymmetrical Digital Subscriber Line", *IEEE Communications Magazine*, May 1990, pp 68-73.

[6] S. Dixit, "Data Rides High on High-Speed Remote Access", *IEEE Communication Magazine* Vol. 37, No. 1, pp. 130-141, January 1999.

[7] A. Azcorra, En. J. Hernadez-Valencia, J. Berrocal, "IP/ATM Integrated Services over Broadband Access Copper Technologies", IEEE Communications Magazine, pp. 90-97, May 1999.

[8] Luis Rodrigues, "ATM and ADSL for High Speed Internet Access", IBC-Asia "High Speed Internet Infrastructure '97", November 1997.

[9] ADSL Forum, "ATM over ADSL Recommendation", ADSL Forum, March 1999.

Broadband Access, WDM Metro and Network Management
D.W. Faulkner and A.L. Harmer (Eds.)
IOS Press, 2000 © AKM

Evaluation of Access Network Modifications Required by the Introduction of VDSL

Harri Mäntylä
Helsinki University of Technology, Laboratory of Telecommunications Technology
e-mail: harri.mantyla@iki.fi

Abstract. A simplified access network (AN) model is used to study a few upgrade options with different fiber penetration levels. The model gives exact values for cable lengths and for the required number of fiber nodes in each case. The area of the model corresponds to an area covered by one local exchange in the current network. In the studied cases 1, 7, 25 and 96 fiber nodes are placed into the area. The average length of the remaining copper loop shortens correspondingly from 1800m to 1000m, then to 500m and finally to 150m. The main components included in the following cost analysis are cable infrastructure costs, costs related with the ONU site, ONU equipment costs and VDSL modem costs. VDSL service take rate is taken into account mainly in the VDSL modem costs that are directly proportional to the number of VDSL subscribers. Other costs are less affected by the take rate and contain therefore higher risks since these costs have to be invested before any VDSL subscribers exist. Rough cost estimates are given to the components and a comparison of the total cost of the considered topologies with different take rates is made. The results show that introducing VDSL requires big investments as could be expected. VDSL take rate has a vital effect on finding the best way to upgrade the original network. Topologies with higher fiber penetration level become more and more attractive as the VDSL take rate increases. Especially the calculations of the cost of capacity make the topology with the highest fiber penetration level look very attractive after certain take rate.

1. Introduction

Copper pairs, connected to practically every household, form one of the most common access networks (AN) in the field of telecommunications. Digital subscriber line (DSL) technologies offer a possibility to utilize part of this huge existing copper base for broadband connectivity. Very high speed DSL (VDSL) offers the most transmission capacity, planned to operate over very short copper loops of only a few hundred meters. To be able to use VDSL major changes to current network topology are required. Part of the existing copper AN must be replaced by fiber as the remaining copper loop length must be shortened significantly. Active nodes called optical network units (ONUs), containing fiber endings and DSL access multiplexers (DSLAMs) have to be placed close to subscriber locations, meaning placing of active electronics in a possibly hostile environment.

In this paper four upgrade options are presented with the help of a general AN model. Different topologies are compared by applying rough cost analysis. The main goal is to find a solution on how to upgrade the present copper AN to a broadband VDSL network in the best possible way.

2. Copper Access Network

2.1 Current Topology

Present copper ANs consist of passive twisted copper pairs running from local exchanges (LE) to subscribers. Typically the length of the feeder cabling running from LE to crossconnection site is significantly longer than the distribution cabling from crossconnection to subscriber. Different sections of the subscriber line may consist of different types of cables. For example in BT's network 0.32 – 0.9mm gauges are typical [1]. In the feeder cabling copper pairs are usually combined to big cable bundles of 0.4 or 0.5mm pairs. Closer to subscribers 0.5 or even 0.6mm pairs are typical. In the urban areas most of the crossconnection sites are located inside buildings, while in the areas with lower population density small distribution points are often located in street cabinets or poles.

2.2 Access Network Model

The SYNTHESYS model of [2] was chosen as a base for the used AN model. This polygon based model is shown in Figure 1. It was chosen because of its simplicity and general applicability. From the model it is possible to calculate exact cable lengths and the number of required fiber nodes (ONUs) in different upgrade options.

In Figure 1 one segment of the whole polygon, presenting the whole LE area, is enlarged. The subscribers are connected to distribution points, letter F in the figure. Like in real AN also here individual segments are gathered together at flexibility points – B, C, D and E in the figure – and brought in big cable bundles to the LE location, presented by letter A in the model. Subscribers that are located near the LE, area A_1 in the model, are connected directly. After fixing R and α the rest of the link lengths can be derived from geometry [2]. Table 1 shows the values used in further calculations.

2.3 Loop Length Distribution

LEs are typically located in the middle of densely populated areas. This leads to number of subscribers with short loop lengths being higher than those with long loops. An approximation of the cumulative distribution of subscriber's loop lengths in Helsinki area (including urban, semi-urban and rural areas) is presented in Table 2. The average length is about 1800 m and 90 % of subscribers have loops shorter than 3500.

Table 1 *Chosen values for the AN model.*

Parameter	Value
R	3000 m
α	60°
AB, BC, CD, CE	866 m
AF, BF, CF, DF, EF	564 m
$F \rightarrow Sub$	15 – 293 m
$A \rightarrow Sub$	579 – 3456 m

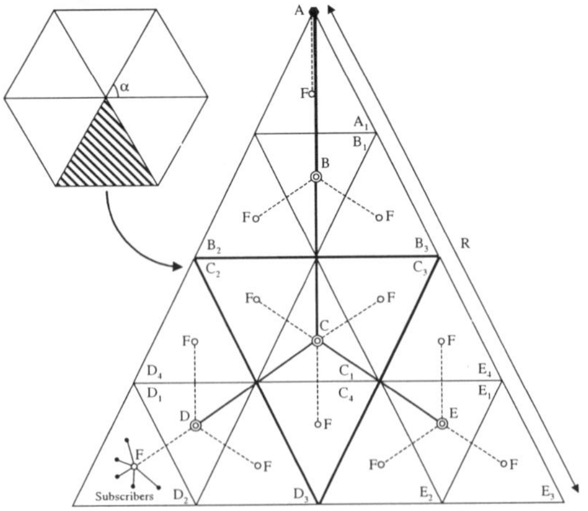

Figure 1 *The SYNTHESYS model [2].*

Table 2 *Rough loop length distribution in Helsinki area.*

loop length [m]	% of subscribers reached
1000	30
2000	60
3500	90
5000	100

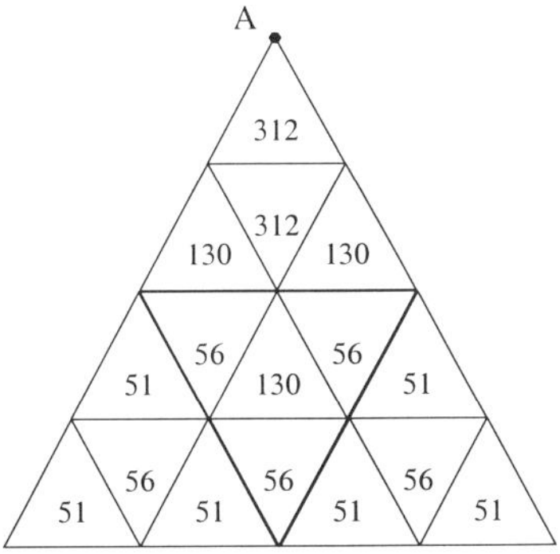

Figure 2 *Real distribution applied to place 1600 subscribers into one segment of the AN model*

After fixing the dimensions of the AN model the resulting loop length distribution depends on how the subscribers are placed into the model. The number of subscribers at each distribution point represented by the smallest triangle in the model has to be decided. Also the way the subscribers are distributed inside these distribution areas has to be determined in order to calculate the exact loop length distribution.

Figure 2 shows 1600 subscribers placed into one segment of the AN model according to the distribution of Table 2. As can be seen the number of subscribers close to the LE, A in the figure, is significantly higher than the number of subscribers near the border of the area. The resulting average loop length is 1834m, quite close to the real value.

2.4 Upgrade Options

In the current network, fiber is brought to the LE (FTTLex). In the AN model this is presented by Fiber to A topology. If the topology of current AN is studied it seems that the most suitable places for fiber nodes would be the crossconnection sites. In the AN model these are presented by B, C, D and E. In the following four different upgrade scenarios with increasing fiber penetration are presented, Fiber to A and B (FTAB), Fiber to A and C (FTAC), Fiber to A, B, C, D and E (FTABCDE), and Fiber to F (FTF). The cumulative distributions of remaining copper loop lengths in different upgrade scenarios are shown in Figure 3. The average and maximum loop lengths are shown in Table 3.

There might be problems with placing new equipment to desired locations because often there is no extra space available, making it necessary to build expensive new housings. Here it is assumed that fiber nodes may be placed into or in the immediate vicinity of the desired locations and later when some rough cost calculations are made a cost for these sites is included as one cost component.

Fiber to A

This is the current topology that makes it possible to offer VDSL only to subscribers whose loop lengths are short enough. If 1500m is assumed as maximum VDSL distance

Table 3 *Ave. and max. remaining loop lengths.*

Topology	Cu Ave [m]	Cu Max [m]
Fiber to A	1834	3456
Fiber to A and B	1137	2590
Fiber to A and C	1059	1724
Fiber to ABCDE	523	857
Fiber to F	154	293

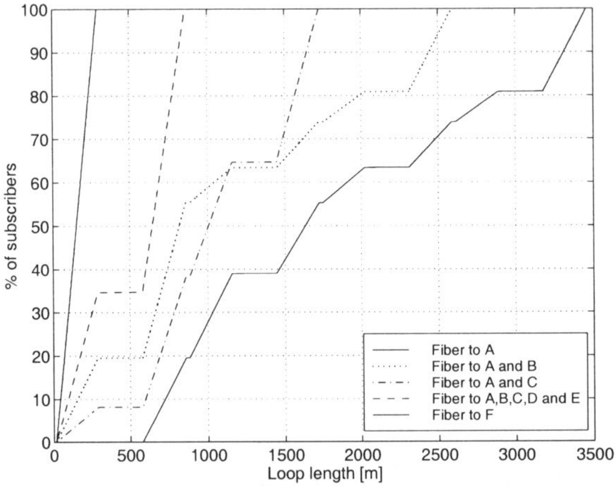

Figure 3 *Remaining copper loop length distributions of studied topologies.*

then 42% of all subscribers are inside this limit. Rest of the subscribers could be reached with for example ADSL technology, which is better suited for longer loops.

Fiber to A and B or Fiber to A and C

The first evolution step could be taking Fiber to either B or C. In both cases 7 ONUs are needed, one in each segment of the AN model and one at A. Figure 3 shows that FTAC results in slightly more potential VDSL subscribers, while FTAB results in more subscribers with short loop lengths indicating more transmission capacity.

Fiber to A, B, C, D and E

This is the first topology examined here that makes it possible to offer VDSL service to all subscribers in the LE area The number of needed ONUs increases to 25 and the remaining loop length is dropped to about half compared to the previous topology.

Fiber to F

The last topology examined here takes fiber to each distribution point of the AN model. Number of needed ONUs increases significantly to 96. Also the remaining loop length is shortened significantly to less than 10% of the original. All subscribers can be offered a high quality VDSL connection.

3. Capacity of VDSL

Each VDSL subscriber can be offered a certain maximum transmission capacity limited by the characteristics of the twisted pair cable and possible constraints set by VDSL system. Maximum capacities of all VDSL subscribers in an ONU area can be added up to upper limit of the possible total capacity that needs to be handled by one ONU. Fiber cables installed into the AN should be able to transmit this estimated maximum capacity.

3.1 Capacity of VDSL over Single Copper Pair

Some varying estimates have been suggested of VDSL capacity as a function of loop length. The one presented in [3] was chosen to be used in further calculations. The symmetric capacity values are presented in Figure 4 limited to expected VDSL operating range of 1500m and max. data rate of 34 Mbits/s [4,5].

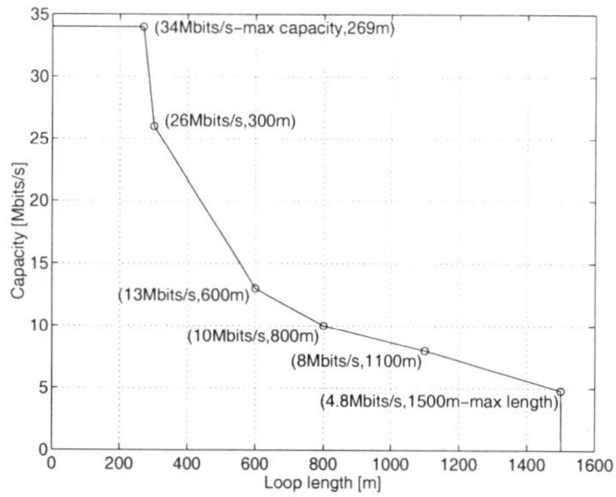

Figure 4 *Symmetric transmission capacity of VDSL.*

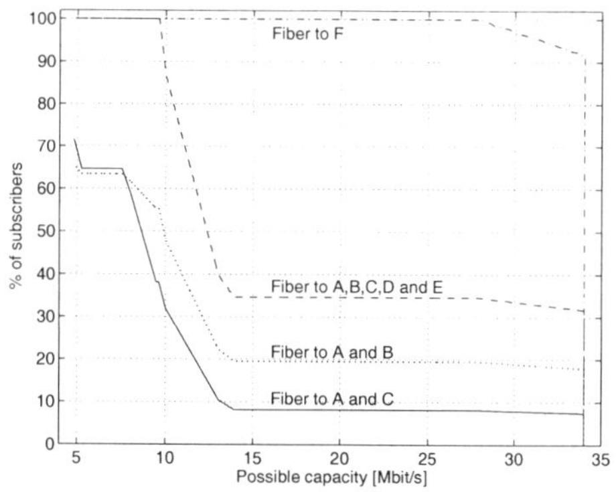

Figure 5 *Capacity offered to certain percentage of the subscribers*

Table 4 *Maximum total and average capacities with 10% take rate.*

	VDSL subs	Cap. total [Mbit/s]	Cap. ave. [Mbit/s]
Fiber to A and B	628	10957	17.4
Fiber to A and C	686	8391	12.2
Fiber to ABCDE	960	18304	19.1
Fiber to F	960	32384	33.7

3.2 Capacity of VDSL Network Covering the Area of AN Model

By combining the cumulative copper loop length distributions of different topologies, presented in Figure 3, and the VDSL capacity from Figure 4, the maximum capacities of examined topologies are obtained as a result. Table 4 shows the maximum total capacity and average maximum capacity per VDSL subscriber for different topologies with an assumed 10% VDSL take rate. The average capacity actually used by VDSL subscribers may be significantly lower depending on the used services.

Figure 5 shows the increase of capacity as a percentage of subscribers who can be provided with certain capacity. For example service level of 10 Mbit/s can be offered to 48% of subscribers in FTAB, 32% in FTAC, 87% in FTABCDE and 100% in FTF.

As the number of ONUs is raised a remarkable increase in transmission capacity is achieved. Especially the upgrade to FTF topology with 96 ONUs means this, but also the upgrade from 7 ONUs of FTAB and FTAC to 25 ONUs of FTABCDE shows a significant improvement of the service level and also the coverage of the VDSL service.

Comparison of FTAB and FTAC shows FTAB offering significantly more transmission capacity than FTAC. Because of this FTAB could be recommended. The reason for the difference in capacity is the big subscriber number located at the vicinity of points A and B. FTAB provides these subscribers with short loops and high transmission capacities. FTAC reaches more subscribers, but offers only low transmission capacity to many of them.

4. Cost Analysis of the VDSL Network

4.1 Cost Components

The total cost of the VDSL upgrade consists of several different components. Four VDSL specific cost components were selected to be included in further analysis. The main costs of the upgrade are tried to be included in them. The costs are roughly estimated for comparing studied topologies. These rough estimates are shown in Table 5 and Table 6.

Table 5 *Rough cost estimates for cable, ONU site and VDSL modem.*

Cabel [Fibers]	Euro/m	ONU Site	Euro		Euro
48	17	LE	10000	**VDSL modem (NT&LT)**	600
24	15	AN	20000		
12	13				

Table 6 *Rough cost estimates for ONU equipment.*

ONU Size	50	250	500[1]	750[2]	1000[2]	1250[2]	1500[2]
Euro	10000	15000	20000	25000	30000	35000	40000

[1] Only ONU(A) and ONU(C) in the AN model with take rate up to 30%.

[2] Only ONU(A) in the AN model with take rate up to 30%.

Cable Infrastructure

Cable costs include the cost of new fiber and the installation costs. The costs depend on meters of new cable, cable type, area type and number of fibers taken into use. Fiber cable cost and the cost of taking installed fiber into use form a stable part of the costs. The biggest cost fluctuations are involved with digging and installation work costs that are strongly influenced by the area type and especially on the availability of ducts [3].

Although the cable infrastructure forms a significant part of the total costs it may be considered to contain only low risk. The installed fiber will most likely be used anyway, assuming the long term goal to being FTTH network. In order to decrease the future costs it would be wise to dimension the installed fiber according to estimated future needs.

ONU Site

Cost of ONU housing and installation are part of the ONU site costs. Because of the importance of the site cost it was decided that this should be handled separate from other ONU costs. Also if future development of costs is compared it is most likely that prices of ONU equipment will go down while the cost of ONU site most probably does not get any cheaper.

ONU Equipment

ONU equipment costs are formed by the costs of equipment needed for building an ONU into the ONU site, meaning rack, shelves, necessary cards and backup powering equipment.

VDSL Modems

VDSL modem costs include NT and LT equipment and the installation work. VDSL modem costs differ from other costs presented here in a very important way. As other costs need to be made at the time of building the VDSL network, with only predictions of the possible number of VDSL subscribers, the VDSL modem costs actualize after VDSL connection has been requested by the customer. This means considerably lower risk if compared to the other cost components.

4.2 Total Cost

Figure 6 shows the total cost of upgrading to different topologies. It also shows how the costs are formed of different components. It can be noted that the proportion of high risk cable, site and ONU costs are very high at low take rates.

4.3 Cost per VDSL Subscriber

In Figure 7 the cost per VDSL subscriber is shown for studied topologies with different take rates. As could already be expected from the total costs the FTF topology has by far the biggest cost level per subscriber especially in case of low take rate where the high total costs are divided by a low number of connected VDSL subscribers.

4.4 Cost of Capacity

The cost of capacity for studied topologies is shown in Figure 8. As the take rate increases the cost of capacity decreases and as shown in the figure the decrease rate differs significantly in different topologies. In FTF change in the cost of capacity is the most remarkable. Already at surprisingly low 10% take rate the cost of capacity in FTABCDE topology is cheaper than in FTAC.

If the main criterion for the topology selection process were to minimize the cost of capacity, this result indicates that the FTAB topology has very low cost per capacity level if compared to the other options. The cost difference to FTABCDE and FTF starts to be at acceptable level at 20 – 25% take rates. The cost level of FTAC is surprisingly high, other upgrade topologies seem to be a lot more attractive.

4.5 Summary of the Cost Analysis

The results show that the VDSL take rate has a vital effect on finding the best way to upgrade the present network. FTAB topology, with 65% coverage, seems very attractive as a starting point, especially if the cost per capacity is examined. FTAB also provides the lowest cost per subscriber, slightly less than FTAC. At low penetration rates the ratio of low risk VDSL modem costs is the highest in FTAB topology. FTABCDE could be thought of at 15 – 20% take rates. FTF could be considered at 25 – 30% take rates.

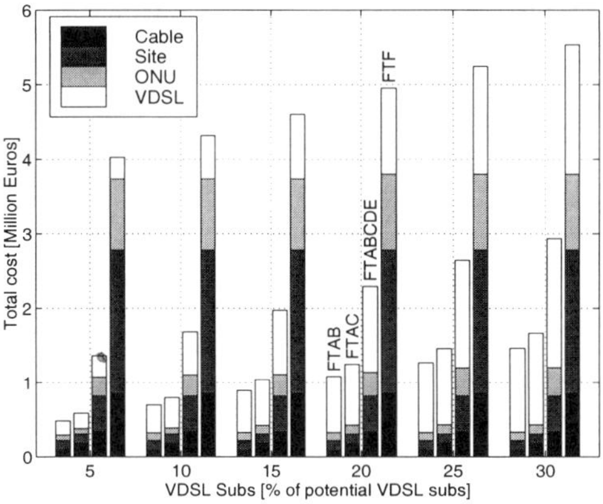

Figure 6

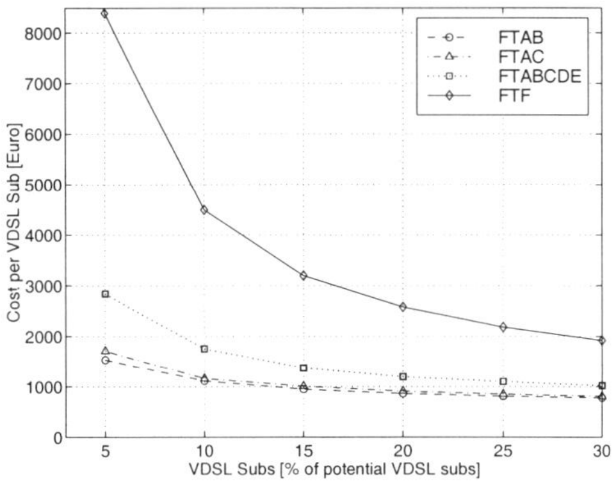

Figure 7 *Cost per VDSL subscriber with different take rates.*

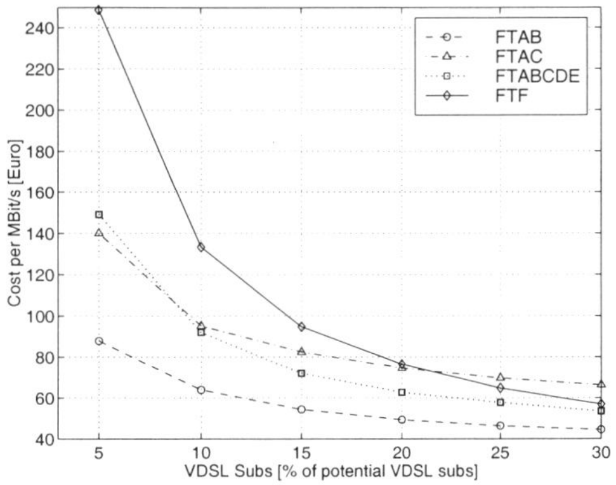

Figure 8 *Cost of capacity with different VDSL take rates.*

5. Conclusions

In this paper different upgrade options of the present copper based AN with VDSL were studied. By utilizing the existing network with VDSL some savings are achieved, although big investments are still required as was shown.

Instead of analyzing existing LE areas and real samples of the AN, a general geometric model of the AN was used. The model makes the analysis of the upgrade options more straightforward, remaining loop lengths are known exactly and the effect of different topologies may be freely calculated. The size of the modeled area was chosen such that could also exist in reality. As also the network topology of the model is similar to that of real networks, it may be expected that similar results will be obtained from real cases.

As a general rule it can be said that as the service take rate increases, topologies with high fiber penetration level become more attractive. This is what could be expected, as there are more connected subscribers among whom the investments are divided.

The calculations of the cost of capacity at different take rates gave interesting results. Above certain take rates the otherwise by far the most expensive 96-ONU Fiber to F topology starts to look very attractive, providing quite low values for the cost of capacity. This suggests that when the take rate is high enough, it might be feasible to take the fiber to a distance of only a couple hundred meters from the subscribers.

Acknowledgement

This work is part of a research joint project of Helsinki University of Technology, Laboratory of Telecommunications Technology and Tampere University of Technology, Telecommunications Laboratory, funded by Technology Development Center of Finland (TEKES).

References

[1] G. Young, *et al.*, "Broadband multimedia delivery over copper," *BT Technology Journal*, vol. 13, no. 4, Oct. 1995, pp. 78-96.

[2] I. Welling *et al.*, "Geometric models," *Broadband access networks: introduction strategies and techno-economic evaluation*, edited by L. A. Ims, London, UK, Chapman & Hall, 1998, pp. 63-73.

[3] L. A. Ims *et al.*, "Costs of upgrading the residential telecommunications infrastructure to broadband," *Proc. IEEE 1998 Global Telecommunications Conference (Globecom '98)*, Sydney, Australia, November 8-12, 1998, pp. 3153-3158.

[4] ANSI T1E1.4, Draft Technical Document, Revision16, *Very-high-speed digital subscriber lines, system requirements*, September 1, 1999.

[5] ETSI Technical Specification, TS 101 270-1 V1.1.5, *Transmission and Multiplexing (TM); Access transmission systems on metallic access cables; Very high speed Digital Subscriber Line (VDSL); Part 1: Functional requirements*, Jun. 1999.

Broadband Access, WDM Metro and Network Management
D.W. Faulkner and A.L. Harmer (Eds.)
IOS Press, 2000 © AKM

Assessment of IP Network Interruptions on End-System Applications

L. Aguirre-Torres, C. Phillips*, L. Sacks, P. Lane
University College London, Dept. of Electronic & Electrical Engineering, Torrington Place, London WC1E 7JE, UK
**Queen Mary and Westfield College, Dept. of Electronic Engineering, Mile End Road, London E1 4NS, UK*

Abstract. This article intends to provide a quantitative assessment of the performance of adaptive and non-adaptive applications when carried over an optical backbone network. Specifically, the study examines the impact of protection at the optical layer, focusing on Optical Shared Protection Rings for the transport of IP based premium services. Using *entropy* as a measure of disruption in the network, an "unbiased" analysis of the results is provided focusing on the QoS degradation.

1 Introduction

The rapid growth in Internet traffic has motivated the need to provide greater network capacity and high-speed access. The advent of data-rich applications, multimedia services and virtual networks have triggered an unprecedented demand for robust, high capacity networks for both the access and carrier backbone.

Packet-based optical backbone networks for high-speed data services are currently under development as a direct response to these demands. Particularly, optical fibre transmission technology has evolved from lower rate multi-mode fibre links to single-mode amplified fibre links capable of carrying multiple 10 Gbits/s channels per fibre. Likewise optical amplification in conjunction with Dense - Wavelength Division Multiplexing (D-WDM), have resulted in great capacity expansion, cost reduction and operations simplification [6]. These two major developments may be combined to provide considerable transport bandwidth, allowing the network topology to be dynamically reconfigured in response to changes in traffic conditions effectively increasing the overall system performance and capacity.

Currently, Internet Protocol (IP) services are provided using a wide range of electronic equipment. A typical network may include as many as three or four different electronic multiplexing and switching layers. The optical layer has been appended to the bottom of the multi-layer protocol stack as a straightforward extension of existing networks. This multitude of layers, however, produces bandwidth inefficiencies, as packets have to be encapsulated and sometimes fragmented as they are passed down the protocol stack. This also adds to the latencies of connections as extra processing, framing and aligning has to be performed, possibly hindering the provision of quality of service assurances. This worsens when the different layers are unaware of each other, causing duplication of network services.

A reduction in the number of layers in the protocol stack can lessen costs and overheads, certainly maximising the useful transport bandwidth. This could also minimise complexity in provisioning fault detection, network planning and restoration.

2 Optical Networking

High demand for increased network capacity has pushed core network service providers to find suitable architectures, capable of supporting bandwidth beyond tera-bit-per-second rates.

The increased demand for bandwidth has forced a reassessment of network designs and the development of network topologies based on D-WDM [6].

The way information is physically encoded onto the transport layer involves a number of trade-offs between utilisation of existing infrastructures, investment in new technologies, and scalability of the network. Initial systems employed purely single carrier information transport based on Time Division Multiplexing (TDM) it became apparent that increased capacity could be economically implemented with a D-WDM overlay on the existing TDM network structure.

2.1 Protection and Restoration

Various failures, such as fibre cuts, hardware and software failures, can occur causing the disruption of network services. Some failures may affect a whole multiplexed section, triggering independent recovery mechanisms. If these different protection schemes operate in isolation from each other, uncoordinated behaviour can lead to further degradation. This primarily affects those applications employing closed-loop feedback mechanisms such as Transmission Control Protocol (TCP).

Protection and restoration are two methods networks use to recover from failures. Protection refers to hardware based, pre-planned, fast failure recovery, and is generally limited to simple topologies like rings. Restoration, on the other hand, refers to software based, dynamic slower recovery. Protection becomes an especially important feature in the wavelength domain because of the potential for large data loss during a failure event. In this context, Optical Add Drop Multiplexors (OADM) networks using an Optical Shared Protection Ring (OSPR) architecture may potentially be more desirable compared with other survivable networks, as they allow the protection bandwidth to be shared.

2.2 Optical Shared Protection Rings

Optical Shared Protection Rings (OSPRs) provide a method for protecting selected traffic streams by offering them redundant transport paths between Optical Add Drop Multiplexors (OADM) nodes. However, the protection path is not exclusively reserved for a particular *premium* traffic stream. Instead, it forms a shared resource that is normally used for transporting *extra* traffic (ET). In the event of any single inter-node span failure, the protection path is capable of transporting the premium data between any two nodes, displacing the ET. As a result, OSPRs offer cost savings over traditional 1+1 protection schemes whilst yielding a negligible availability degradation.

The OSPR network architecture comprises a number of OADM nodes inter-connected by two concentric bi-directional fibre rings. One fibre-pair, referred to as the *working fibre*, normally carries the premium traffic. Conversely, the second fibre-pair, called the *protecting fibre*, normally carries the ET. Under normal operating conditions, the protecting fibre carries ET on the protecting fibre whilst the premium traffic is transported along the working fibre pair. However, when a failure is detected the premium traffic is moved onto the protecting fibre and the ET is discarded.

OSPRs provide two primary types of failure protection mechanism, referred to as *span* and *ring* switching. With span switching, the loss of integrity of a particular working fibre span causes the OADM nodes at either end of the failed link to attempt to signal each other on the protecting fibre span that also directly links the two nodes. If the protecting fibre is operational, the ET currently using this link is discarded and the premium data that would hitherto been directed along the failed working fibre, is now re-directed onto the protecting fibre. When a failure arises that causes a loss of integrity to both the working and protecting fibres of a span between two OADM nodes, a ring switch is attempted. Under these circumstances, signalling messages from the node either side of the failure are transferred around the ring along the protecting fibre in the opposite direction to the point of failure. Provided all the remaining protecting fibre spans are only being used to transport ET, then all the intervening nodes discard this ET and the are used to forward the premium traffic between the two nodes adjacent to the failure.

Figure 1 illustrates a conceptual view of an OADM made up of four bi-directional transport interfaces, two on every direction – East and West – each capable of carrying up to N wavelengths, and four single-wavelength bi-directional tributary interfaces. In each direction, one transport interface is used to convey high protection-priority premium data, whilst the second interface conveys the ET.

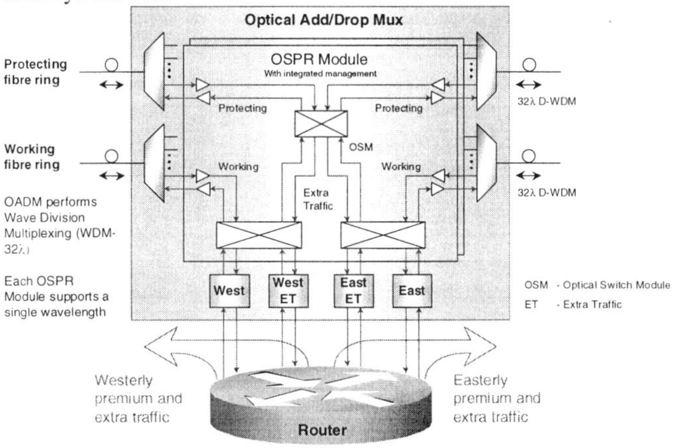

Figure 1: Optical Add / Drop Multiplexor Structure

3 Traffic Characterisation

Transport protocols operating across packet based networks can implement congestion control to avoid congestion collapse. Congestion collapse occurs when the network is so overloaded that it is only forwarding retransmissions, and most of them are delivered only part way before being discarded. Congestion control reduces the offered load on the network when it becomes congested.

3.1 Adaptive Traffic Characterisation

Transmission Control Protocol (TCP) is an adaptive transport protocol with congestion control that constitutes about 90% of Internet traffic [1]. It achieves congestion control by managing the transmission window size, specifying the amount of data that can be sent unacknowledged at any one time, and the retransmission timer value, that defines how long to wait before re-sending unacknowledged data. The TCP congestion control algorithm uses packet loss as an indication that there is congestion somewhere in the network. This loss manifests itself as either a timeout event or the receipt of duplicate acknowledgements.

The flow control mechanism is comprised of two parts [5]. Firstly, the sender's effective window must be increased to take advantage of any available bandwidth, as it becomes available. This employs a congestion avoidance mechanism to ensure that the increasing data transmission is handled by slowly, but continually, increasing the size of the send window. The aim is to take advantage of available resources, but not so fast that the network overloads. In particular, the increase should be sufficiently slow so that feedback from the network or remote end of the connection arrives before there is significant risk of increased network congestion. Secondly, the sender's effective window should be rapidly and significantly reduced once congestion is detected.

3.2 Non-Adaptive Traffic Characterisation

Non-adaptive applications refer to those without a closed-loop feedback mechanism to adapt to changing conditions in network traffic, routing, load, etc. In this study, it refers to those applications running on top of UDP, such as real-time voice services.

4 Performance Modelling

The complexity of the chosen network scenarios preclude the use of analytical methods. Hence, the work involved the use of *OPNET Modeler*[1], a network-modelling tool to develop a simulation model capable of representing the functionality and performance of IP networks together with the salient features of the underlying OSPR technology.

The research confined itself to the performance impact of protection events on premium services. In addition the work focussed on scenarios where all the ET is directed exclusively onto the protecting fibre under normal conditions and discarded when protection events arise causing the premium traffic to be re-directed onto the protecting fibre. There is no interaction between the premium and ET flows. The study also envisioned scenarios where the protected traffic flows would be present in relatively high quantities, resulting in typical Offered Load figures of around 60%. This allowed the work to focus on scenarios where TCP flows are present in quantities that enable them to expand their congestion windows to a sizeable value under normal operating conditions.

Several scenarios were considered that reflect the anticipated usage of Internet services by three types of user: the large corporate user, the Small - Medium Enterprise (SME) and residential users via an Internet Service Provider (ISP). Additionally, the services being considered were: voice, email, ftp and http. The percentages of the total traffic for each of these services for the three scenarios are shown in Table 1.

	Voice	email	ftp	http	Access Rate
Large Corporate	30%	20%	10%	40%	155 Mb/s
Small/Medium Enterprise	30%	15%	5%	50%	2 Mb/s
Residential (ISP)	40%	15%	5%	40%	128 kb/s

Table 1: Percentages of Total Traffic

To facilitate a comparative assessment between the data obtained from each of the scenarios, these were normalised in that the total offered load remained fixed at 60% of the inter-OADM link capacity, set to STM-1, STM-4 and STM-16 rates. This provided an opportunity to dimension the traffic generators using the total packet arrival rate as a starting point. Furthermore, these three scenarios were mapped into a network model consisting of an OADM ring with four nodes, each one representing a core router implementing Open Shortest Path First (OSPF) as the dynamic routing protocol. Each router is considered to be part of an OSPF routing area expanded to provide the number of traffic generators required in every scenario.

In order to assess the performance of the different scenarios during a protection event, a failure was introduced in the form of a fibre cut causing the diversion of traffic onto a neighbouring span or in the opposite direction along the ring. The main parameters varied during the different simulation runs, were those considered as to having greater impact on the behaviour of end-system applications. The distance between nodes was set to be the same between each pair of OADM nodes, varying between 100 and 300 km. The cumulative switch-over time, composed by different discrete events, was set to vary between 50 ms and 400 ms. Finally, the inter-OADM link speeds were varied between STM-1 and STM-16.

[1] OPNET is a registered trademark of OPNET Technologies Inc.

5 Analysis and Discussion

The work culminated in an extensive series of tests. The results take into account the effects of aggregation and the impact of protection events, while maintaining a fixed offered load and systematically varying the distance, the number of concurrent connections and cumulative switch-over time.

5.1 Adaptive Traffic Behaviour

Within this paper the authors confine themselves to describing the aggregate traffic behaviour in order to provide a quantitative unbiased estimate of the disruption in the network in response to a link failure. A metric to quantify such disruption is introduced in the form of Entropy and Directed Divergence.

Different simulation sets were run with different seeds, probing the aggregate TCP session delay amongst other relevant statistics. The shape of the probability density curve, in every case, indicates a normal distribution; thus being statistically characterised by the mean and the variance. These two parameters, however, represent single numerical values from the information in a random sample, and therefore subject to sample variability. Therefore a confidence interval is employed.

For each one of the three scenarios, the mean and the variance were used to create a confidence interval in which the aggregate TCP session delay should fall with high degree of confidence as shown in Figure 2.

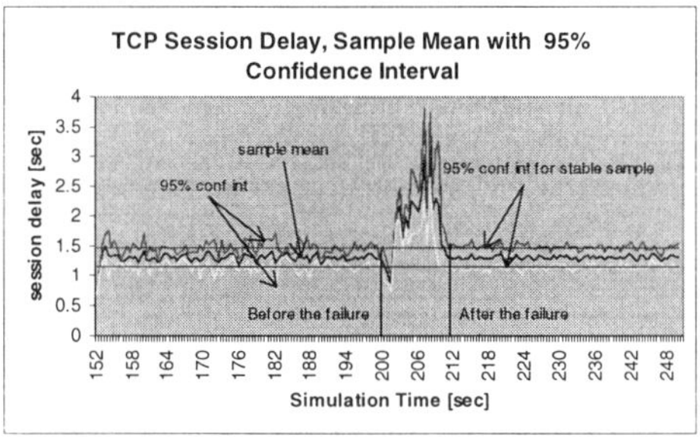

Figure 2: TCP Session Delay with Confidence Interval

Figure 3 illustrates how, for the Large Corporate case, the values of the TCP session delay fall into the confidence interval defined for the stable case, providing an unbiased estimate of the time elapsing between the moment the failure arose and the moment when the network returns to a stable state.

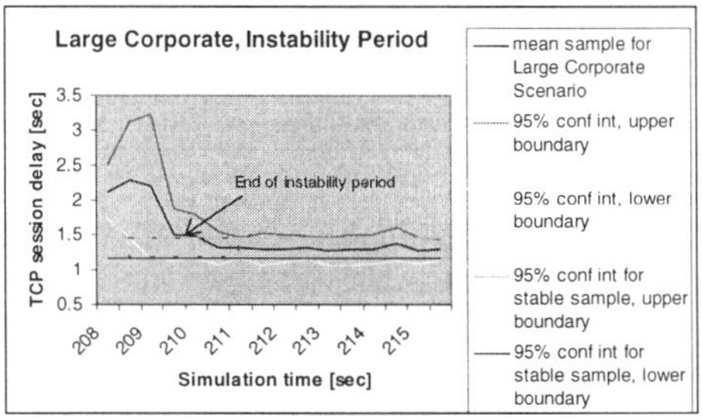

Figure 3: Large Corporate, Instability Period

Likewise, a similar procedure was followed to estimate the maximum disruption period for the remaining two scenarios, the Small Medium Enterprise and the Residential ISP.

In Table 2, it can be appreciated that, even though the switch-over event is less than half a second for every case, it has greater impact on the aggregate TCP session delay. It is clear that for every case the period of instability is different, increasing as the number of concurrent connections is reduced[2].

Scenario	Instability period
Large Corporate	between 9 and 11 seconds
Small/Medium Enterprise	between 7 and 9 seconds
Residential/ISP	between 5 and 6 seconds

Table 2: Instability Period

Even though it would seem obvious that after a period of instability the network stabilises and starts operating more consistently, it is required to statistically define whether or not there is significant difference between the mean and the variance before and after the failure. A Null hypothesis and a rejection area had to be defined and tested for every case. After the testing there appeared to be no significant difference between the means of the TCP session delay sample sets before and after the failure. That is, the network indeed returns to a stable state after a greater period of instability immediately following the protection event.

5.1.1 Entropy as a Measure of Disruption

From simple observation, it is possible to see that a greater disruption occurs immediately after the failure occurs, it is however hard to measure the amount of disruption caused by the protection event and the cumulative switch-over time. The concept of Entropy is introduced to provide a quantitative measure of the disruption in the network.

According to the maximum-entropy principle [3]: *"Given some partial information about a random variable, we should choose that probability distribution for it, which is consistent with the given information, but has otherwise maximum uncertainty associated with it."*

In other words, unless the outcome is known *a priori* and always equal, there is some uncertainty associated with every outcome which is somehow related to both the *a priori* distribution (relative entropy) and the spread of our set of possible outcomes. Thus the entropy measure will be maximum when there is maximum uncertainty arising when the outcomes are equally likely (and different), and minimum (and possible zero) when there is no uncertainty about the outcome – there is only one outcome certain to occur.

Alternatively, entropy can be expressed in terms of divergence or directed divergence [2]. This can be interpreted as a form of distance or dissimilarity between two probability measures, although being more mathematically rigorous, it is more a measure of distortion as considered in information theory, which is a generalisation of the notion of distance.

Shannon suggested in 1948 a measure of entropy, directly inherited from thermodynamics and previously suggested as a measure of the information in a random signal. Like Shannon's measure of entropy, this new measure of disruption has to be consistent with Laplace's principle of insufficient reason that unless there is information to the contrary, all outcomes should be considered equally likely [3]. In this study, after a failure occurs, the behaviour of the aggregate traffic would varies depending on the state of the transmission and, in the case of adaptive applications using TCP, whether they are in slow start or congestion avoidance phase. It also depends on the number of unacknowledged segments lost during the switch-over time. We could say that for a period of time, it would be practically impossible to forecast the behaviour of the end-system applications.

---

[2] In order to maintain a normalised fixed offered load for the three scenarios, the number of concurrent connections is increased as the per-connection access rate is reduced (see Table 1).

If the behaviour of the network during stability periods is known, it is also possible to associate it with a prior distribution. Then another measure of entropy, called Bayesian Entropy, would be defined by the expression below [3].

$$B(p) = -\sum_{i=1}^{n} p_i \ln \frac{p_i}{\alpha_i (\alpha_i)_{\min}} = -\sum_{i=1}^{n} p_i \ln \frac{p_i}{\alpha_i} - \ln(\alpha)_{\min}$$

Moreover, from Information Theory, the first term in the Bayesian Entropy expression, can be defined as the Directed Divergence [2] of a distribution P from the a *priori* distribution A, leading to the following expression:

$$B(p) = -\sum_{i=1}^{n} p_i \ln \frac{p_i}{\alpha_i} - \ln(\alpha)_{\min} = -I(P:A) + \ln\left(\frac{1}{\alpha}\right)_{\min}$$

It can be appreciated that if A is associated to the collection of outcomes from the experiment when no failure exists, and P corresponds to the values obtained from an experiment where disruption, in the form of i.e. greater TCP session delays, exits, the Directed Divergence from A to P would be minimised, maximising the Bayesian Entropy expression. Therefore, the greater the Bayesian Entropy, the greater the disruption and the less the certainty of the possible outcomes.

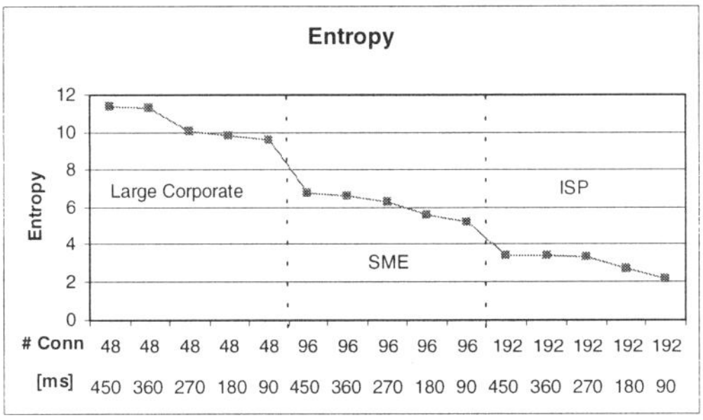

Figure 4: Entropy as a Function of the Number of Connections and Switch-Over Time

Figure 4 shows how the Entropy changes when both the cumulative switch-over time and the number of connections are systematically modified. After assessing the extent of the relationship between both the number of connections and the cumulative switch-over time, an indication of a weak positive relationship between the cumulative switch-over time and the Entropy was found. This contrasts with a strong negative relationship between the number of connections and the degree of disruption in the network. Hence, if the number of concurrent connections is increased, maintaining a fixed offered load and therefore reducing the transmission window size, the disruption in the network would be also reduced as a direct consequence of short-lived transmission flows. Moreover, the cumulative switch over time affects the performance of the network, becoming more evident when the number of connections is reduced, as each connection is likely to have a greater transmission window, therefore allowing for more segments to be lost during the protection event.

5.2 Non-Adaptive Traffic Behaviour

Voice Over IP (VoIP) services were characterised as running over UDP with no retransmissions of lost or corrupted data. Therefore the volume of lost data is a simple function of the switch-over time. This loss of data is inevitable since the OSPR functionality provides 1:1 protection. However depending on the mechanisms employed in the higher layers of the protocol stack, this loss may or may not have an impact on the quality of the service received by the end user.

The effect of this loss of data is not as significant as one might at first expect. This is because VoIP services employed to date typically use a playback buffer to minimise the effects of the variable transmission times of the packets as they cross the network. Both the size of this playback buffer and how the system reacts when the buffer empties also varies between implementations.

It can therefore be determined that the OSPR switching will only have a perceived impact to the end service user if this results in the playback buffer emptying, referred to as buffer under-run. When the buffer empties the impact will relate to the control mechanism employed in the playback buffer. For example this might require the buffer to refill to half capacity or another predetermined level before recommencing the playback, or to start playback on receipt of the first data packet, possibly at a temporary reduced playback rate. This latter mechanism could result in further buffer under-runs if the network is congested causing large variations in packet delay.

6 Summary

For UDP based services with no transport layer feedback mechanism, such as VoIP, the switch-over delay results in a service interruption of approximately the same duration though this can be ameliorated by the operation of the playback buffer at the receiver.

TCP performance is strongly affected by switch-over delay with parameters such as session delay worsening as the switch-over time increases due to the reduction in both the congestion window and the slow start threshold. For a given switch-over time, decreasing the number of concurrently active connections increases the disruption they experience resulting from the switch-over event.

It was also found that the magnitude of the disruption becomes noticeable worse when the switch over time exceeds 200 ms, which is partially attributable to the granularity of the retransmission scheme in practice at the transport layer along with the particular delayed acknowledgement policy. A switch-over time of less than 200 ms provided an acceptable degree of protection for premium services in the selected scenarios, however, further research is needed to ascertain the sensitivity of this figure to different TCP implementations.

Acknowledments

The authors greatly appreciate the support of the many people from Nortel Networks Limited who contributed their time and energies to improving this paper. They include Malcolm Duff, Mike Kelly, Kevin Warbrick, Peter Atterton and Ashraf Khan. However, the views expressed in this paper are personal to the authors and do not necessarily represent those of the people we acknowledge, their employers or our employers.

References

[1] K. Thompson *et al.*, "Wide-Area Internet Traffic Patterns and Characteristics", MCI Telecommunications, IEEE Network, November/December 1997, pp.10-23.

[2] R. Gray, "Entropy and Information Theory", Springer-Verlag, 1990.

[3] J. Kapur, "Maximum Entropy Models in Science and Engineering", Wiley Eastern Limited, 1989.

[4] Schmitt, Samuel A., "Measuring uncertainty: an elementary introduction to Bayesian statistics", Addison-Wesley , 1969.

[5] W. Stevens, "TCP Slow Start, Congestion Avoidance, Fast Retransmit, and Fast Recovery Algorithms", RFC2001, January 1997.

[6] D. Al-Salameh *et al.*, "Optical Networking", Bell Labs Technical Journal, January 1998.

Broadband Access, WDM Metro and Network Management
D.W. Faulkner and A.L. Harmer (Eds.)
IOS Press, 2000 © AKM

Implementation and Management of the KomNet DWDM Metropolitan Network

Klaus-Dieter LANGER, Joachim VATHKE, Godehard WALF
Heinrich-Hertz-Institut für Nachrichtentechnik, Einsteinufer 37, D-10587 Berlin, Germany

Gert EILENBERGER, Lars DEMBECK, Ulrich BRONIECKI
Alcatel SEL AG, Holderäckerstrasse 35, D-70499 Stuttgart, Germany

Ralf-Peter BRAUN, Andreas GLADISCH, Sabine SZUPPA, Frank TETZLAFF
Deutsche Telekom Innovationsgesellschaft mbH, Goslarer Ufer 35, D-10589 Berlin, Germany

Lutz BERSINER, Hans LAUSEN, Erhart LEDERER, Günter NEUMANN
Robert Bosch GmbH, Robert-Bosch-Strasse 200, D-31139 Hildesheim, Germany

Harald BOCK, Detlef STOLL, Alexander RICHTER, Patrick LEISCHING, Georg FISCHER
Siemens AG, Hofmannstrasse 51, D-81359 Munich, Germany

Abstract. Under the framework of the German initiative KomNet, a consortium of industrial partners, cooperating research institutes and a network operator is pushing R&D on key technologies for the implementation of DWDM networks. This paper describes the implementation of an experimental metropolitan network from various viewpoints including network conception and trial topology, multi-vendor interoperability issues, OADM architectures and network operation aspects. Innovative network management concepts being developed in that technical context are discussed.

1. Introduction

The unprecedented worldwide growth of Internet traffic and the migration of the Internet towards the Broadband Internet lead to an ever-increasing performance requirement for the telecommunication networks. Under the framework of the German Federal Ministry of Education and Research (BMBF) research initiative 'KomNet – Innovative Communication Networks for the Future Knowledge Society', numerous R&D projects have been under way since mid-1998 that contribute to implementation options for corresponding transport networks. The projects [1] focus on the development and evaluation of advanced DWDM (Dense Wavelength Division Multiplexing) networking technologies.

Metropolitan area networks based on DWDM technologies are expected to play an important role in the evolution of photonic networks, [2], [3], [4], [5]. Challenging deployment aspects including cost-effective network operation, metro optical networking, multi-vendor interoperability, standardization issues, metropolitan network management systems, reconfigurable optical add-drop multiplexers (OADM), and in-service monitoring of the signal transmission quality are research objectives covered by the KomNet metro network workplan. This paper describes the architecture of the managed experimental metropolitan network and summarizes first achievements.

Long-haul Transmission
- ETDM 40 Gbit/s, OTDM 160 Gbit/s
- NRZ, RZ, QDST Transmission
- WDM *n* x 40 Gbit/s, *n* x 160 Gbit/s
- Bidirect. Single Fiber WDM 2 x 8 x 10 Gbit/s

Metropolitan Networks
- Multi-Vendor Transparent WDM Ring Networks
- 16 x 2.5 Gbit/s ... 80 x 10 Gbit/s

Access Networks
- IP & ATM via HFC and PONs
- SDSL Wideband Internet Access & ISDN
- Advanced xDSL Technology for QoS
- OCDMA Transmission via PONs
- Hybrid Fiber Radio

Network Functionalities
- Network Management
- OAM & Performance Monitoring
- Optical Protection Switching
- Optical Add-Drop Multiplexing
- IP over WDM

Network Elements & Components
- Transmitter & Receiver OEICs up to 40 Gbit/s
- Transceivers (Hybrid & PIC)
- ETDM ICs 40 Gbit/s (Si & GaAs)
- Configurable OADMs
- Fiber Dispersion Compensators
- IC & OEIC Packaging (2.5 ... 40 Gbit/s)

Figure 1: KomNet Topics

2. KomNet - Scope and Field Trial Topologies

KomNet is a national-scale platform, initiated to perform and integrate research and experiments for the next generation Broadband Internet. The KomNet consortium formed by Alcatel SEL, Bosch Telecom, Infineon Technologies, Lucent Technologies, Profile, Robert Bosch, Siemens and T-Nova collaborates in 18 R&D projects with 21 German universities, SMEs and research institutes. The scope of the work is illustrated in Figure 1, and a more detailed description of the program can be found in [6].

Using the existing Deutsche Telekom standard single mode fiber infrastructure, a demonstration communication network interconnecting the experimental metropolitan network located in Berlin with sites in the cities of Darmstadt, Mannheim and Stuttgart, Figure 2, is currently being implemented and will be tested under real life operating conditions.

In close cooperation with T-Nova, the field experiment configuration of the metropolitan network, Figure 3, is being established in the city of Berlin. The trial topology consists of two interconnected DWDM rings controlled by a network management system, together with an interface facility to the long-haul network. To demonstrate the interoperability of multiple vendors, Alcatel and Bosch are jointly realizing one ring network, and Siemens is providing the second one.

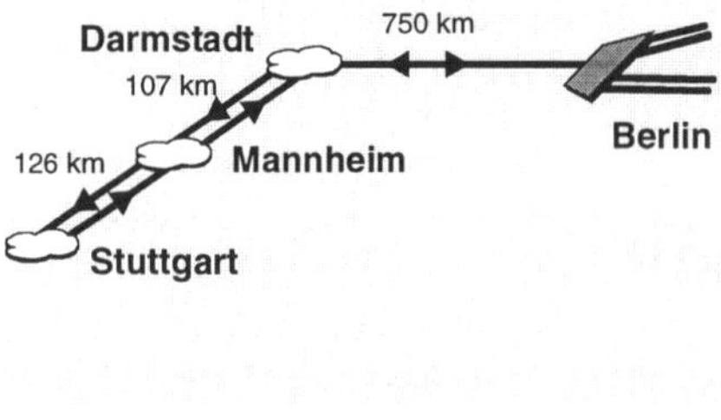

Figure 2: Long-haul field trial topology

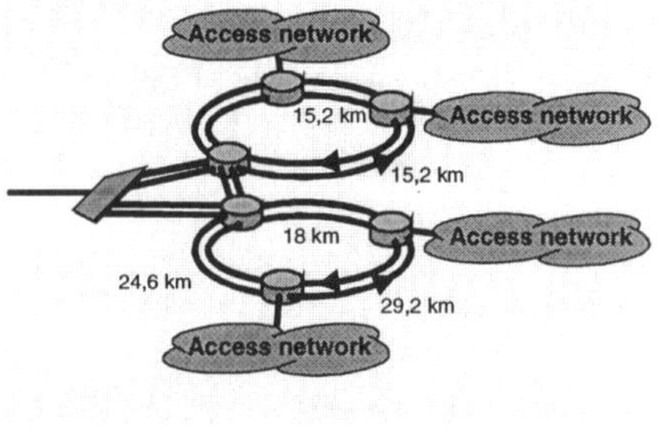

Figure 3: Metro field trial topology in Berlin

3. Metropolitan Network Configuration

The optically transparent demonstration testbed in KomNet consists of two interconnected two-fiber bi-directional line switched rings designed to optionally boost the number of wavelengths in each standard single mode fiber from a nominal of 16 up to 80. Three OADMs will be placed along each ring arrangement to enable the interconnection of various advanced access networks via the ring backbone with a total length of more than 100 km. Different node architectures coexist, including partially equipped hub nodes at the rings connection point and fixed or flexible OADMs of different sizes. Siemens is developing fully reconfigurable OADMs suitable for use with large channel count WDM systems, i. e. up to 80 channels at 10 Gbit/s data rates. Alcatel is contributing opto-electronic nodes that perform routing at the IP and ATM levels while operating in the optical layer at up to 10 Gbit/s and with different optical signal formats (e. g. OCDMA, digital feeder). The design of the network elements includes the development of electronic, optical and opto-electronic (o-e) components such as transmitter modules with integrated wavelength reference and laser modules incorporating integrated electro-absorption modulators. It is intended to assess the key functionalities of this metropolitan network under real operating conditions. Using different types of novel access networks, a mixture of advanced multimedia applications will be demonstrated, ranging from broadband Internet traffic to optically transparent connections over leased fibers.

Comprehensive management and control functions are required for the operation of complex communication networks. Along with the design of a uniform network management architecture, fundamental control and management functions will be developed including optical supervisory channels and optical protection switching schemes.

Fast protection schemes realized by optical hardware are a great asset in general of WDM networks and especially for transparent WDM metro networks, and will thus enable considerable efforts in that field. Optical multiplex section (OMS) protection as well as optical channel (OCH) protection has to be considered. Without precluding OMS protection schemes [7], OCH 1+1 protection has been chosen for implementation within the KomNet metro network. The number of wavelength channels required is summarized e. g. in [8]. OCH 1+1 protection means duplicating the optical signal at the transmitter side using a 3 dB power splitter and transmitting the signals over two separated paths. Protection switching is performed at the receiver side. As long as protection is not active in lower layers, OCH protection offers a lowest-layer measure against fiber break consequences.

4. System Concept and Characteristics of a Configurable OADM

The concept of Siemens' OADM for the field trial is depicted in Figure 4. This functional diagram shows the OADM configured for bi-directional ring traffic. The optical line interfaces (OLI) compensate for the fiber and internal OADM losses by using optical booster amplifiers (EDFAs). The add-drop process is performed by a two-stage add-drop switching unit (OADM-R). The first stage is the add-drop stage which filters the drop channels out of the WDM signal and adds tributary signals into the ring. The second stage is an optical cross connect (OCC) which assigns the dropped signals to the channel access modules (CAM). The OADM-R on the left adds and drops the westbound traffic, on the right a similar unit is related to the eastbound traffic. No separation of working traffic and protection traffic exists in bi-directional 2-fiber rings. The working traffic usually takes the shortest path to the destination node. Thus, a distribution stage (OCC) is necessary to sort the channels from westbound/eastbound traffic to separate working and protection tributary interface arrays. In the system the distribution unit is realized as a 4×4 space

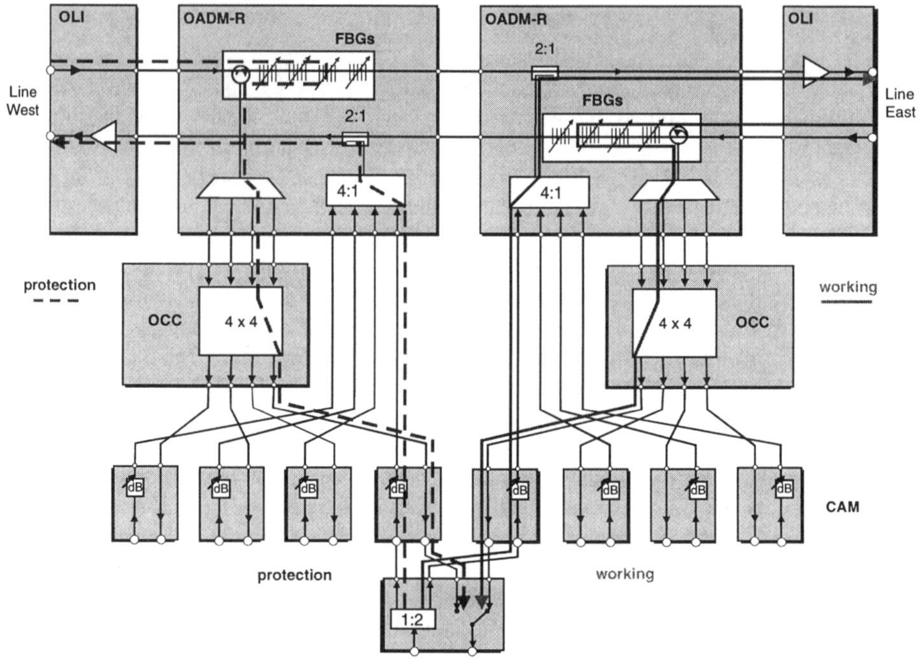

Figure 4: System concept of Siemens' configurable OADM used in the KomNet field trial

switching matrix. Different switching technologies, polymer-based waveguide optical switches and MEMS-based technologies, will be tested during the field trial. CAM units provide tributary interfaces and monitor diodes for optical power level supervision. An array of four 2×1 switches and four 1:2 couplers (3 dB power splitters) per OADM is implemented for optical path 1+1 protection.

The DWDM signal (channel spacing 100 or 50 GHz resp.) entering the add-drop stage from the left ring input is forwarded clockwise by the optical circulator to the next port. The signals to be dropped are reflected by tunable fiber Bragg gratings (FBGs). The DWDM channels can be dropped or passed through by tuning these band reject filters, i. e. if their resonance frequency is tuned to a channels center frequency, this channel is dropped. If it is tuned to a neutral park frequency (between two adjacent wavelength channels), the optical signal will pass the filter [9]. Each signal to be dropped requires an individual tunable filter. The reflected signals are dropped via the circulator. A static optical demultiplexer performs the final separation of the dropped WDM channels. Moreover, it suppresses interchannel crosstalk significantly. Up to 50 dB can be achieved. The add process of four DWDM channels per OADM is realized in a cost-efficient way by using a frequency independent 4:1 coupler and a 2:1 coupler.

The cascading of tunable FBGs at 10 Gbit/s is investigated by means of experimental results and theoretical simulations. Each span consists of an attenuator (linear and nonlinear fiber effects and crosstalk were excluded), an EDFA (21 dB gain) and an add-drop module. The receiver penalty is calculated as a function of the detuning between laser frequency and the channel center frequency for signals that have travelled through a certain number of spans, Figure 5. A more detailed description can be found in [10]. In real world systems, slight variations in the frequency positions of the FBGs drastically reduce penalty peaks, which are attributed to ripple in the group-delay of the FBG, Figure 6. Further results on power penalties at 10 Gbit/s and thermal stability of FBGs are published in [9] and [11]. These results show that tunable FBGs are promising candidates for dropping of

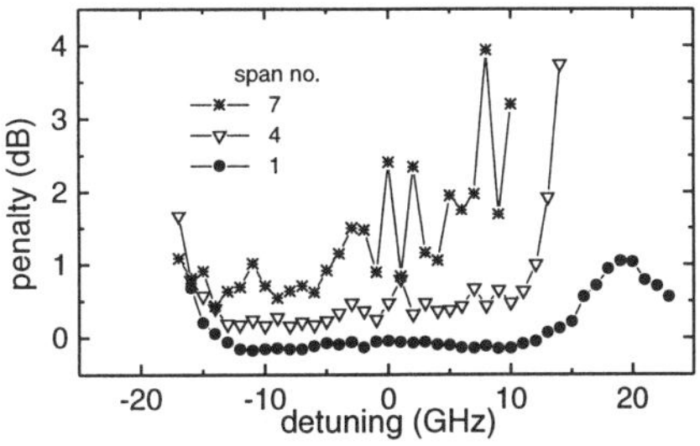

Figure 5: Receiver penalty spectra resulting from the cascading of identical OADM modules

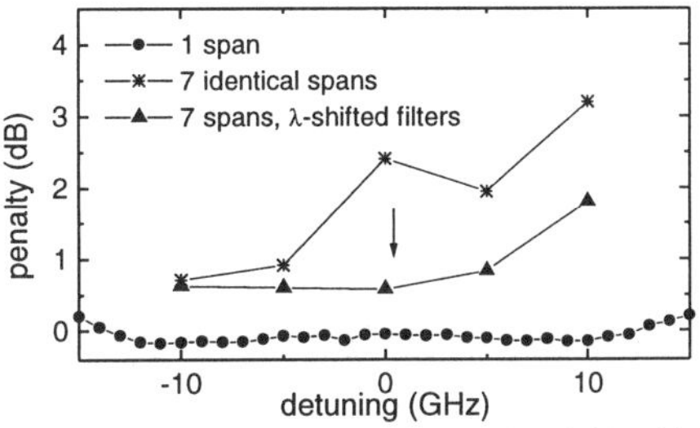

Figure 6: Simulated reduction in penalty obtained by detuning the FBGs' center wavelength within 5 GHz

DWDM signals dynamically. It is expected that the loss of the filters can be reduced to 0.3 dB. Thus it will be possible to cascade 16 filters or more depending on the system concept.

5. Parameters for Performance Monitoring and Fault Management

The specific characteristics of optical Network Elements (NE) have been studied widely in research projects and have already been introduced partly in standardization documents for WDM core networks [12]. In principle, optical NEs in WDM metro networks will apply to similar rules and will deal with the same parameters for performance monitoring and fault management:

- Optical signal power
- Operation temperature of components/modules
- Optical Trail-ID
- Wavelength accuracy and stability
- Optical signal-to-noise ratio

Most of these parameters on optical NE level are hardware related and strictly dependent on the individual realization options. Besides vendor individual variants of optical components and modules, there also exist application specific variants. The WDM metro networks under study in the KomNet program, for instance, will be designed under specific boundary conditions – one of them being the strict low-cost target – which will result in a reduced choice of processes to be implemented inside the atomic functions of the related optical layers, i. e. optical transmission section (OTS), OMS and OCH. The important assumption, that a WDM metro network will consist of a single transparency domain, will have a major impact on the functional definition and the presence of processes.

Despite all these implementation options, the representation of the NE functionalities to the Network Management System (NMS) has to be unique, abstract and must hide equipment related and vendor specific features. Thus, for fault management, the defects on NE level are mapped onto only two alarms related to the Opt. Channel Trails, Opt. Multiplex Trails and Opt. Transmission trails of the WDM layer: Loss_of_Continuity (LOC) and Trail_ID_Mismatch (TIM). At the NE level, a set of Operation, Administration and Maintenance (OAM) data needs to be communicated between adjacent NEs in order to implement fast protection schemes in the optical layers. These signals comprise Forward Defect Indication (FDI), Backward Defect Indication (BDI) and a few auxiliary signals to indicate the protection state and the related Trail ID etc. The detailed hardware related parameter values are also made available to the NMS on demand for maintenance purposes.

For performance monitoring, the degradation of hardware related parameters (anomalies) will cause notification messages to the NMS as soon as programmable thresholds are crossed and are also provided on demand, but unlike alarms they are not processed in real-time, since only off-line maintenance actions will be initiated.

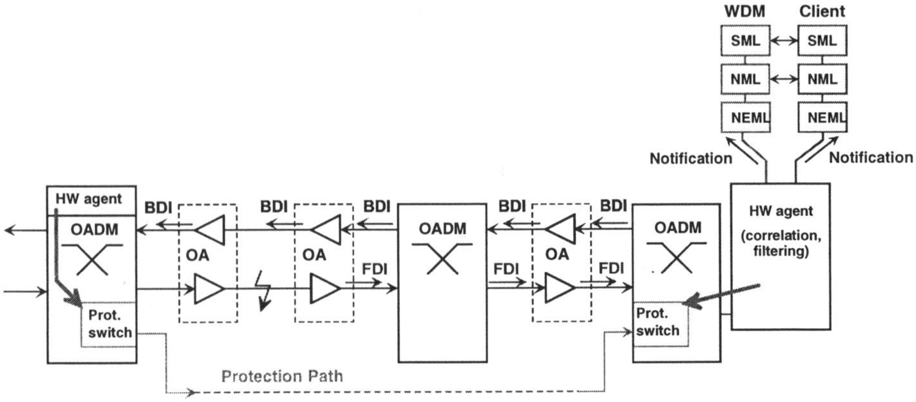

Figure 7: Protection schemes based on FDI/BDI signals

6. Management of Protection Schemes

The protection schemes are controlled on NE level with the help of the said FDI/BDI signals. A NE detecting a defect on any of the WDM layers will issue FDI in forward direction and BDI in backward direction both related to the affected layer. These OAM signals are transported via the Optical Supervisory Channel (OSC), a separate wavelength channel carrying the overhead signals for the WDM layers, directly to the adjacent NE. The FDI/BDI signals are forwarded to the next routing node, where a protection switch is implemented, and trigger the protection actions, Figure 7. This mechanism is not controlled by the management system and is extremely fast (in the range of a few milliseconds).

The OAM signals also include a protection indicator which enables the use of working and protection paths on both ends of the bi-directional connection simultaneously. Since only OCH connection protection is implemented, any failure in lower layers will result in Server Signal Fail (SSF) signals passed to the OCH layer. There, an OCH-FDI is forwarded towards trail termination sink and an OCH-BDI towards trail termination source. For the case of a network connection running over more than one ring interconnected by OADMs, ring-connecting protection is provided by passing the corresponding overhead signals (i. e. FDI/BDI) for the interconnecting channels between the rings. After completion of the protection switching, the management system is notified about the action to update the configuration databases and to eventually initiate maintenance and/or restoration actions.

7. Management Systems Interworking

Most of the client layer systems using the WDM server layer provide their own fault management with protection and restoration schemes. It is obvious that interfering fault management systems with similar latency times and protection switching schemes may lead to unpredictable situations and could even result in unstable oscillating conditions in the network. Since the reaction times of client and server layer management systems – due to exhaustive protocol stacks and long transmission distances – will range several orders of magnitude longer than the hardware based protection schemes, the coordination of protection actions in the Network Management Layer (NML) or Service Management Layer (SML) is nearly impossible.

The only practical way to harmonize the behavior of protection schemes is seen on NE level. Figure 8 shows a specific example of a hybrid o-e routing node in a WDM metro network. The NE consists of an OADM part for routing and add-drop multiplexing of WDM channels (i.e. termination of optical channel trails) and an electronic part dealing with client layer trails. This NE may combine interfering protection schemes in both WDM server and electrical client layers.

The proposed approach consists of an interworking scheme between server and client layer management systems on different levels. On NE level, all necessary OAM data need to be exchanged between server and client layer to enable a consistent and coordinated protection switching action within the required time, instantly in a few milliseconds, which would not be possible otherwise. This interworking scheme has not yet been considered up to now in standardization bodies and will potentially affect existing standards. For the coordinated behavior on network level in terms of restoration and reconfiguration, a second scheme is proposed on NML or SML level with integrated management services to ensure a consistent network view over both server and client layers for various aspects of configuration, fault and performance management.

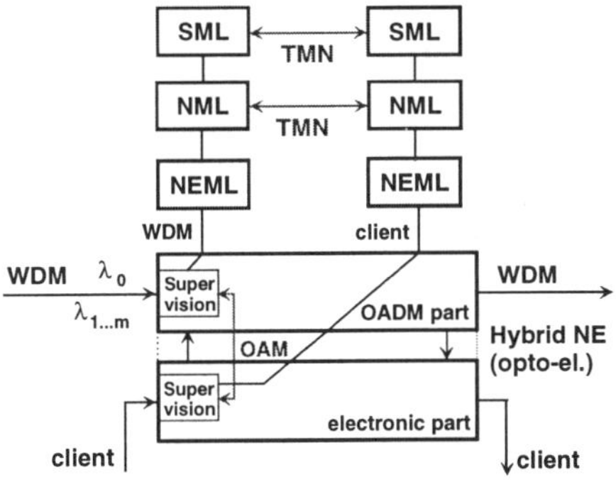

Figure 8: Interworking scheme of management systems

8. Further Aspects on Management of Optical Metro Networks

Metro networks have to meet requirements of both, core and access network, mainly the transport and integration of multi-client formats (voice centric and data centric formats), the increasing demand on capacity, and the seamless manageability of the optical transport network (OTN) within an umbrella network management.

Figure 9 shows a network connection through an OTN via different separated domains, e. g. administrative, vendor, and transparent transmission domains. This allows for individual monitoring and performance measurements to guarantee the performance of the specific sub-trails. For supervising the end-to-end trail, an extension of the functional architecture of OTNs [12] is currently discussed using digital framing (e. g. digital wrapper [13]) and tandem connection monitoring. Therefore o-e conversion and 3R-regeneration are required at the borders between domains e. g. in inter-domain interfaces (IrDI).

DWDM metro networks are considered to be transparent domains meeting the various

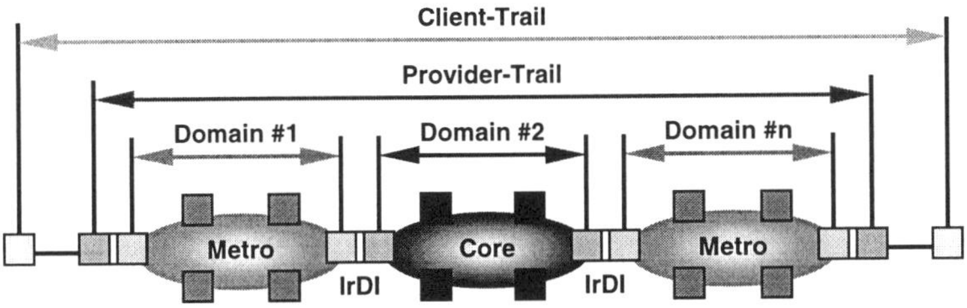

Figure 9: Monitoring and Management of trails within the optical transport network. Domains are connected via inter domain interfaces (IrDI). Boxes correspond to network elements.

requirements and avoiding o-e conversions and 3R regeneration. This efficiently reduces the system costs which is an essential factor to metro networks. Further cost reductions can be achieved by transparent tunneling of the metro network, i. e. by omitting the 3R regeneration in NEs at the interfaces of the metro network. How the transmission performance could be guaranteed in that case is an open issue to transparent networks. In any case, a pro-active trail monitoring and survivability within the metro network domain has to be achieved.

9. Future Work

We have described the concept and the implementation plan of the initial experimental metropolitan network as a part of the KomNet telecommunications testbed. This demonstration network will be in operation from mid-2000 to the end of the KomNet project in 2002, and will serve as research platform to field-test emerging and advanced networking technologies.

Future work will also focus on testing and demonstrating IP-based bandwidth-intensive real-life applications using the entire trial network (metro network segment, novel access networks and ultra-high-capacity long-haul transmission systems). A complementary goal is to develop management architectures and systems for all levels of the WDM metropolitan network and to study issues associated with the multi-vendor network operation including equipment interoperability and the common view of management systems.

10. Acknowledgement

The authors would like to thank their project partners involved in KomNet for fruitful collaboration. The financial support by the German Ministry of Education and Research under contracts 01BP 805, 807, 812, 815, and 816 is gratefully acknowledged.

References

[1] K.-D. Langer, J. Vathke, G. Walf, KomNet - a modular platform to assess optical networking techniques and components. In: A. Bononi (ed.), Optical Networking. ISBN: 1-85233-641-2, Springer Verlag 1999, pp. 48 - 51.

[2] G. Ellinas et al., Architecture considerations in merging multi-vendor WDM rings for the Washington D.C. network. Proceedings of OFC '99, vol. 2 (1999), pp. 190 - 192.

[3] L. Berthelon, AC355 PELICAN: Advanced photonic networking field trial in both access and transport. HORIZON Workshop on WDM networks and systems, the convergence of data and optics, 28.10.1999.

[4] Web site http://www.infowin.org/ACTS/RUS/PPROJECTS/ac354.htm.

[5] G. Bendelli et al., PEGASO OTN Testbed. Proceedings ECOC '99, Vol. 2 (1999), pp. II-114.

[6] Web site http://www.hhi.de/komnet.

[7] A. Gladisch, N. Hanik, G. Lehr, Netze mit optischem Frequenzmultiplex. Der Fernmeldingenieur, Vol. 51, No. 6 - 8 (1997), pp. 68 - 73.

[8] S. Johansson et al., A cost-effective approach to introduce an optical WDM network in the metropolitan environment. IEEE Journal on Sel. Areas in Communications, Vol. 16, No. 7 (1998), pp. 1109 - 1122.

[9] P. Leisching et al., Optical add/drop multiplexer for dynamic channel routing. Electronics Letters, Vol. 35, No. 7 (1999), pp. 591 - 592.

[10] H. Bock et al., System impact of cascaded optical add/drop multiplexers based on tunable fiber Bragg gratings. Proceedings OFC 2000, pp. 296 - 298.

[11] A. Richter et al., Passive temperature compensation of piezo-tunable fibre Bragg gratings. Electronics Letters, Vol. 35, No. 15 (1999), pp. 1269 - 1271.

[12] Architecture of optical transport networks. ITU-T draft recommendation G.872 (ex G.otn).

[13] Lucent Technologies - North America, A proposed implementation for a digital "wrapper" for OCh overhead. ITU-T Delayed contribution D.498 (WP4/15), June 1999.

Multi-Band Hybrid WDM Transmission of Voice, Video & Data Enabled by Zero-Waterpeak Optical Fiber

Stefan HUNSCHE, Atul K. SRIVASTAVA, Bernard R. EICHENBAUM, Marcus A. PACZKOWSKI, James J. THOMAS and Charles L. WOLF

Lucent Technologies, 101 Crawfords Corner Rd, Holmdel, NJ 07733, USA

Abstract: Using combination of coarse (waveband-based) and dense wavelength division multiplexing (WDM), we demonstrate simultaneous transmission of analog video signals (AM-VSB plus QAM) and high-speed digital circuit-based (STM-64) and packet data (Gigabit Ethernet) over 60 km of single mode fiber. Error-free operation with negligible crosstalk between baseband digital and analog signals is made possible by using different transmission windows, covering the whole optical spectrum from 1300 nm to 1550 nm, for different signal types.

1. Introduction

Metropolitan fiber optic networks that link high-capacity backbones and emerging high-speed access solutions, are currently a major focus for global bandwidth growth. A prime requirement for these Metropolitan networks is the flexibility to support a wide variety of applications, services, protocols, speeds, and signals – both legacy and emerging. Wavelengths division multiplexing (WDM) offers the potential of combining dissimilar signals on a single fiber while avoiding costly or inefficient electronic multiplexing schemes. Other than in the long-haul backbone, where sheer capacity is the primary goal, achieved by combinations of high speed time division multiplexing (TDM) and dense WDM, flexibility and system cost are key factors in the Metro area. While specific Metro DWDM systems have started being deployed, cost reduction of these systems still is a major goal, and the competition with traditional TDM solutions is expected to remain fierce for the near future. Coarse or "wide" WDM with optical channel spacing of tens of nanometers, rather than the customary 0.8-1.6 nm used in dense WDM, are currently being developed as cost-effective solutions. Unrestricted compatibility with these emerging systems and all established transmission technologies is absolutely essential for current Metro fiber deployment. The recent development of zero-water peak optical fiber (Lucent Technologies AllWave™ fiber) has opened a new wavelength window for optical signal transmission around 1.4μm. As traditional "single-wavelength" optical interfaces are standardized to operate in bands, e.g. 1260nm--1360nm for the nominal wavelength of 1310 nm, rather than actual well defined wavelengths, opening a new transmission window not only provides more capacity, but also an important new degree of freedom to mix and match various services on a single fiber plant.

We have constructed and tested a system demonstration that multiplexes and transmits the technically most challenging combination of analog and high speed digital over a common infrastructure. This experiment explores the possibilities of the new zero-water peak fiber for CATV operators striving to enhance traditional one-way broadcast with targeted and two-way digital services. Figure 1 shows both the attenuation spectrum and dispersion curve for AllWave fiber and conventional single mode fiber. The attenuation peak at 1385nm (schematically indicated by the dotted line) that is present in conventional single mode fiber is completely eliminated in the manufacturing process for AllWave fiber. The attenuation in the 1.4 μm band is essentially determined by the Rayleigh scattering limit, with the attenuation around 1.4 μm being lower than at 1.3 μm, so that the full spectrum from 1.2 to 1.6 μm is usable for optical data transmission.

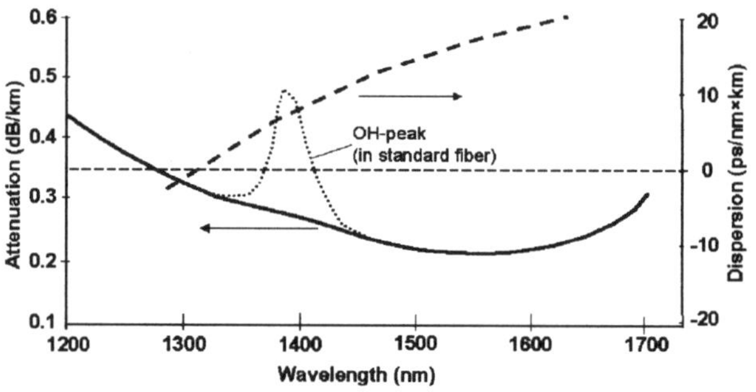

Fig. 1: Attenuation and dispersion of zero-water peak (AllWave™) fiber.

2. Experiment

One technical challenge of using WDM technology to transmit simultaneously different services on the same fiber is posed by cross-talk among different WDM channels due to optical non-linearities. To suppress impairments of dense WDM (DWDM) transmission in long-haul systems, a certain minimum amount of fiber dispersion is required. The optimum dispersion in the C-band (1530-1565 nm) of erbium-doped fiber amplifiers (EDFA) is provided by non-zero dispersion shifted fibers such as Lucent's TrueWave RS, which allows to transmission of a maximum number of high-speed DWDM channels at the lowest system cost. However, if one or more of the wavelengths carries an analog CATV AM-VSB signal, the need for cross-talk suppression is strongly increased because of the high noise sensitivity of this analog signal format. Therefore, transmission of several WDM channels carrying baseband digital data and analog CATV signals is generally not practical within a single wavelength band.

Opening the "fifth window" for optical transmission allows the placement of separate signals in different bands that have sufficient spectral separation to avoid the generation of cross talk by non-linear effects such as stimulated Raman scattering (SRS)[1]. In addition, the favorable, low but nonzero, dispersion level (6-7 ps/nm-km) renders the 1.4 μm band highly suitable for high-speed (10 Gb/s) DWDM transmission. High-speed NRZ data transmission on AllWave™ fiber in 1.4 and 1.55 μm bands was previously demonstrated over an 80 km distance[2]. In the present paper we report simultaneous error-free transmission of Gigabit Ethernet data frames at 1.3 and 1.4 μm, high-speed (10 Gb/s baseband NRZ format) telecom traffic over several DWDM channels in the 1.4 μm band, QAM data over another 1.4 μm wavelength, *and* transmission of AM-VSB video signals in 1.55 μm band on the same fiber with a CNR >50 dB over a distance of 60 km of AllWave fiber.

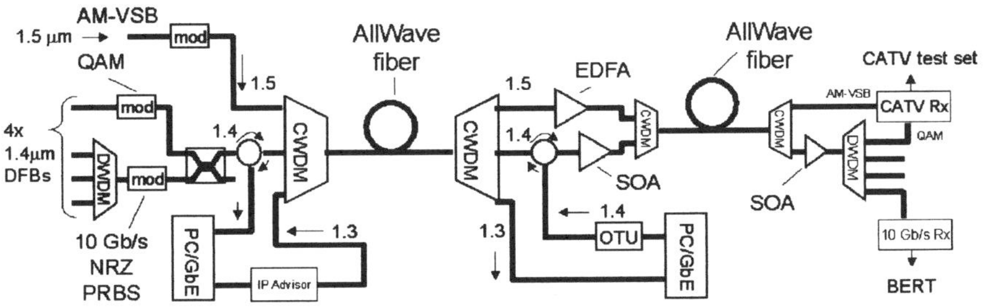

Fig. 2: System set-up

A schematic overview of the experimental setup is shown in figure 2. The system is built around two spans of AllWave fiber, 20 km and 40 km in length, respectively. Thin-film filter based coarse WDM (CWDM) devices are used to combine the 1.3, 1.4 and 1.5 μm band signals into one fiber, and split the signals after transmission for amplification or detection, respectively.

In the 1.5 μm band, we use a 77 NTSC-channel (55.25 – 535.25 MHz) Matrix Systems CATV carrier source to modulate a Scientific Atlanta Prisma 6475-3 transmitter. The optical modulation depth (OMD) on the transmitter is 3.0 % per channel, and the optical output power 8.9 dBm. After the first 20 km of fiber and band splitting by a CDWM, this signal is amplified by a Scientific Atlanta Prisma 6476-16 EDFA, transmitted over a second span of 40 km of AllWave™ fiber. After this fiber span and another CWDM the 1.55 μm CATV signal is detected by a Scientific Atlanta Prisma 6971 receiver and analyzed with a Hewlett-Packard CATV analyzer to characterize the signal quality by carrier-to-noise ratio (CNR), composite second order (CSO) and composite triple beat (CTB) measurements.

The optical signals in the 1.4 μm band consist of four DWDM channels between 1390 and 1391.6 nm with a 100-GHz channel spacing. After multiplexing with a silica wave-guide-router (DWDM), three of these channels are modulated with a Mach-Zehnder (MZ) modulator at 10 Gb/s using a 2^{23}-1 NRZ pseudo-random bit sequence (PRBS). The fourth channel is modulated with a second MZ optical modulator driven by a Scientific Atlanta D9467 QAM modulator that creates a 6 MHz DAVIC 64-QAM channel centered at 750 MHz. This fourth channel was not used during use of the Matrix Systems input but is used otherwise to quantify QAM transmission quality in the presence of 46 channels of AM-VSB broadcast programming. The QAM modulated optical signal is combined with the 10 Gb/s stream in a 3 dB coupler. The total output power of the four DWDM channels going into the 20 km AllWave span, after passing through a circulator and the CWDM, is –5 dBm. These 1.4 μm band signals are amplified in a Uniphase semiconductor optical amplifier (SOA) having 50 nm gain bandwidth and a maximum gain of 15 dB, recombined with the CATV channel into the second AllWave span. A second SOA is used as a pre-amplifier before demultiplexing the four 1.4 μm channels in a second waveguide router and detection of the PRBS bit stream with a 10 Gb/s receiver and a bit error-rate test set (BERT). The QAM channel is detected simultaneously with the same CATV receiver as the AM-VSB signal, and characterized by error-vector magnitude (EVM) and modulation error ratio (MER) measurements taken with a Hewlett-Packard 89441A vector signal analyzer.

The channel at 1.3 μm establishes the downstream part of a Gigabit Ethernet (GbE) between two PC workstations in a client/server configuration. The IP/Ethernet packet traffic over this link is analyzed by a HP internet advisor. The downstream 1.3 μm signal is derived directly from the SM output of the internet advisor's GbE interface, while we use an optical translation unit (OTU), consisting of a receiver and a directly modulated DFB laser to shift the upstream GbE traffic coming from the client PC into the 1.4 μm band. This upstream signal is coupled into and out of the 20 km of AllWave fiber using two circulators. For benchmark tests such as throughput and latency measurements on the GbE channel, we use the internet advisor to generate and receive IP data packets at a line rate of 1.25 Gb/s, bypassing the PCs and looping the downstream signal directly back via the OUT, obtaining a total link distance of 40 km.

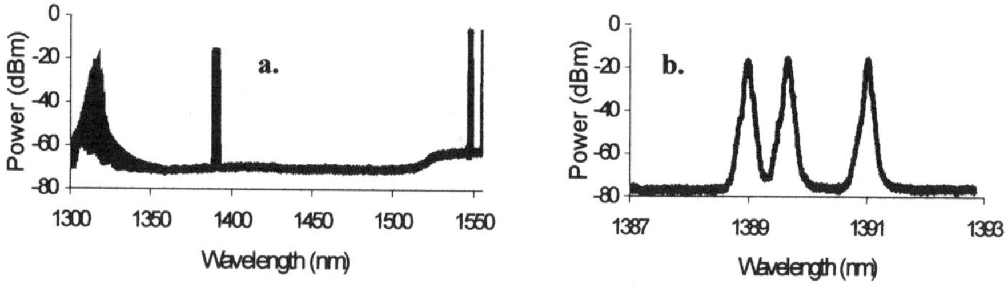

Fig 3: 1.3/1.4/1.5 micron WDM channel spectra.

In summary, a downstream 1.55 μm broadcast CATV AM-VSB signal, three downstream DWDM 1.4 μm band 10 Gb/s NRZ signals, one downstream QAM channel over a fourth 1.4 μm DWDM wavelength, one upstream 1.4 μm band 1.25 Gb/s LAN signal, and one downstream 1.3 μm 1.25 Gb/s LAN NRZ signal share a single 20-km AllWave fiber. After amplification, the downstream broadcast 1.55 μm wavelength and the four 1.4 μm band DWDM wavelengths share an additional 40 km of AllWave™ fiber.

3. Results and Discussion

A spectrum of all the downstream channels in 1.3, 1.4 and 1.55 μm regions is shown in figure 3a and the spectrum of the 1.4 μm channels, with channel 3 turned off, is shown at higher resolution in figure 3b (the upstream 1.4 μm channel is not shown). The CNR measurements on the1.55 μm channel showed nearly 3dB degradation for the entire frequency range, after transmission over 60 km fiber, relative to back-to-back measurements. Results for four frequencies, 55.25, 247.25, 319.25 and 535.25 MHz are shown in figure 4. The CNR for the entire frequency range stays above 51 dB, which is typical for commercial broadcast-fiber feeder input to a fiber node. The chief concern was whether or not the presence of the NRZ transmissions would affect the transmission quality of the 1.55 μm transmission through SRS or any other mechanism. As the data in figure 4 show, we saw no effect. The results were unaffected by the presence of the NRZ-modulated wavelengths in the 1.4 μm band. Since the separation between these two signals is about 22 THz, the Raman gain is only 6-7 % of the peak value at 12-14 THz separation[3]. Therefore no cross-talk is created due to SRS. Results of CSO and CTB measurements are shown in Table 1. For all frequencies CTB is between –68 to –69 dBc. Except at the highest frequency (535.25 MHz), the CSO is between –66 to –69 dBc. Again we see that the CSO and CTB values are unaffected by the presence of 1.4 μm band signals. These values are well within the CATV node input target of -65 dBc. The CSO on the higher frequency channel could be degraded due to self-phase modulation.[4]

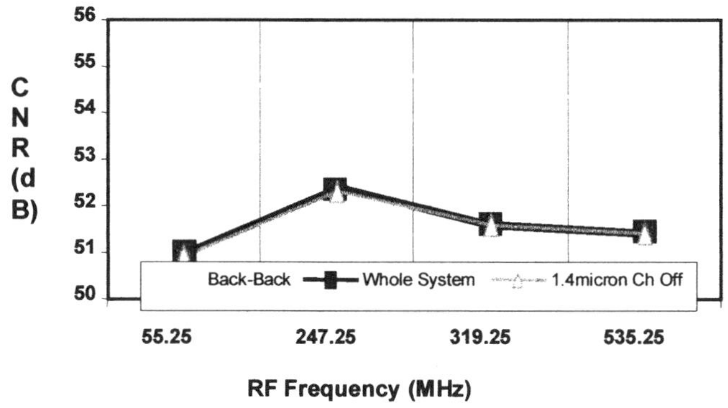

Fig 4: Carrier-to-noise ratio comparison for the analog signal.

Bit-error-rate measurements on the three 10 Gb/s 1.4 μm channels showed error free performance with the bit-error-rate curves similar to those reported earlier[2]. The receiver sensitivity for 10^{-9} BER was -24 dBm with negligible penalty over the back-to-back configuration. Also, we did not observe any BER impairment due to the presence of the AM-VSB signal at 1.55 μm. When conventional single mode fiber with 0.4 dB/km loss at 1385 nm was substituted in place of Allwave™ fiber in the 40 km span, the 1.4 μm band bit error rate increased to greater than 10^{-5}. The QAM channel over 60 km of AllWave fiber and two SOAs typically ran with an EVM below 2.7% and a MER of 29 dB, which corresponds to a BER of approximately 10^{-10} before forward error correction.

Table 1

Frequency (MHz)	OMD (%)	Back-to-back CSO (dBc)	CTB (dBc)	Whole System CSO (dBc)	CTB (dBc)	1.4μ Channel Off CSO (dBc)	CTB(dBc)
55.25	3.26	-69.8	-68.0	-67.3	-68.2	-66.8	-67.9
247.25	3.19	-69.1	-68.7	-69.3	-69.1	-69.5	-69.1
319.25	3.12	-67.7	-69.1	-66.1	-69.6	-65.7	-69.9
535.25	3.01	-67.0	-66.9	-58.7	-68.7	-58.2	-68.5

For testing of the GbE link, the HP Internet Advisor was configured to send and receive IP packets in Gigabit Ethernet mode which simulates transmission of data frames between two servers or routers at 1.25 Gb/s rate. Frame throughput as a function of the frame size, up to 1500 bytes is shown in figure 5. No significant frame loss was observed up to this frame size and the performance was identical to that measured under back-to-back condition. The sensitivity of the Internet Advisor receiver at 1.25 Gb/s was –20 dBm. At this power no frame check sequence (FCS) errors were observed. The latency, which was independent of frame size, was measured to be 207 μs, close to the theoretical value of 200 μs (based on 40 km round trip).

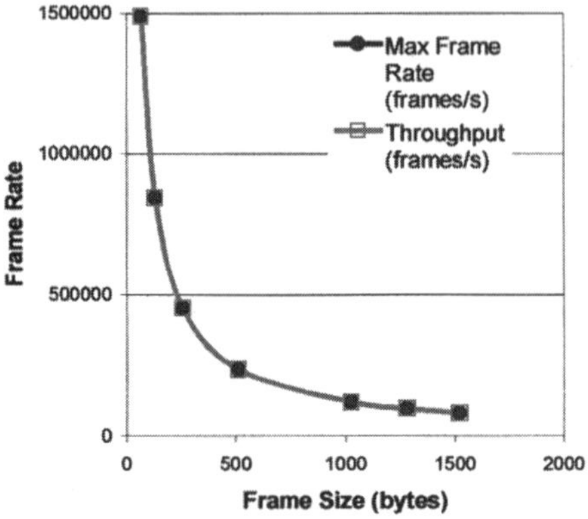

Fig 5: IP/Gigabit Ethernet frame throughput as a function of frame size.

4. Conclusion

We have demonstrated the first impairment-free transmission of both analog and various digitally modulated signals on a single fiber by using AllWave™ fiber. The signals consist of a bidirectional Gigabit Ethernet channel at 1.3 and 1.4 μm, telecom traffic at 3x10 Gb/s at 1.4 μm, QAM at 1.4 μm, and CATV AM-VSB signals at 1.55 μm. The demonstrated ability of zero-water peak fiber to support a most heterogeneous mix of optical signals directly translates into a maximum flexibility for service providers to upgrade their networks and add new services without installing new fiber. These considerations become even more important as fiber moves deeper into the access plant. In particular the overlay of bi-directional data services on broadcast CATV services will offer tremendous revenue opportunities at dramatically reduced cost since all services can be transmitted over a common fiber plant. While conventional fiber requires separate fibers to avoid signal impairments, the same goal can be achieved by spectral separation of analog and digital signals over a single strand of AllWave fiber. This reduces the total amount of fiber required in the network, as in the number of fibers needed to access a small fiber node, and therefore enables significant system cost savings.

Acknowledgment

This demonstration was made possible by the generous technical support from Hewlett Packard (now Agilent Technologies), JDS Uniphase, E-Tek, and Scientific Atlanta. We gratefully acknowledge their contributions.

References

[1] Z. Wang, A. Li, C. J. Mahon, G. Jacobson, I. Bodtker, Photonic Technol. Lett. Vol. 31, pp. 1538 (1995).

[2] A. K. Srivastava et. al, Proc. of Optical Amplifiers and their Applications , Post-deadline paper PD-3, 1998.

[3] G. P. Agarwal, Fiber-Optic Communication Systems, John Wiley, New York pp. 380 (1997)

[4] M. R. Phillips, T. E. Darcie, D. Marcuse, G. E. Bodeep, and N. J. Frigo, IEEE Phot. Technol. Lett.," vol. 3(5), pp. 481 - 3, May 1991

Broadband Access, WDM Metro and Network Management
D.W. Faulkner and A.L. Harmer (Eds.)
IOS Press, 2000 © AKM

Advanced IP classification techniques for a hybrid routing node in a DWDM metropolitan network

Carlos Macián[1], Peter Domschitz[2]

macian@ind.uni-stuttgart.de, pdomsch@rcs.sel.de

Abstract

Although the progress in the optical technologies will put an unheard-of transmission capacity at the user´s disposal in the near future, to fully take advantage of that bandwidth the nodes in the network have to be able to provide QoS and Value Adding Services (VAS) to those same users at wire-speed. One of the most critical aspects, which is being researched inside the high-speed routing activities within the GIGA-RING project (itself part of the german initiative KomNet), is packet classification in a hybrid routing node. In this paper, we present a novel approach that overcomes the scalability issues of this question by means of a hierarchy of filter databases based on the semantics of the rules. We also demonstrate how our model helps the nodes support QoS and VAS with an example of outsourcing Virtual Private Network (VPN) services.

1. Introduction

Within the German initiative KomNet (Innovative Communication Networks for the Future Communication Society, see [1]), a research project called GIGA-RING has been established to develop and demonstrate the benefits of future broadband access networks. Metropolitan area networks (MANs) are designed to close the gap between an optical core network and a flexible scenery of last-mile networks. Wavelength division multiplex (WDM) technologies are used for transmission as well as for routing among the access systems attached to the MAN via optical access nodes. A ring-formed optical transport network with hybrid routing nodes for the metropolitan area has been designed and set up, and will be exhaustively tested within a field trial. The GIGA-RING project consists of two activities: One part is optic-oriented, focusing on the research and demonstration of the potential of new architectural and management aspects for DWDM (Dense Wavelength Division Multiplex) optical access networks. Subject of the other project activity is an innovative, universal but cost efficient electronic routing node, which is able to process the bandwidth offered by the DWDM-based metropolitan network. The electronic switching and routing functions are projected for data rates up to some Gbps, covering all network protocol layers up to the ATM and IP protocol level. Therefore, this hybrid routing node is designed for a flexible interconnection of existing and future broadband last-mile networks using various technologies. This paper is focusing on the advanced, QoS-oriented IP functions of the node, including support for virtual private networks (VPNs).

The immense bandwidth that the DWDM technology promises to deliver cannot be fully taken advantage of without network nodes that are able to process packets at that same speed. Ideally, those nodes should perform their task without abandoning the optical domain, but although there have been constant improvements aiming thereto, it seems that hybrid routing nodes, performing part of the packet processing optically and part electronically are still the only viable alternative in the immediate future. In recent years the design of routing nodes suffered a strong revolution, with researchers as well as manufacturers swifting towards more distributed approaches which take the processing burden out of a central unit and distribute it among the network cards [1], [3], [4]. That, together with the substitution

1.Corresponding author. Postal address: University of Stuttgart, Institute of Communication Networks and Computer Engineering (IND), Pfaffenwaldring 47, D-70569 Stuttgart, Germany.
2.Alcatel Corporate Research Center, Stuttgart, Germany.

of the bus as switching plane against solutions more amendable to parallel transmission of packets or cells, like shared memory or crossbar switches, have taken the routers to the point of delivering throughputs of hundreds of Gbps and beyond (see [5] and [6] for just two examples).

Existing implementations typically achieve high speeds at the cost of flexibility, which makes them optimal for the routing of best-effort traffic, but stretches them to their limits as soon as QoS or Value Adding Services (VAS), like Virtual Private Networks (VPNs), come into play. One of the tasks which more strongly unveil this fact is the classification of packets. Since the first step towards being able to deliver QoS or VAS is the ability to discriminate the packets to be able to serve them differently, it is crucial that routers can perform classification at wire-speed. In this paper we will describe a novel approach to this problem, which shall enable routers to classify packets based on any number of header fields at high speeds. We will furthermore show how such a design can improve the flexibility of the nodes and help them to provide VAS. For the sake of brevity, we will restrict ourselves to the description of the main properties of our model, deferring most details and numerical results for a future paper.

The remainder of the paper is structured as follows: Section II presents our hybrid routing node; section III describes the packet classification problem and the previous work done to solve it; section IV presents our model and analyses its properties, and section V concludes the paper by providing a summary and looking into the future of our research.

2. The GIGA-RING hybrid routing node

A global trend is the increasing demand for fast IP networks in all environments, but specially in the access area, where a strong backlog exists. New technologies like WDM have had little impact in this segment so far, but that is set to change: The aggregate traffic from emerging high-speed last-mile infrastructures (xDSL, HFC, WLL, ...) calls for powerful optical technologies.

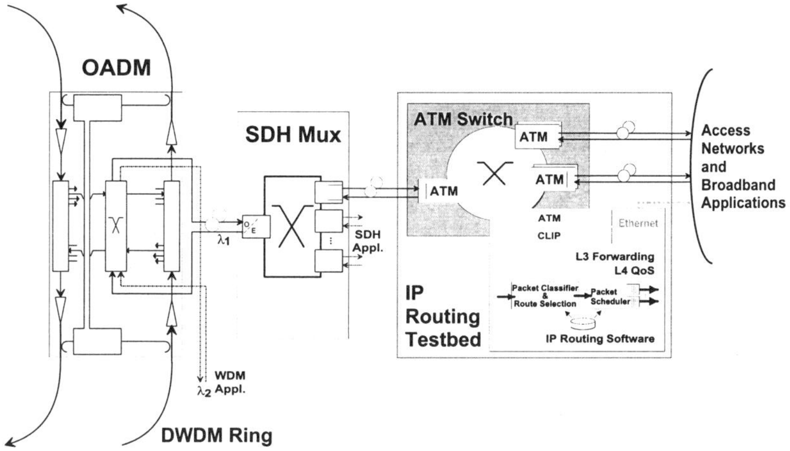

Figure 1: The GIGA-RING hybrid routing node

Hybrid opto-electronic routing nodes are expected to be key elements of future MANs. The GIGA-RING research activity addresses this challenge by analysing the overall system requirements, structures and concepts without focusing on a specific implementation of functional models. Rather, it is intended to point the way for future system developments.

Until mid 2000 a first demonstrator for the hybrid routing node will be supplied, combining an optical add/drop multiplexer (OADM, in quasi-fixed configuration with protection) and an electronic gigabit switching node, whose fundamental structure is depicted in Fig. 1.

The electronic part is characterized by the function blocks of the ATM and particularly the IP layer. The SDH multiplexer works as a connecting link, offering synchronous data streams extracted from a wavelength channel selected by the OADM to the higher layers. The architecture of the switch/router is designed for high performance and optimized for the interoperation of ATM and IP, following the idea of establishing a flexible platform for the prototypical implementation and examination of advanced concepts and protocols in the metropolitan area. These new concepts will be evaluated in a field trial involving fast IP applications running over an ATM core delivered to real users.

3. The packet classification problem

The packet classification problem, aiming at deciding what kind of treatment a packet deserves based on a corpus of filters or rules, is a well-known multidimensional matching problem. In a way, it can be seen as an extension of the IP Routing Lookup problem, in which the treatment of the packet (simply, to which output port to send it) is decided by looking at just one header field (the destination address) and looking for the matching rule (the longest matching prefix) in a destination prefix database. That much more simple operation was supposed to be one of the main bottlenecks in the new router architectures, since it is a very time-consuming task. Recently, new matching algorithms have been proposed which solve or at least leverage that problem (see for example [13] for a good survey on the subject).

In today´s routers, the broader question of packet classification was seen as a minor problem, because the filtering was performed usually only for security purposes (firewalling), which contain only a small number of rules to permit communication with known partners, and blocking all the rest. Classical search algorithms are run over that filter database every time a packet arrives, to see if it is conform with the security policy. The problem is, those algorithms do not scale well with the number of rules and/or with the length of the header segment to compare, and both variables can be expected to increase strongly in the near future.

The delivery of QoS or VAS in the IP world is centred around the concept of *flow*. If we assume that the proportion of traffic requiring QoS is going to grow strongly in the future, mainly due to the broad dissemination of real-time multimedia applications and the willingness of the people to pay a little bit more for a better service, the importance and the difficulty of distinguishing and separating flows to deliver individual QoS guarantees will greatly increase. This is the approach taken by the Integrated Services Architecture [7].

Another proposal, Differentiates Services, tries to overcome the scalability question that arises in Integrated Services by aggregating traffic in a reduced set of classes [8]. But even in that case, there are points in the network (the edges) where the (de-) aggregation of the traffic has to take place, involving once again a potentially huge number of individual flows which have to be individually classified.

Furthermore, the inclusion of VAS like VPNs (and, in the future, QoS-enabled VPNs) makes it even more difficult: Think of a scenario in which a customer outsources its VPN to its Internet Service Provider (ISP). The ISP has to guarantee that the traffic of different VPNs will not interfere, and possibly deliver different degrees of QoS to the different traffic types (real-time, non-real-time, ...) that every client might define. In such a case it might be necessary to very accurately analyse the content of the packet header to decide first to which VPN it belongs, and then to which traffic class and priority level.

The number of rules or filters needed in the routers for classification is directly proportional to the number of flows to be considered. Add to the increasing number of rules that the identification of a typical flow could require the analysis of its destination address, source address, source port, destination port and Type Of Service (TOS) Byte, making a total of some 102 bits spanning five header fields. Not all the algorithms can cope with such a broad key to match. Clearly classification algorithms have to scale well both with the number of rules and the length of the key (i.e., the header chunk to be matched).

We do not claim the size of the filter database to be going to grow unboundedly in the near future, but we do claim that current solutions can not cope with realistic scenarios, which are not that far away. Recall the previous example: Let´s say that the ISP has 20 VPNs to support. Every VPN has defined 200 addresses or prefixes spread around the globe with which a communication shall be allowed (other delegations, partners, providers, etc.). Every client wants at least three different levels of QoS (say, for real-time, non-real-time and best-effort traffic). For security reasons, they set up different rules for different services (HTTP-Access, FTP and the like). Say, 10 different services. That makes 20*200*3*10 = 120,000 rules, which even for this quite restricted example is already well two orders of magnitude over actual databases [9]. Notice that this problem is independent of the way the (possibly aggregated) traffic is transported over the network, since it still has to be classified and separated at the edges.

Since the classification problem is the multidimensional case of the well-studied IP Address Lookup problem, the most common approach in the literature has been to try to extend the solutions for the former to the later. Most proposals use evolutions of the binary tree search algorithm over the whole length of the key, like the grids-of-tries [11]. Others have try to apply recursivity to the problem, like in [9] or [10]. Others try to decompose the multidimensional problem in *K* unidimensional ones, where *K* is the number of header fields to be matched, and then try to find a filter that matches the concatenation of the results of the individual classifications [11]. A geometric approach has also been taken in [12].

In all the cases there are three main variables to consider to judge the quality of a classification algorithm: The speed at which a matching filter is found (in the average as well as in the worst case), the memory requirements of the chosen solution and the time that it takes to make an update in that structure (i.e., to add or remove a filter or rule).

To the best of our knowledge, none of the afore mentioned solutions can minimize all three variables. Most of them try to find a good compromise of the first two at the cost of very expensive updates, with some rare exceptions (see [13]). In our opinion, that is a dangerous approach, because, as we asserted already, the basic unit to be classified is the flow, and a flow is intrinsically dynamic. Recent measurements show that it is not unusual for any core router today to support 100K flows simultaneously at any one moment, with update rates of several hundred flows per second. If we envision such orders of magnitude as the possible evolution of edge routers in the future, we clearly see that to support fast update rates is mandatory. In our proposal, we strongly address this issue by trying to concentrate related rules in contiguous positions in our data structure, a feature that we name the ***locality*** of the rules.

Most of the proposals neglect that locality and disperse related rules over the whole data structure, or even replicate them in various positions. We believe that the cause is that up to now no attention was paid to the ***semantics*** of the rules. The diverse rules are not independent from each other, but grouped in blocks with „similar meanings", i.e. which address a same function: Enforcing a security policy, discriminating a certain service (HTTP-Access, for example), differentiating among VPNs, etc. That is to say, we could take advantage of that semantic relationship among rules -which is purely logical- and transform it in a physical relationship by locating them together in our classifier. Furthermore, this semantic grouping is recursive: There is a subgroup of rules which address a certain service for traffic corresponding to a certain VPN, etc. (see Fig. 2). Said another way: There is a natural ***hierarchy*** concerning the rules in a packet classifier, which has never been taken into account before.

Another important point related to this is that not all packet headers give the same amount of information concerning the group the arriving packet belongs to, and the cost to retrieve that information is also different. We call this property the ***discrimination capability*** of every header field, for it tells us how useful this header is to distinguish or discriminate among the various groups the packet could belong to and how high is the cost of doing so. Obviously, the discrimination capability varies depending on the scenario. For example, recall the VPN outsourcing service provided by a certain ISP that we mentioned before. In that environment, the Source Address of a packet delivers an almost unambiguous information about its VPN, which the Destination Address might not so easily do. Furthermore, the Source Address provides that information with a much simpler operation, since the number of possible source addresses (20 in our example, if every VPN has only an address prefix) is naturally in an edge node much smaller than the number of possible destination addresses (200 * 20 = 4,000 in the example).

By paying attention to the semantics of the rules we can target a hierarchy of groups of rules, which has the field with the highest discrimination capability at the root, the second highest in a further step, and so on. Although, as we said, this hierarchy and the fields which will be considered at every step are application-dependent, we believe that a near-optimal combination can be pre-computed at boot time by analysing the structure of the initial filter set. Its exact content will change due to the updates, but it is reasonable to assume that its main characteristics (like the number of VPNs, or the number of permitted destinations) will not be severely affected, assuring the resiliency of the chosen structure with time. In this paper we will nevertheless centre only on VPN-support, which we think is a very important scenario nowadays, bringing together the two principal novelties of future networks: QoS and VAS.

4. The hierarchical model

A question which always arises when designing a packet classifier is the choice between a hardware or a software implementation. Although software implementations have classical advantages, like the flexibility or the re-programmability, to achieve the throughputs that are to be expected, it is our opinion that only hardware solutions are viable. Moreover, the idea of hierarchy introduced on the previous section is highly amendable to pipelining, as we will see. An adequate design also permits to take advantage of another classical feature of hardware implementations: the use of parallelism. Hence, we decided to optimize the design for a hardware implementation from an early stage.

A rough representation of our model can be seen in Fig. 2. In the presented example, a hierarchy of

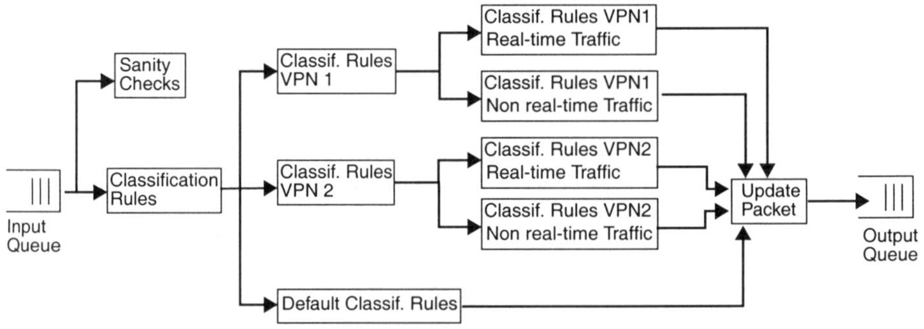

Figure 2: Example of a 3-step hierarchical classifier supporting 2 VPNs plus default traffic

three steps has been depicted for a simple scenario with just two VPNs and two types of traffic: real-time and non-real-time.

The fact that the classification shall be performed at the input of a router to avoid wasting valuable resources (like bandwidth in the switching backplane or space in the output queues), permits the coupling of two independent functions inside the packet classifier: The verification of the correctness of the packet (the so-called „sanity check“[1]) and the classification itself. These two functions can easily be performed in parallel, serving the former as a validation for the later. Shall the packet not be considered "sane", then independently of the classification result it shall be discarded.

Our approach consists in starting with a rough discrimination of the filter group whom the packet might match against, and progressively bound the subgroup with every further step. In other words: We establish our hierarchy along the line of maximum negative slope of discrimination capability. Ideally only

1.In fact, the sanity check performs various tests over the packet: That the (source, destination) address exists, that the time to live has not reached zero, that the length of the packet makes sense, etc.

one header field is checked every time, although the model itself does not preclude the use of two or more fields in any one step, if it is found to be more suitable. As an example, a combination of destination port and value of the TOS-Byte might serve to define the priority and service of the packet (Service: Real-time, Priority: Highest), and since the number of possibilities is tightly bounded (in our example, we supposed only 3 priorities and 10 services), it might be more efficient to couple them.

In every step only one field is examined, allowing the use of the new algorithms developed for the unidimensional case (IP Address Lookup[1]). The fields considered important for the classification are the five mentioned before: Source and destination addresses and ports, plus the TOS-Byte (or the DS-Field for Differentiated Services). The point in the hierarchy at which they are examined could depend on the concrete scenario. For our example, it is clear that the source address can almost unambiguously define the VPN and should therefore be used on the first step.

After a number of steps, depending on the characteristics of the concrete filter database, it will not be possible to efficiently further refine the classification. At this point, a list containing all the filters which correspond to this subgroup will be used to perform the last step of the classification. The packet is then confronted with the original multidimensional matching problem, albeit in a much more restricted context, i.e., the number of rules to compare against has been greatly reduced. In this way we overcome the scalability problems that most of the existing algorithms have when the size of the rules databases grows beyond some hundreds or thousands of elements.

At every step, depending on the characteristics of the original database, a different matching algorithm can be used, to take advantage of their different properties. Some algorithms are better suited for longer keys, for example (like a combination of source address and source port: 32 + 16 = 48 bits), while others scale better with the number of rules, albeit only when the length of the key is kept small. At the same time, to achieve the high classification rates we aim at, we treat every step in our hierarchy as a logical pipeline. In this way we can perform a classification every clock cycle, independently of the algorithms used.

It should be noted that our structure situates the rules in its complete form (the combination of values for the five header fields) only at the end of the hierarchy. In the previous steps only a list containing the values of one of the fields is stored. Moreover, in that list only the values which actually are found in existing rules will be stored: If we define just three different priorities in the TOS-Byte, like in our example, it makes no sense to have a list with all the 255 possible values. By locating related rules together we make it simpler to trace where a new rule shall go (or an old one shall be removed from). This increase in the locality of the filters thus highly simplifies the update of the database: To add or remove a rule, only the corresponding small table at the end of one of the branches of the hierarchy must be updated, plus its ancestors, if the adding (or removing) of that rule introduces (or deletes) a value not yet present on them (a value that no remaining rule possesses). In this way, the update rate that can be achieved is highly improved.

Consider the example depicted in Fig. 3, which represents the classifier resulting from the rules in the table next to it. For clarity, we have used names instead of numbers for the different fields. We consider an scenario with three VPNs, seven possible destinations (distributed between providers -Prov- and delegations -Dele-), four different services and just two priorities. The chosen hierarchy derives from the analysis of the ruleset. As stated, the header field with the highest discrimination capability is the source address, being therefore placed at the root. The next steps are different for every VPN. If a packet defined by <VPN1, Dele2, Srv2, Prio2> arrives at the classifier, it will follow the highlighted path prior to achieve classification. At every point, the way to take will be selected by examining the chosen header field, except in the last step, where two fields have to be checked. The packet has been compared against much smaller tables, and in all cases but one only one dimension had to be checked.

Let's consider the addition of the rule <VPN1, Dele5, Srv2, Prio1>. Since all the prefixes were already

1.Although a rule can be specified as: a range of values, a prefix, a suffix and an exact value, all of them can be transformed into prefixes, as explained for example in [11].

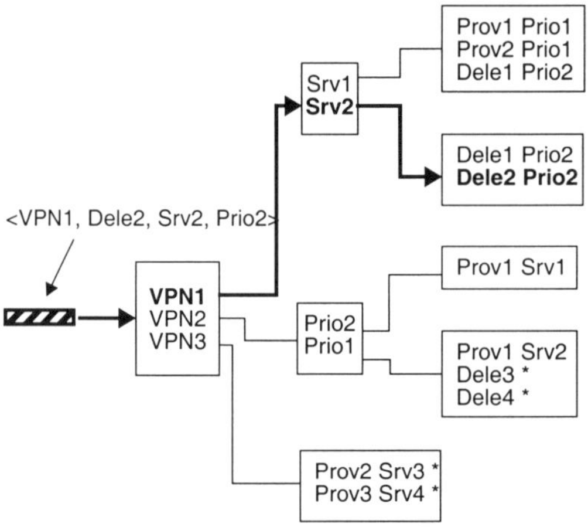

Src Addr	Dst Addr	Dst Port	Priority
VPN1	Prov1	Srv1	Prio1
VPN1	Prov2	Srv1	Prio1
VPN1	Dele1	Srv1	Prio2
VPN1	Dele1	Srv2	Prio1
VPN1	Dele2	Srv2	Prio2
VPN1	Dele1	Srv1	Prio2
VPN2	Prov1	Srv1	Prio2
VPN2	Prov1	Srv2	Prio1
VPN2	Dele3	*	Prio1
VPN2	Dele4	*	Prio1
VPN3	Prov2	Srv3	*
VPN3	Prov3	Srv4	*

Figure 3: Classification example

present in the corresponding tables but Dele5, only the last table has to be updated as stated in Fig. 4. If, on the other hand, the filter <VPN2, Prov1, Srv1, Prio2> had to be removed, since there are no more rules corresponding to the value Prio2 for VPN2, the second and third tables in that branch have to be updated as expressed in Fig. 4. In fact, being the last table now empty, it is simply removed.

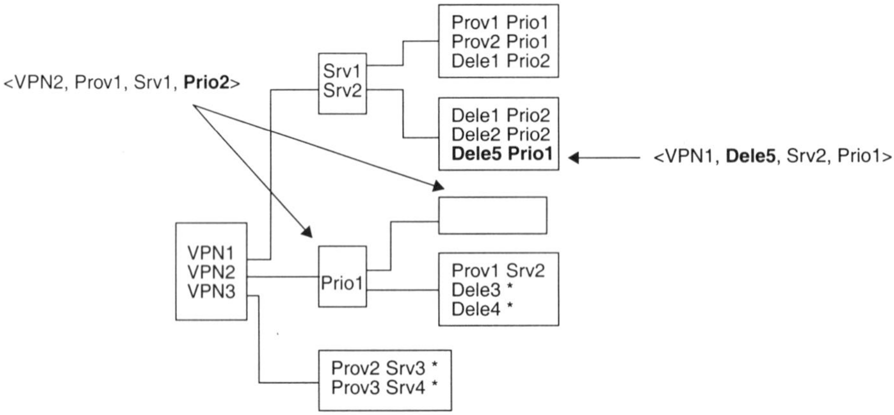

Figure 4: Addition and removal of filters

It is difficult to have access to real filter databases, and the ones which are available are far too small to serve as a testbed for our experiments. That is why for the validation of our model we had to create our own databases by analysing the contents of some smaller databases from the routers at the Institute of Communication Networks and Computer Engineering and extrapolating their contents to the sizes we expect to find in the near future. At the same time, we extended the functions of the filters to account for the introduction of emerging QoS-aware and value adding services in the networks.

At the moment of writing, we have started the measurements of our model by means of emulation in a software platform based on Linux as well as by simulation. First steps towards implementing a prototype have also been taken. As a most interesting scenario, we are considering the case depicted in this paper, where an ISP is offering VPN services to many companies at the edge of a core network. Several configurations (depth and structure of the hierarchy, matching algorithms employed along it, etc.) are being tested. The results of those measurements, as well as a detailed description of our prototype and an analysis of the effects of the security issues typically present in VPNs (like encryption, authentication and tunnelling), which have also been taken account of in our model, are deferred for a future paper.

5. Conclusion and further work

The change of paradigm in the IP world towards the inclusion of new QoS-aware and value adding services presents severe challenges to the network nodes. One of the most critical examples is the classification of packets at wire-speed. The algorithms developed to solve this multidimensional matching problem have generally tried to achieve high-speed and an acceptable storage requirement at the cost of very low update ratios. In this paper we have presented a classifier which scales well thanks to its hierarchical approach based on the semantics of the rules; performs classification at high-speed through the use of a hardware implementation of the hierarchy, the assumption that fields with a better discrimination capability should be checked first, and by keeping tables small and generally unidimensional; and presents a favourable structure to achieve high update ratios by maximizing the locality of the rules in its tables. We have furthermore demonstrated its virtues in a VPN-scenario, for it brings together the two principal novelties of future IP networks: QoS and VAS.

The exhaustive testing of our model by means of simulation and emulation in software will give path to its implementation in hardware. Meanwhile we are working on new classification algorithms specially designed to take advantage of the characteristics of our model.

The described work has partially been funded by the German Federal Ministry for Education and Research (BMBF) through the project GIGA-RING (01 BP 815/3) within the KomNet program [1]. The authors alone are responsible for the content of the paper.

References

[1] http://www.hhi.de/komnet

[2] Partdrige, C. et al.: A 50 Gb/s IP Router, IEEE/ACM Transactions on Networking, vol. 6, n . 3, June 1998

[3] Keshav, S. and Sharma, R.: Issues and trends in Router design, IEEE Communications Magazine, May 1998

[4] Kumar, V.P. et al.: Beyond best-effort: Router architectures for the differentiated services of tomorrow's Internet, IEEE Communications Magazine, May 1998

[5] NX64000, Nexabit Networks, http://www.nexabit.com

[6] SP2400, NEO Networks, http://www.neonetworks.com

[7] Herzog, S.: RSVP extensions for policy control, IETF RFC 2750, proposed standard, January 2000

[8] Blake, S. et al.: An architecture for differentiated services, IETF RFC 2475, informational, October 1998

[9] Gupta, P. and McKeown, N.: Packet Classification on Multiple Fields; SIGCOMM'99, Harvard University

[10] Gupta, P. and McKeown, N.: Packet Classification using Hierarchical Intelligent Cuttings; Hot Interconnects VII, August 1999, Stanford University

[11] Srinivasan, V. et al.: Fast and Scalable Layer Four Switching; SIGCOMM'98, Vancouver

[12] Lakshman, T. and Stiliadis, D.: High-Speed Policy-based Packet Forwarding Using Efficient Multi-dimensional Range Matching; SIGCOMM'98, Vancouver

[13] Tzeng, H. and Przygienda, T.: On fast address-lookup algorithms, JSAC, vol. 17, nr. 6, June 1999

Broadband Access, WDM Metro and Network Management
D.W. Faulkner and A.L. Harmer (Eds.)
IOS Press, 2000 © AKM

A WDM-BASED MULTIPLE CONTROL CHANNEL LIGHTWAVE ARCHITECTURE

Ioannis E. Pountourakis
Department of Electrical and Computer Engineering
Division of Computer Science
National Technical University of Athens
157 73 Zographou Athens GREECE
e_mail: ipount@cs.ntua.gr

Abstract: In this paper we study a single-hop network for a passive star topology, where each station communicate with one another by assigning a fixed wavelength to each station. A station transmits its packets to another station by transmitting over its dedicated wavelength. All these dedicated wavelengths are multiplexed by WDM into a wultiwavelength broadcast system. To enable the destination station to know which packets from the WDM system are destined for it, we introduce the Multichannel Control Architecture(MCA) [2]. The MCA with an appropriate Network Interface Unit(NIU) at each station gives the solution to the problem of the electronic processing bottleneck that the single common shared control channel for pretransmission coordination introduces [1].
We present and analyze a new protocol architecture based on Poisson approximations statistics for finite population and examine the receiver collision phenomena [3-4].

1. NETWORK ARCHITECTURE MODEL AND ASSUMPTIONS

The system under consideration as Figure 1 shows is a wavelength division multiplexed broadcast passive star network. The bandwidth of the optical fiber is divided into v+N channels each using a different wavelength, $\lambda_{c1},...,\lambda_{cv},\lambda_{d1},...,\lambda_{dN}$, to serve a finite number N of stations. Each station is assigned with a fixed wavelength $\lambda_{di}\{i=1,2...N\}$ for transmitting data packets. For propose of signaling we introduce the MCA at wavelengths $\lambda_{c1},...,\lambda_{cv}$ which is shared by all stations for coordination of data packet transmission. The network system can be described as a $[CC]^{v}-TT-FR^{v}-TR$. It means that there are v control channels and each station k has a tunable transmitter tuned at each of $\lambda_{c1},...,\lambda_{cv}$. Also the same tuneable transmitter can tuned at its dedicated λ_{dk} .The outcoming traffic from a station is connected to one input of the passive star coupler. The NIU of every station uses v fixed tuned receivers one for each control channel and one tunable receiver to any of data channel $\lambda_{d1},...,\lambda_{dN}$. The incoming traffic to a user station is splitted into v+1 portions by a 1x (v+1) WDMA splitter as Figure 1 indicates. In our analysis the access method to MCA is based on the ALOHA protocol. The transmission time of a fixed size control packet is used as time unit (control slot) and the data packet transmission normalized in control slot time units is L (L>1) which is called data slot. The control packet is consisting of the transmitter address, the receiver address. The access method follows the "tell and go" policy, which is suitable, when the round trip delays are much longer than the packet transmission times. By this strategy a station wishing to (re) transmit a data packet informs the destination about its intention without waiting any more. This policy has the advantage of decreasing the waiting time before data packet transmission. A station will hear the result of the transmission of its control packet by listening to the star coupler multichannel system since it operates as a broadcast medium. In the receiving mode if a station sees its address announced in a control packet, immediately adjust its receiver to the transmission wavelength channel (station) which is specified in the control packet for packet reception. We say that the tunable receiver of a station Z is active if it is tuned receiving a data packet from a data channel $\lambda_{dk}\{k = 1,2...N\}$. If a data packet is addressed to station Z and the receiver is active, the packet is rejected. This phenomenon is called receiver collision [2-4]. The stations participating in unsuccessful transmissions defer their retransmissions for a random time until successful retransmission. We assume uniform distribution of retransmissions. The random time delay introduced between two consecutive retransmissions of backlogged stations is uniformly distributed from 1 to K time data slots. We also assume that the total offered traffic from new generated and retransmitted packets obeys to Poisson statistics.

2. ANALYSIS

2.1 Case I

A station desiring to communicate with an other, choose at random a control channel λ_{ci} at the instant of its generation or at the end of the random time chosen for retransmission and sends first the control packet and after that the corresponding data packet over its own dedicate λ_{dk} data channel. Taking into account that each station choose randomly a control channel independently from the others in a complete unsynchronized manner to transmit following the unslotted ALOHA protocol, the vulnerable time period for the control packets lasts 2 time units.

Let

G = the mean offered traffic over v-channel control system in a time unit period, that obeys to Poisson statistics.

$G_j = G/v$, the traffic to j_{th} control channel j∈{1,2,...,v} given that each control channel is chosen with equal and constant probability $p_j = 1/v$.

P_{suc} = the probability of one Poisson arrival in a control channel during a time unit. Then

$$P_{suc} = \frac{G}{v} e^{-2G/v} \tag{1}$$

A_w = random variable representing the number of control packets in a time unit $0 \le A_w \le v$.

Thus the probability of finding $A_w = k$, control channels everyone with one Poisson arrival during a time unit, obeys to binomial probability low.

$$\Pr[A_w = k] = \binom{v}{k} P_{suc}^k (1 - P_{suc})^{v-k} \tag{2}$$

S_c = the average successful transmission rate of control packets in a time unit in steady state.

$$S_c = E\{\Pr[A_w = k]\} = \sum_{k=1}^{\min(M,v)} k \Pr[A_w = k] = Ge^{-2G/v} \tag{3}$$

We define P_f as the probability of successful (re) transmission of a control packet over the control multichannel system. From (3) we take

$$P_f = S_c / G = e^{-2G/v} \tag{4}$$

G_d = the offered traffic per data slot per station.

$$G_d = \frac{G}{N} L \tag{5}$$

S_{ds} = the average successful transmission rate of data packets with destination a given station during a data slot time.

$$S_{ds} = G_d P_f = e^{-2G/v} \frac{G}{N} L \quad (6)$$

The throughput reduction induced by receiver collision is related with the possibility a receiver to be active. Thus we examine what will happen if a data packet which starts transmission at time instant t_r and has as destination a station Z who belongs to the set 1,2,...M. It is obvious that if a data packet has started transmission between $t_r - L$ and t_r will have activate the receiver of station Z and a receiver collision occurs [3],[4]. In the opposite case, the Z station tuneable receiver is idle and the data packet will be received correctly. Thus the vulnerable period time lasts L time units. It is evident that any other data packet transmission, starting between t_r and $t_r + L$, will find the station's Z receiver active and will be aborted, causing receiver collision. In other words the collision window interval is extending from t_r to $t_r + L$. The multichannel nature of the control communication system allows more than one successful headers transmission from different control channels to denote the same destination of data packets during the same time unit. Thus the station receiver is tuned only to one of incoming transmissions and rejects the others. In this case, there are different policies to select which packet is received correctly at destination. As an example we can say that the data packet that is transmitted on the lowest data channel number win the competition. The probability that station Z is destination of a data packet is $1/M$ (for sake of simplicity of the analysis we assume that a station may sent packets to itself). We define the following events for the vulnerable time period $t_r - L \le i \prec t_r$:

C_j, the successful transmission of a control packet in j_{th} control channel j$\in$\{1,2,...v\}, starting transmission in the $i_{th} - 1$ time unit of the vulnerable period.

B_j, the data packet starting transmission in the i_{th} time unit, whose corresponding control packet has been transmitted over j_{th} control channel at $i_{th} - 1$ time unit has as destination a given station, let's say Z. Taking into account the independence of the of the C_j and B_j events for each time unit, we define A as the event that no one data packet has as destination the station Z during the vulnerable period.

$$A = \bigcap_{i=1}^{L} (\overline{\bigcap_{j=1}^{v} C_j \bigcap B_j}) \quad (7)$$

We define as P_{cor}, the probability of correctly reception of a data packet from station Z starting transmission at the given time instant t_r. This probability is equivalent with the probability $P(A)$. From (7) we get,

$$\begin{aligned} P_{cor} = P(A) &= \prod_{i=1}^{L} (\prod_{j=1}^{v} \{1 - P(C_j \cap B_j)\}) \\ &= \prod_{i=1}^{L} \{1 - P(B_j / C_j) P(C_j)\}^v \\ &= \prod_{i=1}^{L} \{1 - \frac{1}{N}\frac{G}{v} e^{-2G/v}\}^v \\ &= \{1 - \frac{1}{N}\frac{G}{v} e^{-2G/v}\}^{vL} \end{aligned} \quad (8)$$

We define as throughput in steady state, the fraction of data packets that are received correctly at destination by a station during a data slot time.

$$S_N = S_{ds} P_{cor} \quad (9)$$

Using approximations in the above equation, we take

$$S_N \approx \frac{L}{N} Ge^{-2G/v} \exp[-\frac{G}{N} Le^{-2G/v}] \tag{10}$$

2.2 DELAY

Let D = the interval between the time a data packet is generated and the moment that has been successfully received at destination. Data packet delay D, is composed from two parts as follows:

$$D = D_r + L + 1 \tag{11}$$

D_r = the delay from packet transmission to beginning of the successful reception at destination.

F_r = the probability a packet successful reception at destination in steady state. From Eq. (10) we get

$$F_r = e^{-2G/v} \exp[-\frac{G}{N} Le^{-2G/v}] \tag{12}$$

$Q_r(n)$ = the probability a packet successful retransmission after n trials. Assuming that the probability of success is the same on any try, n has a geometric distribution, i.e.

$$Q_r(n) = F_r (1 - F_r)^{n-1} \tag{13}$$

R = the average number of trials for successful transmission of a packet. So

$$R = E[n] = \sum_{n=1}^{\infty} nQ_r(n) = 1/F_r = e^{2G/v} \exp[\frac{G}{N} Le^{-2G/v}] \tag{14}$$

$D_b = (K+1)/2$, the average delay between successive retransmissions. The average packet delay D, is

$$D = E[D_r] + L + 1 = (R-1)(L+1+D_b) + L + 1 =$$
$$\{e^{2G/v} \exp[\frac{G}{N} Le^{-2G/v}] - 1\}\{L + 1 + (K+1)L/2\} + L + 1 \tag{15}$$

2.3 Case II

The main problem of the previous protocol is the inefficient use of control channels. In this protocol case the time axis is slotted with the size of the control packet. All control packets are synchronized in a parallel system constituting the control multislot. A station k desiring to communicate with an other, choose at random a control channel i, waits the beginning of the next multislot and sends first the control packet(header) over the λ_{ci} channel. After the end of control packet transmission, the station transmits the corresponding data packet over its own dedicate λ_{dk} channel. The vulnerable period, for destination conflicts, lasts L-1 time units. The notations used here are consistent with those used in the previous protocol. So

$$P_{suc} = \frac{G}{v} e^{-G/v} \tag{16}$$

$$P_f = \frac{S_c}{G} e^{-G/v} \tag{17}$$

$$S_N \approx \frac{L}{N} G e^{-G/v} \exp[-\frac{G}{N}(L-1)e^{-G/v}] \tag{18}$$

For the delay analysis we introduce

D_w = the interval between the packet is generated till the beginning of the next multislot. So we get,

$$\begin{aligned}[D] &= E[D_w] + E[D_r] + L + 1 = (R-1)(L+1+D_b) + L + 1.5 = \\ &\{e^{G/v} \exp[\frac{G}{N}(L-1)e^{-G/v}] - 1\}\{L+1+(K+1)L/2\} + L + 1.5\end{aligned} \tag{19}$$

3. NUMERICAL RESULTS

If we set the first derivative of Eq. (10) with respect to G equal to zero, we find the optimal value of G that maximize the throughput.

$$G_{opt} = \frac{v}{2}$$
$$S_N(\max) = L\frac{v}{2N} e^{-[1+\frac{v}{2N}L\frac{1}{e}]} \tag{20}$$

If the above equation is differentiated with respect to v, can be found the optimal of v for fix values of N and L.

$$v_{opt} = \frac{2N}{L} e \tag{21}$$

Substituting the above equation into (20) we take

$$S_N(\max) = \frac{1}{e} \tag{22}$$

The above value is independent of L.

For the case II we take

$$G_{opt} = v$$
$$S_N(\max) = L\frac{v}{N} e^{-[1+\frac{v}{N}(L-1)\frac{1}{e}]} \tag{23}$$

$$v_{opt} = \frac{N}{L-1} e \tag{24}$$

$$S_N^{'}(\max) = \frac{L}{L-1}\frac{1}{e} \tag{25}$$

For L=2, we get the maximum possible value of $S_N^{'}(\max) = \frac{2}{e}$

4. CONCLUSIONS

The novel concept of the Multichannel Control Architecture and the distribution of control information over a number of control channels, in conjunction with the fixed tuned receivers, minimize the headers processing requirement at each station. Each station fixed tuned receiver has only to process packet headers at a rate divided by a factor v (fixed tuned receivers number) compared with the single control channel communication system. In this way a network designer can overcome the limit of the maximum processing rate of the electronic interface of a station.
We have presented and analyzed a new protocol architecture in which the "tell and go" strategy has been adopted for the random access to the communication network. The proposed schemes except for the reducing of the electronic processing bottleneck improves the performance measures by avoiding data channel collision improving any more the efficient use of optical bandwidth.

References

[1]Pierre A. Humblet, Rajiv Ramaswami, Kumar N.Sivarajan ,"An Efficient Communication Protocol for High-Sped Packet Switched Multichannel Networks," IEEE J. Select. Areas Communic., vol.11 No.4 pp 568-578 May 1993.

[2]Ioannis.E.Pountourakis "A Multiwavelength Control Architecture for Electronic Processing Bottleneck Reduction in WDMA Ligtwave Networks " Computer Communications vol. 22 Number 15-16 September 1999 pp. 1468-1480.

[3]Ioannis.E.Pountourakis " Throughput evaluation of multichannel slotted Aloha-type protocols with receiver collisions" Telecommunication Systems vol.5, No.4, July 1996 pp 413-419.

[4]Ioannis. E. Pountourakis, " Performance Evaluation with Receiver Collisions Analysis in Very High-Speed Optical Fiber Local Area Networks using Passive Star Topology", J. Lightwave Technology, vol.16, No.12 pp 2303-2310 December 1998.

Broadband Access, WDM Metro and Network Management
D.W. Faulkner and A.L. Harmer (Eds.)
IOS Press, 2000 © AKM

The Bundled Full Service (Broadband) Residential Market: Structure, Strategies and Solutions

J.A. Quayle, E.E. Harstead & P.C. Becker
Lucent Technologies, 600 Mountain Avenue, Murray Hill, New Jersey, USA

Abstract. A detailed analysis of the market, operator strategies, and access solutions are brought together to give an insight into what could emerge over the next four years with respect to broadband access. The combined number of residential DSL (Digital Subscriber Line) and cable modem households is likely to exceed 20 million in the US by 2004, roughly 20% of the residential market. Surveys show that 95% of residential customers are interested in long distance and local telephony bundling, and 60% of customers are interested in a bundle of telephony and cable TV. Customer demand is exploding, and bundling of services is becoming a critical success factor.

Incumbent operators have two unique assets that clearly differentiate themselves from the competition, their access network infrastructure, and a workforce to effect change in that infrastructure. APON (ATM Passive Optical Network) used in a FTTH (Fiber to the Home) configuration would appear to be a good long-term solution. Given the advantages of using overhead plant to minimize installation costs, a strategy of targeted incrementalism would appear best for introducing the solution.

However, FTTH technology is not yet a complete solution, issues remain. Key industry groups, such as FSAN (Full Service Access Network) initiative, are actively working to solve and standardize solutions to these issues. There are dramatic changes happening in the telecommunication world, incumbent operators need to make some dramatic decisions to meet the challenges being posed by these changes. Could FTTH finally have its day?

1. Introduction

Many papers have been published on broadband access solutions to the residential market. Why is this paper different? For the first time a detailed analysis of the market, a detailed analysis of operator strategies, and a detailed analysis of access solutions are brought together to give an insight into what could emerge over the next four years.

The paper's main focus is on the US, because of the rapid development of access competition and high Internet usage has resulted in the US leading the race towards the converged network. The converged network is where all network services and transport layers are supported by a unified solution for the provision of information, communication, and entertainment applications. However, this analysis is extended to the rest of the world and differences are highlighted. For example, in Europe and Japan the maturity of mobile telephony means fixed-mobile convergence will initially play a more important role than in the US.

For many years ILECs (Incumbent Local Exchange Carrier) have investigated and experimented with broadband access. However, there was little competitive or market drive to encourage them to make a large pro-active investment in broadband. However, there are emerging some fundamental changes in the ILEC's environment, described in this paper and briefly summarized in Figure 1.

Regulation and competition are making significant changes to the access network competitive environment. Demand from customers is growing and becoming increasingly sophisticated, associated industries are making demands on the connectivity to customers, further increasing demand. All these factors are converging upon the incumbent operator, which will potentially force ILECs to make a re-active investment to provide a full service bundle (broadband).

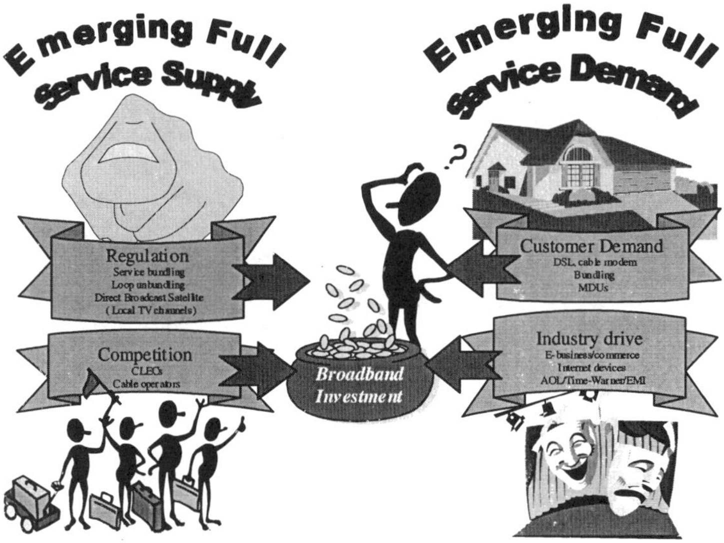

Figure 1. Emerging Broadband Environment

2. Market Analysis

The combined number of residential DSL (Digital Subscriber Line) and cable modem households is likely to exceed 20 million in the US by 2004, roughly 20% of the residential market, see Figure 2. Both ILECs and cable operators are aggressively preparing their access networks for the deployment of DSL and cable modems, respectively. By 2002 about 80% of US households will be passed by either DSL or cable modem. Thus, broadband connectivity is set to explode in the US market. About 20% of ILEC customers in the US are fed by DLC (Digital Loop Carrier), a node remoted from the central office. Open access at the DLC is not resolved which complicates DSL deployment for ILECs.

Cable is predicted to have the greater share. In 1999 over 1 million customers were using cable modems. This market lead will continue. In other markets, e.g. UK, the incumbent does not want to loose potential market share by letting cable operators take the lead. Hence the more aggressive rollout of DSL compared to the cable operators' cable modem plans. However, the over-riding fact is broadband Internet access is set to explode.

Some predictions show DSL-lite dominating residential DSL deployment. This is dependent upon ILECs supporting DSL-lite, which currently does not appear to be the situation. This is partially due to the operational process concerns. However, if a CLEC aggressively supports DSL-lite, which they are likely to because it avoids a truck-roll, ILEC will be forced to respond. The residential market dynamic has changed radically, changes are happening in months rather than decades, hence increased uncertainty in the solution choice.

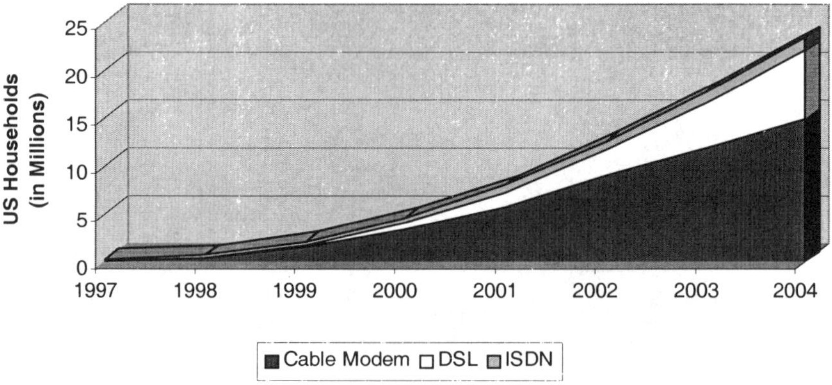

Figure 2. Residential Broadband Demand

Examining the residential market shows a large SOHO (Small Office Home Office) segment, about 40% of households, see Figure 3. This is a relatively loose definition of SOHO, which encompasses after hours working, telecommuters, secondary income generating and primary income generating. This segment is important because they potentially need high-speed communications from the home. Generally SOHO customers are the ILECs' residential cash cow, with high monthly telephone bills, and are likely to be the early adopters of DSL and cable modem technology. The number of SOHO households could increase over the next two years to 50% of the market. These customers have a higher than average propensity to take a full service bundle.

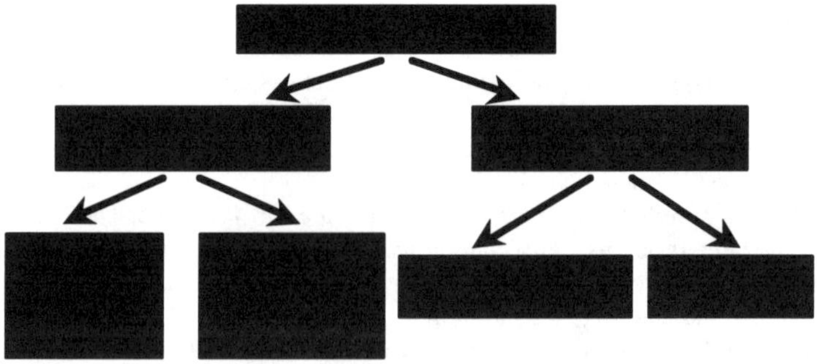

Figure 3. Work at Home Categories

An early battlefield in the residential market will be customers living in MDUs (Multi Dwelling Unit). There are about 21 million MDU households in the US. About 50% of those households are in MDUs of 50+ households. MDUs of this size are likely to have competitive access by 2001. This market will be the first battleground for operators competing with bundled full service packages of voice (local distance and local), video (cable TV) and data (Internet access).

A key feature of residential customers is their interest in bundling of services. Today in Bell Atlantic's operating region there is significant competition in the bundling of long distance and local telephony, and recently Bell Atlantic was allowed to enter the long distance market. These are dramatic changes in the telephony market. Surveys show that 95% of residential customers are interested in long distance and local telephony bundling.

The next service residential customers would like bundled is cable TV, with over 60% of customers interested in a bundle of telephony and cable TV. This fact explains the cable operators' drive to expand their service capability to data and voice. Some ILECs are responding to their customers' desire for the TV & telephony bundle by reselling DBS (Direct Broadcast Satellite). Recently DBS has become a more significant competitor to cable as it can now carry some local programming. However, CLECs (Competitive Local Exchange Carrier) can also resell DBS, so this is not necessarily a unique competitive advantage for ILECs.

In summary, customer demand is exploding, and bundling of services is becoming a critical success factor.

Not described in detail in this paper is the supply-side drive. There is much consolidation in the computing, communications and entertainment industries, e.g. the AOL/Time-Warner/EMI merger, Microsoft's investments in cable operators. Many TV channels have simultaneous WWW (WorldWide Web) events synchronized with their programming. For local access providers bundling is essential to satisfy the supply-side drive.

Outside the US Internet access is growing, in some markets the lack of access competition is restricting the rate of change, however, the regulatory trajectory indicates access competition will increase. The relative strength of cable operators, incumbent telephony providers and second operators are different. However, the key message is full service bundling will become important once competition truly occurs in the local loop.

3. Operator Strategies

In examining the costs and revenue flows in providing a full service bundle, Figure 4. Cable operators are moving into a more profitable business, whilst ILECs are being forced into a less profitable business. This will put the ILECs in a challenging environment. They will be attacked from all sides. From cable operators with their telephony and TV bundle, and from CLECs using unbundled copper plant and potentially bundling DBS, see figure 5.

Five ILEC strategies are assessed in Table 1, and summarized below.

- Build out a video network:
  - Revolution (build out a new access network everywhere which is capable of offering video and high speed Internet access).
  - Targeted incrementalism (target the build of a new access network to areas where video and Internet take-rates will be high).

- Use DBS for video:
  - Resell DBS and package on one bill, using DSL for high speed Internet
  - Focus on customization by providing many tailored packages of services.
- Don't do video. Focus on voice and Internet services (move up the 'value add' stack).

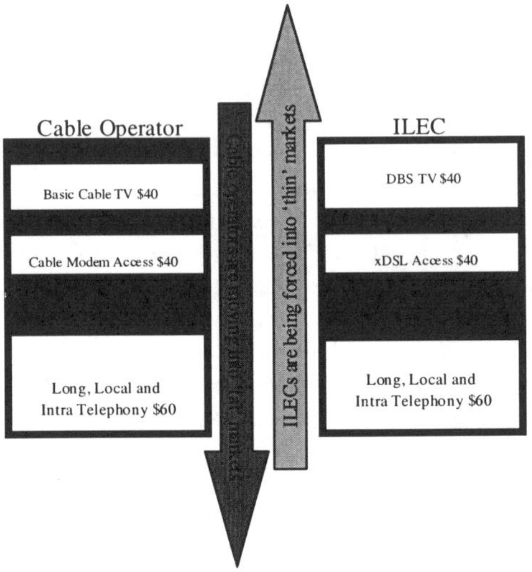

Figure 4. Fat and Thin Markets

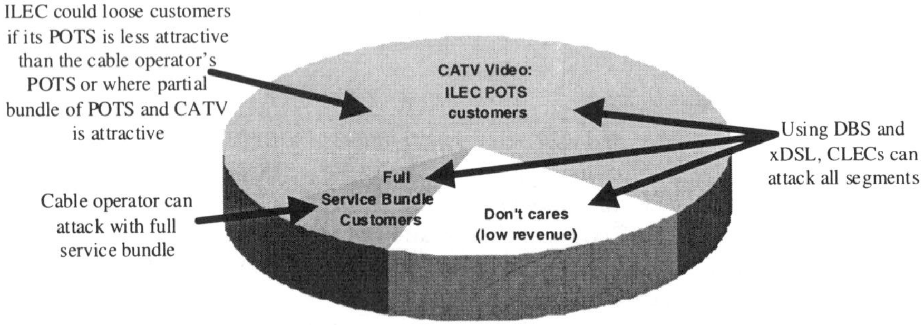

Figure 5. ILECs will be attacked on all sides

Table 1. Possible ILEC Strategies

Revolution: Changeout or overlay a new infrastructure to provide a full service bundle generally aiming for about 30% penetration	Examples	USWest with VDSL SNET with HFC (abandoned)
	Advantages	Competes with and differentiates from cable operators Potentially removes aging copper plant
	Disadvantages	High cost Broadcast TV service provision can be complex, e.g. requiring digital STBs for every TV in the home. Operational issues are not yet resolved
Targeted Incrementalism: Changeout plant in a targeted way for customers who want a full service bundle aimed at very low average penetration, with possibly high density pockets	Examples	Current ILEC ADSL deployment targeted in cable modem deployment areas Bell South with FTTH (First Office Application)
	Advantages	Lower cost at low volumes compared to the revolution strategy Competes with and differentiates from cable operators Natural choice for new build, e.g. new housing developments
	Disadvantages	Operational problems, as potentially requires many interventions in the access plant If penetration starts to ramp up, then costs will be high. Broadcast TV service provision can be complex, e.g. requiring digital STBs for every TV in the home
DBS Reseller: To meet the full service bundle requirements provide a DBS service bundled with POTS and xDSL bill	Examples	SBC and Bell Atlantic
	Advantages	Low cost, does not require intervention with the access plant Enables some differentiation from cable operator.
	Disadvantages	No additional revenues, just stops loss of customers. Local programming issues for some areas. Possibly expensive for the customer as it requires aerials and dishes on the roof and a digital set top box for each TV.
Tailored Packages: ILEC provides a suite of innovative communication service bundles, targeted at different user groups, e.g POTS (long, local, intra), xDSL, Internet, cellular & conferencing for the 'mobile professional who also works at home'	Examples	
	Advantages	Low cost No intervention with the copper plant Enables differentiation from CATV operator
	Disadvantages	Same DBS issues as 'DBS Reseller' strategy Requires operator to have sophisticated billing and customer service operations Open to significant competition from IXCs other CLECs This is the long term residential plan of many CLECs
Focus on Voice and Internet Services: ILECs focus on Internet service provision and a range of value add voice/data services	Examples	
	Advantages	Low cost No intervention with the copper plant
	Disadvantages	ISP is not a core area of expertise Highly competitive

4. Access Solutions

An important access solution for the ILEC, and perhaps their only chance to be successful in the emerging broadband environment, may be the APON (ATM over Passive Optical Network) used in a FTTH (Fiber To The Home) configuration, see reference 1. FSAN (Full Service Access Network) initiative has been fundamental to the specification of this solution, reference 2. Unless ILECs adapt quickly to meet the challenges of the emerging competitive environment they will have difficulties maintaining their current performance.

APON provides a flexible future-proof way to deliver multiple services, within a managed service level agreement, to customers that maximizes the utility of a service providers physically assets, because of its point to multi-point topology. The flexible way capacity is provisioned over the APON, allows almost instant zero-touch provision of services and efficient use of scare access network capacity. In moderate to high volume scenarios APON provides the benefits of point to point fiber at a significantly lower cost.

Currently ILECs are maximizing their copper assets by using DSL. This is providing effective competition to cable modems. However, ILECs have an additional asset, the infrastructure (ducts and poles) and workforce to maximize the value of their local access. Creating a competitive access platform to compete in the full service bundle market will be for some ILECs their only survival route.

The domain of FTTH deployment can be divided between overbuild and new build. Like ADSL, overbuild is where a copper 'lifeline' POTS network already exists, and an overlay system need only supply broadband features (data and video) and additional non-lifeline telephony to subscribers. Aerial plant is the most practical place to start overbuild. Hence a targeted incrementalism strategy, focused upon specific deployment areas with a high percentage of aerial plant would be a good place to start deployment.

Powering issues are a major operational difficult for new build. Maintenance of ONT batteries will be a major operational cost; batteries would be required for lifeline services. However, growth in mobile connectivity means people's perceptions and needs for lifeline service are changing. Lifeline POTS will always be required from some customers, however as personalization and mass-customization increases, the ubiquitous need for lifeline POTS could be questioned.

The best solution for broadcast TV services is to deliver the full stack of channels to the customer's home. The service content must be at a minimum equivalent to the CATV operator. The broadcast service can be delivered with digital channels that are converted into analog in the home by a STB or residential gateway (a STB that serves several TVs over the in-home coax distribution network), or as a full analog multiplex. The solution that will win out has yet to be determined. This is an active study item within the FSAN initiative.

5. Conclusions

In summary, customer demand is exploding, and bundling of services is becoming a critical success factor.

ILECs are in a difficult evolutionary position, bundling of services will be vital to their future success. It will become increasingly difficult for ILECs to differentiate themselves from the competition. CATV operators already have a broadband pipe to the home; this is the competitive technology threat that is keeping ILECs awake at night. There is no 'golden bullet' access solution; there are a range of strategic options available to ILECs.

ILECs have two unique assets that clearly differentiate themselves from the competition, their access infrastructure (ducts and poles), and a workforce to effect change in that infrastructure. As in interim measure using DBS and bundling it with other ILEC services appears to be a preferred approach for many ILECs. However, there are a range of issues, that potentially limit the effectiveness of such an approach. Hence APON used in a FTTH configuration would appear to be a better longer-term solution. Given the advantages of using overhead plant to minimize installation costs, a strategy of targeted incrementalism would appear best.

However, the FTTH technology is not yet a complete solution, issues remain on low cost network terminations, how broadcast TV services are supported, how lifeline POTS services are supported and installation procedures of fiber in the customers premises. However, key industry groups, such as FSAN, are actively working to solve and standardize these problems. There are dramatic changes happening in the telecommunication world, ILECs need make some dramatic decisions to meet the challenges being posed by the changes. Could FTTH finally have its day?

6. References

[1] E.E.Harstead, et al., ATM PON as a Business Access Technology, with Migration Path to FTTH, ISS 2000, May

[2] J.A.Quayle, Achieving Global Consensus for Broadband Access through the Full Service Access Network Initiative, NOC 1999

Broadband Access, WDM Metro and Network Management
D.W. Faulkner and A.L. Harmer (Eds.)
IOS Press, 2000 © AKM

STM Shared Access System for High-Speed IP Communications

Yukihiro FUJIMOTO, Akihiro OHTAKA, Kizuku YAMAKI and Noriki MIKI
NTT Access network service systems labs.
1-6 Nakase, Mihama-ku, Chiba 261-0023, Japan

Abstract. This paper describes high-speed access system based on synchronous transmission mode – passive double star (STM-PDS) system for computer communications. Proposed system uses bandwidth-sharing technologies that enable multiple users to transmit their data packets 10Mb/s at maximum by sharing a 10Mb/s bandwidth path. By using these technologies, we developed the STM shared access system and evaluated the system performances and inter-operability with high-speed backbone networks.

1. Introduction

High-speed access networks have been discussed for decade and wide variety of access systems has been proposed. The main service by using this kind of network is for high-speed data communication represented by the Internet access. Corresponding to this aspect, Cable modem systems and xDSL are widely deployed in access network today [1][2].

Passive optical networks (PON) have been deployed for residential customers to provide telephony and Cable TV services, however, optical-fiber access systems, especially fiber-to-the-home (FTTH) for high-speed communications services is still costly except for large-scale business customers[3].

This paper describes high-speed access system based on synchronous transmission mode – passive double star (STM-PDS) system for computer communications. Proposed system uses bandwidth-sharing technology that enables multiple users to transmit their data packets 10Mb/s at maximum by sharing a 10Mb/s bandwidth path. By using this technology, we developed the STM shared access system and evaluated the system performances and inter-operability with high-speed backbone networks.

2. STM-PDS Access System

NTT have been deploying Pai-system [4], which is a fiber-rich access system for telephony services based on the STM-PDS system [5][6], in the access networks since 1998. The installation cost of the Pai-system is less than that of a conventional metallic system because each optical network unit (ONU) accommodates multiple lines (Fig. 1). Since optical fiber cables have been installed in distribution areas, the access network is easily upgraded to FTTH just by running a fiber from the closest access point. Therefore the best use of Pai-system platform is the viable approach to provide economical high-speed communications services.

For preparing transmission paths for providing high-speed communications services, it is common that to designate paths for all ONUs and to guarantee transmission speeds. This method is generally used to provide leased-line services. However, having high-speed paths for multiple ONUs is not economical because high-speed transmission systems are required for every path whether a path is continuously in use or not. The other method is to share a high-speed transmission path among multiple users. With 32 ONUs sharing a 10-Mb/s path, each ONU can communicate at nearly a 10-Mb/s transmission speed during light signal traffic. Under heavy traffic condition, each ONU can communicate around 300Kb/s at least (Fig. 2).

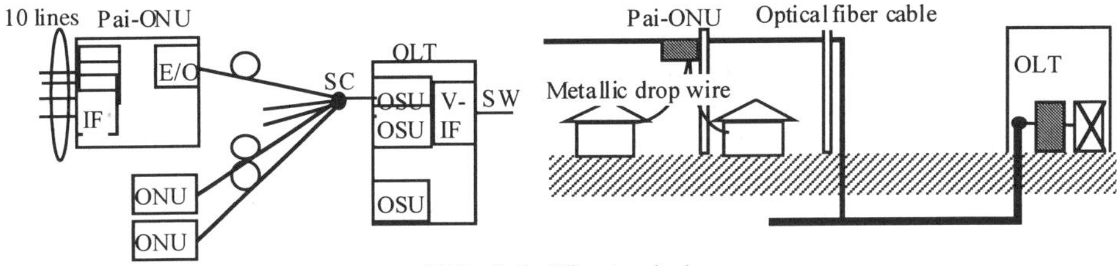

OLT: Optical line terminal
ONU: Optical network unit
OSU: Optical subscriber unit
SC: Star coupler

Figure 1. Pai-System

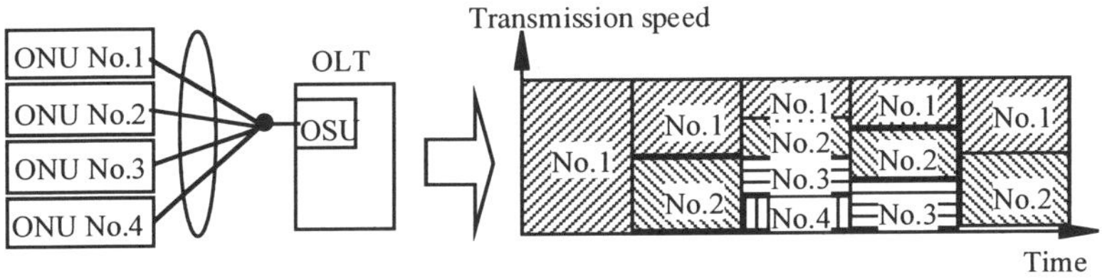

Figure 2. Bandwidth sharing

3. STM Shared Access System

To develop STM shared access system based on the existing STM-PDS system (Pai-system), the transmission frame should be changed for the bandwidth sharing, however, the transmission frame structure must be to ensure compatibility with the existing STM-PDS system, over-modification of the transmission frame must be avoided. Pai-system uses time-division-multiplexing (TDM) for downstream, time-division-multiple-access (TDMA) for upstream, and time-compression-multiplexing (TCM) for bi-directional communication. The pointer information for the downstream channel's receiving position and the upstream channel's sending position for individual ONUs are contained in the downstream block headers. Each ONU uses an authorized channel for telephony services, and this channel is regularly assigned to individual ONUs [7][8].

3.1 Transmission Frame for Bandwidth Sharing

The transmission frame for bandwidth sharing consisits of existing channels and data blocks for bandwidth sharing (Fig. 3). Existing channels are not modified, so ONUs can recognize the authorized channel position as in the conventional method. The data-block length for bandwidth sharing is flexible, ranging from 1 to 255 B (B: 64 Kb/s).

For downstream transmission, ONUs that share bandwidth read the pointer information contained in the header of the downstream data blocks. This pointer indicates the starting positions and the length blocks for sharing. The downstream data block also contains data packets with ONU ID codes. In the STM-PDS system, downstream data blocks are broadcast to every ONU, then each one reads the indicated data blocks and retrieves packets with its own ID.

For upstream transmission, each ONU must know how many data blocks it is permitted to use and the sending positions for the upstream data blocks. When the ONU receives a data packet from a terminal, ONU measures the data packet size. Then, the ONU requests the use of upstream data blocks. In the next burst period, the ONU reads the header of the downstream data blocks containing ONU IDs for authorized ONUs and the positions for send-out. This header information indicates which ONU has permission to use a specific upstream data block for sharing. Next, the ONU places the data packet in the indicated upstream data block or blocks (Fig. 4). When several ONUs request use of upstream data blocks simultaneously, the controller schedules the order of permission for each ONU, attempting to keep an equal level of accessibility among ONUs [9].

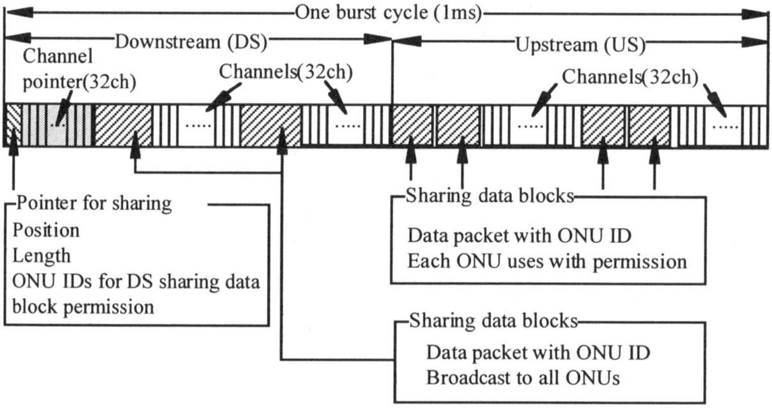

Figure 3. Transmission frame of STM shared access system

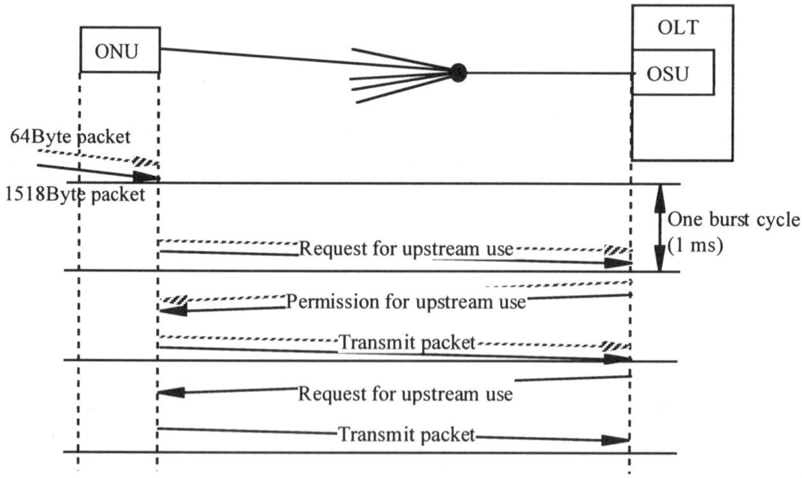

Figure 4. Sequence of upstream transmission

3.2 System Configuration

Fig. 5 shows the system configuration of STM shared access system. Each optical subscriber unit (OSU) can accommodate 32 ONUs and one Router-IF can control 8 OSUs and 256 ONUs. 10 Router-IFs are equipped on an optical line terminal (OLT). Then one OLT drives 2560 ONUs.

The ONU has a 10 Base-T interface so that PC or LAN equipment in the customer premises could be connected to the ONU directly. The Ether-IF has functions of media access control (MAC) address bridging of Ethernet packets and protocol filtering to avoid sending needless packets to the OSU. The Router-IF in the OLT has 4-10Base-T/100Base-TX ports to ensure the inter-operability with switches, routers and other LAN equipment. The main functions of the Router-IF are to build VLAN configuration and bridging.

Once a Ethernet packet passes through the ONU and the Router-IF, the MAC address is stored in the both memory tables. If the Ethernet packet with same MAC address is received by the ONU or the Router-IF, they forward to proper port or ONU. The combination of a Router-IF and ONUs acts like a Layer-2 switch. The system has no layer-3 functions at the time.

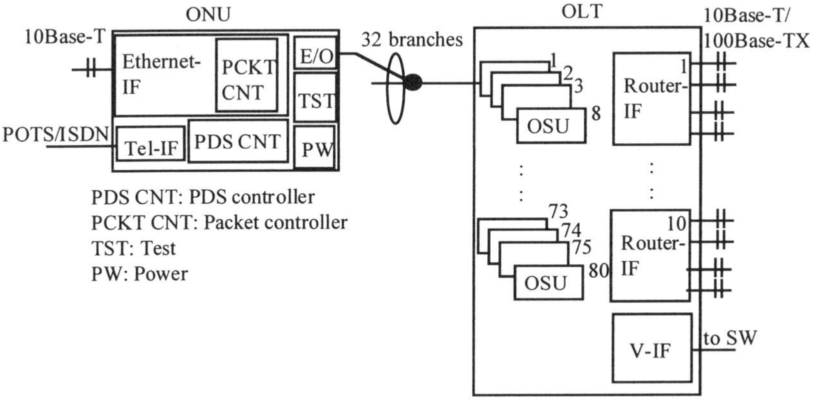

Figure 5. STM shared access system configuration

3.3 VALN function

The STM shared access system is not only an optical transmission system just supporting bridge and repeater functions, but also has two virtual LAN (VLAN) functions (Fig. 6).

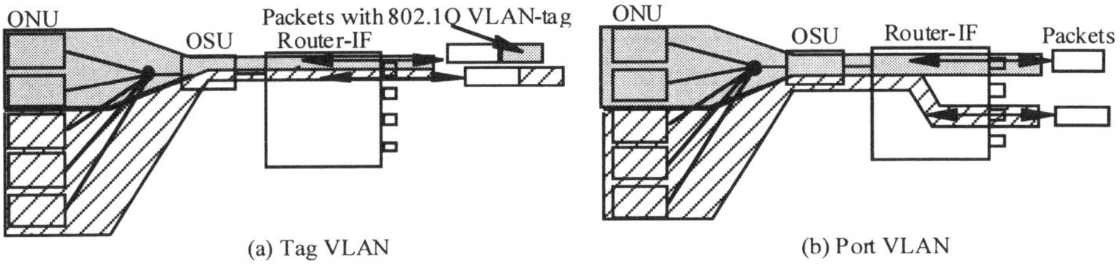

Figure 6. VLAN functions

The one is the Tag-VLAN function. This function uses IEEE 802.1Q VLAN tag [10] and same VLAN-tag is assigned to the same group. The VLAN-tag is attached on the Ethernet packets at the ONU. Logically, it is possible to form 256 VLANs because one

Router-IF manages 256 ONUs. The Router-IF identifies the packets by the VLAN tag number and forward to proper port and ONUs. The VLAN tag will help the core-network closed user group (CUG) function.

The other is the port-base VALN function. This function is to map ONUs to a Router-IF port. The VLAN configuration is managed by the table, which contains the relation between ONU IDs and the port number of the Router-IF. By using this function, transparent transmission is possible and users can form VLAN with IEEE 802.1Q VLAN tag in their groups. However, available VLANs of the Router-IF are limited by the number of Router-IF ports.

In both cases, each VLAN is regarded as a logically independent LAN and communication between VALNs and only communication in the same VLAN is allowed.

4. System performance

We measured the transmission latency and transmission througput of the STM shared access system under the 10Mb/s bandwidth sharing condition.

4.1 Transmission latency

We measured the latency by changing the number of ONUs that send the data packets simultaneously (Fig.7). The length of test packet is 1518-byte, which is the maximum length of Ethernet packet [11][12]. The latency of one ONU, which other ONUs are not sending the data packets, is around 5 ms. To transmit the packets from the ONU to the OLT, it required 3 ms (three burst: one for requesting and two for transmit). To complete the reception of the 1518-byte packets by the Ether-IF in the ONU, it took around 1 ms, as estimated by the bit-time equation (1).

Receiving Time = Packet Size * 8 bit * Bit Time .... (1)
Bit Time: 100 ns (10 Mb/s Ethernet)
Packet Size: bytes.

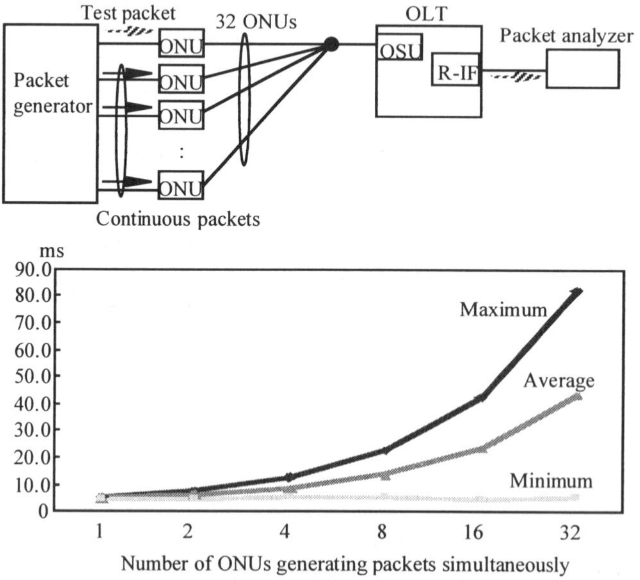

Figure 7. Transmission latency

To receive the packets at the analyzer, the amount of is also needed. Given these times and including the timing variation mentioned before, the actual results are also close to the estimated ones.

In case of increasing the number of ONUs, the latency is varied because the latency depends on the waiting time. For instance, 32 ONUs request the use of the bandwidth simultaneously, the transmission latencies are expected from 5.6 to 82.3 ms. The minimum latency is that ONU could get the permission in the next burst period, therefore it is independent of the number of ONUs. The maximum values are that the ONU is the last ONU to get the permission.

4.2 Transmission throughput

We measured the data throughput of each ONU by changing the number of ONUs that sent data packets (Fig. 8). The data throughput decreased in heavy traffic due to the multiple. However, the minimum transmission speed was ensured because the bandwidth was equally shared by the ONUs. For instance, if the sharing bandwidth was set to 10 Mb/s and 32 ONUs used it, each ONU could transmit at a minimum rate of around 300 Kb/s. The size, appearance and frequency of data packets generated by the terminals would fluctuate, so greater throughput would be expected. Using this system for LAN access, the transmission speed would exceed the 10 Mb/s-Ethernet LAN because minimum transmission speed is ensured and accessibility for ONUs is equal in heavy traffic.

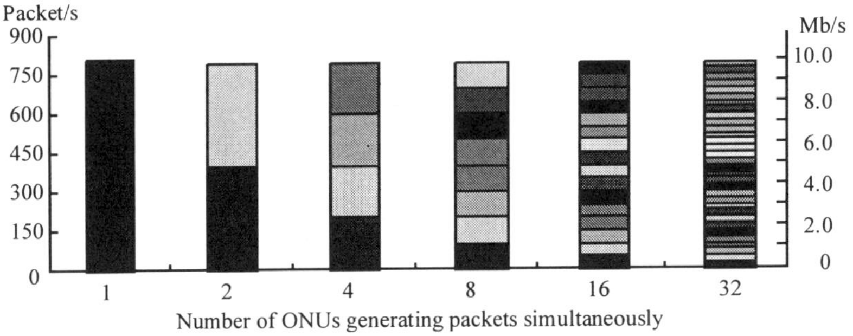

Figure 8. Transmission throughput

5. Applications

The shared-bandwidth method is useful for high-speed computer networks. We shows some examples.

5.1 Communitiy and SOHO networks

By using shared access system, community networks and inter-SOHO networks are formed easily, because any protocols on the Ethernet are transpealently transmitted in the shared access system (Fig. 9). Conventional these kinds of networks are formed by leased lines and the shortage of the line capacity causes the bottoleneck of the network. Furthermore, customers have to install and set up bridges or routers to connect the leased lines. The ONU of the shared access system has a 10 Base-T interface, customers can connect their LAN equipment or PC directry with CAT-5 cable.

If the scale of network is not so large, less than 254 PCs in the network for instance, router-less operation is also possible. The network is seemed as a sub-network and each PC can share or transfer files under the same as 10 Mb/s LAN environment.

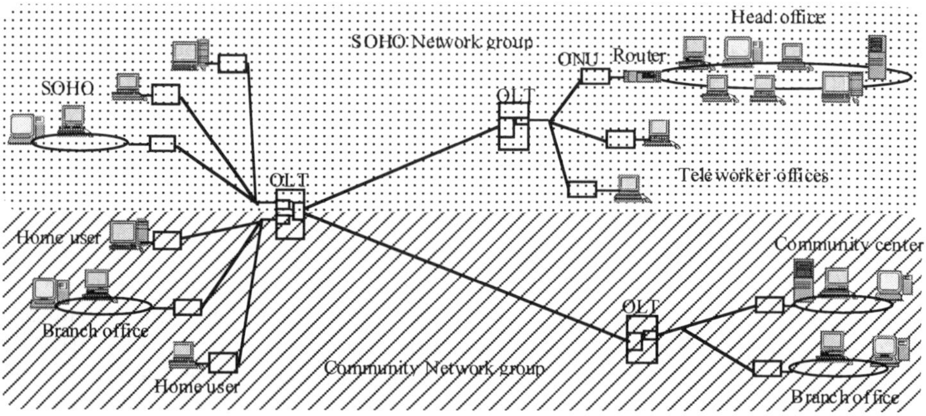

Figure 9. SOHO/Community networks

5.2 High-speed Internet Access

High-speed Internet access have become the most required. STM shared access system will be also viable for the Internet access services, because of its transmission speed and affinities of LAN equipment. By using VLAN fanctions, we can form ISP subscriber groups (Fig. 10). If the ISP supports closed user grouping (CUG), nation-wide VLAN froming would be possible. For subscriber authentication, PPP over Ethernet (PPPoE) [13] would be useful, because there is no IP-layer termination between user's terminals and ISP access servers. The OLT of shared access system supports the fowarding mode that is not multcastiing to other ONUs, PPP packets are not captured by other ONUs. The authentication and access proceedures much depends on the scale of the number of subscribers and the configurations of core networks. We will start these tests from this April.

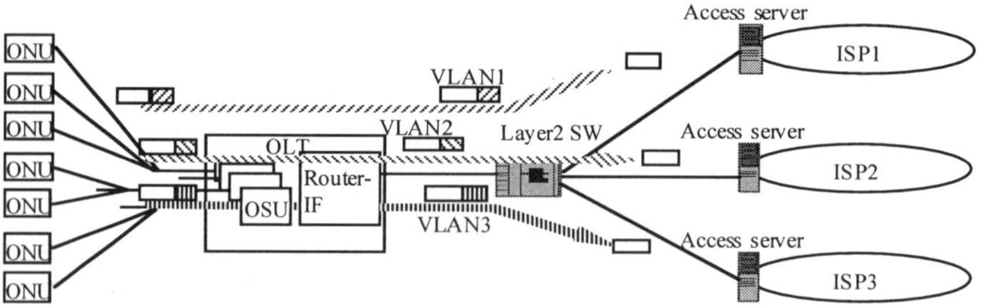

Figure 10. High speed Internet access example

6. Summary

We described the STM shared access system for high-speed data communications access. This system is compatible with the existing STM-PDS system, which is able to provide telephony service at a low cost, and the system is able to provide high-speed IP communications services by using bandwidth sharing technique. The performance of the system has been evaluated and the system could be applied as a high-speed access for IP communications.

References

[1] S. Dixit, "Data Rides High on High-Speed Remote Access", IEEE ComMag., Jan 1999.

[2] S. Yamano, " The Range of Passband QAM Based ADSL in NTT's Local Networks", IEICE Trans. Comn., Vol. E78-3, No. 9, 1995.

[3] L. A. Ims, et al., "Economics of Broadband Access Network Upgrade Strategies: The European Perspective", Access Network Mini Conf., GLOBECOM '98, pp.1-8, 1998.

[4] Y. Shibata, et al., "New 'Fiber To The Near Home System' Using PDS Technologies", Access Network Mini Conf., GLOBECOM '98, pp.77-81, 1998

[5] K. Okada, et al., "Fiber-Optic Subscriber System for Point to Multi-point Transmission Architecture", Proc. ECOC'92, 1992.

[6] N. Tamaki, et. al., "Studies on Design of Optical Receiver for TCM/TDMA Passive Double Star System", IEICE Trans. B, Vol. J82-B, No. 12, pp. 2266-2274, 1999.

[7] Y. Fujimoto, et al., "Fiber Optic Access System and expectations for LSI Technologies", International Symposium on Key Technologies for Future VLSI Systems, pp30-33, 2000.

[8] Y. Fujimoto, et al., "A Transmission Method Using Bandwidth-Sharing for Multi-grade High-speed Access Services", Broadband Access Conference 99, pp15-21, 1999.

[9] N. Miki, et al., "Optical Access System Suitable for Computer Communication", Access Network Mini Conf., GLOBECOM '98, pp.82-87, 1998.

[10] IEEE 802.1Q "IEEE standards for Local and Metropolitan Area Networks: Virtual bridged local area networks".

[11] RFC 793 "Transmission control protocol."

[12] RFC 889 "Internet delay experiments."

[13] RFC2516 "A method for transmitting PPP over Ethemet".

Broadband Access, WDM Metro and Network Management
D.W. Faulkner and A.L. Harmer (Eds.)
IOS Press, 2000 © AKM

Design of an Adaptive Access Network for IP Communications

Kazunari Irie, Hirokazu Nakagawa, and Norihisa Ohta

NTT Information Sharing Laboratories
3-9-11 Midori-Cho Musashino-shi, Tokyo 180-8585 JAPAN

Tel: +81 422 59 4982
Fax: +81 422 60 7815
E-mail: irie.kazunari@lab.ntt.co.jp

Abstract

As the Internet and enterprise intranets become more popular, more time-critical and mission-critical applications will be being run on IP networks in addition to non-real-time applications. Supporting such applications requires QoS schemes that control the bandwidth and transmission delay time. New access networks are desired where users can dynamically control the bandwidth through easy interfaces.

We propose an adaptive access network scheme that can provide virtual IP leased circuits to users on demand. The design objectives are to provide guaranteed bandwidth IP communication services to home and business users in a particular region. It supports conventional applications such as those using FTP as well as applications that use RSVP. We have studied two kinds of access system, ATM-PDS and carrier-shared systems, for use in the fiber-optic infrastructure. The transmission rate is controlled according to the number of carrier signals within the latter system. Policy-managed time-slot control, or bandwidth assignment to users, in ATM-PDS is also designed.

We describe policy-managed access system protocols that provide dynamic and unified control of the bandwidth by linking the IP layer to the lower layers. Java technologies enable interworking between the physical layer and the upper layers. Reservation requests are sent directly to the policy server to control the bandwidth. An experimental system configuration to evaluate the basic operations of the setup protocols is shown. UNIX PCs with multiple network interface cards are used to emulate user and network nodes.

1. Introduction

IP communication services are widely used by business and home users [1]. Many companies have constructed enterprise intranets to connect, for example, their branch offices. However, since guarantees concerning bandwidth availability and transmission delay are generally not provided or specified for conventional Internet and intranet use, best-effort services have been the most common type of service. Increasingly, though, time-critical and mission-critical applications, such as Voice-over-IP, video-conferencing, and financial applications, run on IP networks. Supporting such applications requires QoS schemes that control the bandwidth and transmission delay time. Bandwidth control is particularly critical in terms of video transmission quality and file transfer throughput. To date, bandwidth-shared systems have been studied mainly for access networks wherein only a minimum bandwidth is guaranteed for each user. New access networks are desired where users can dynamically control the bandwidth through easy interfaces.

We propose an adaptive access network scheme that can provide virtual IP leased circuits to users on demand. The design objectives are to provide guaranteed bandwidth IP communication services to home and business users in a particular region. This scheme can ensure that the bandwidth required for the transmission of large data files or video communication is provided. Because the bandwidth is set between user nodes using WWW client applications, the system supports conventional applications such as those using FTP that are not provided with RSVP [2].

We have studied two kinds of access system, ATM-PDS and carrier-shared systems, for use in the fiber-optic infrastructure. Policy-managed time-slot control, or bandwidth assignment to users, in ATM-PDS is designed. When the policy server receives a bandwidth assignment request from a user, it checks the user information data and sets up the user and network nodes. After the setup sequences, the user can transmit data at the reserved rate. Carrier-shared systems within which the transmission rate is controlled according to the number of carrier signals are also proposed. IP data is carried over multiple frequency signals or sub-carrier signals. In this case, either an FDM-type configuration or a WDM-type configuration is possible. The two types of systems are compared.

We also describe policy-managed access system protocols that provide dynamic and unified control of the bandwidth by linking the IP layer to the lower layers. Java technologies enable interworking between the physical layer and the upper layers. Reservation requests are sent directly to the policy server to control the bandwidth. An experimental system to evaluate the basic operations of the setup protocols is now being developed. The system configuration is described in which UNIX PCs with multiple network interface cards are used to emulate user and network nodes.

2. Access Network System

2.1 Shared Access System

The system objectives are to provide guaranteed bandwidth IP communication services to users on demand. We have developed a scheme that uses a policy server to manage network resources as well as user information and that dynamically controls the bandwidth in the lower layers of the access networks accommodating the users. The typical access system configuration is shown in Fig. 1.

Passive Double Star (PDS) systems [3] have been developed to introduce the fiber infrastructure into access networks cost-effectively. Both ATM and STM transmission schemes are applicable to PDS systems. For example, STM-shared systems [4] are proposed where the spare time slots are shared by many users. These systems provide IP communication services at a rate from 500 kbit/s to 10 Mbit/s cost-effectively. The ATM interface, on the other hand, was specified by FSAN (Full Service Access Network) and standardized as G.983.1. The ATM-PDS [5] is more promising for future systems that support multimedia applications because it provides a transmission rate of 150 Mbit/s for upstream and 150 or 600 Mbit/s for downstream and is suitable for high-speed IP communication services.

For alternative transmission schemes in the PDS systems, we propose carrier-shared systems within which the transmission rate is controlled according to the number of carrier signals. IP data is carried over multiple frequency signals or sub-carrier signals. Two types of systems are studied: a WDM-type configuration where multiple wavelengths are used, and an FDM-type configuration where a single wavelength is used for each of upstream and downstream signals in an access line.

Some mechanism will be required to rapidly set up the bandwidth corresponding to the user request in the above-mentioned shared-type access systems wherein the network resources are shared by unspecified users. We have designed an access network control protocol that can rapidly set up the user and network nodes.

2.2 ATM-PDS System

(1) System Configuration

The ATM-PDS system configuration and its upstream and downstream transmission frame formats are shown in Fig. 2. The network node (OLT: optical line terminal) accommodates subordinate user nodes (ONUs: optical network units) by an optical fiber via a star coupler. A downstream frame consists of 56 slots of 53 bytes while the upstream frame

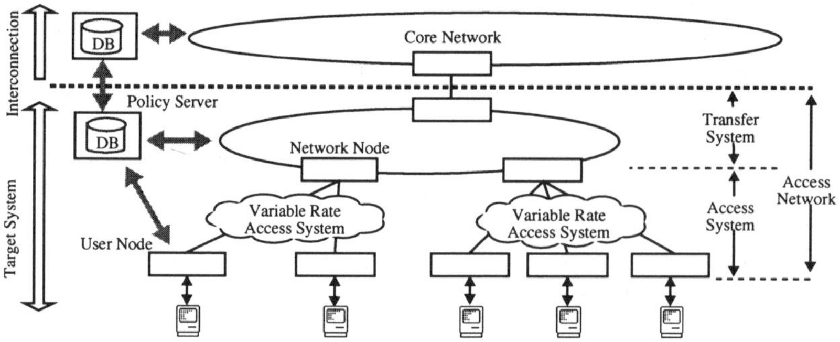

Fig. 1 Adaptive Access Network System

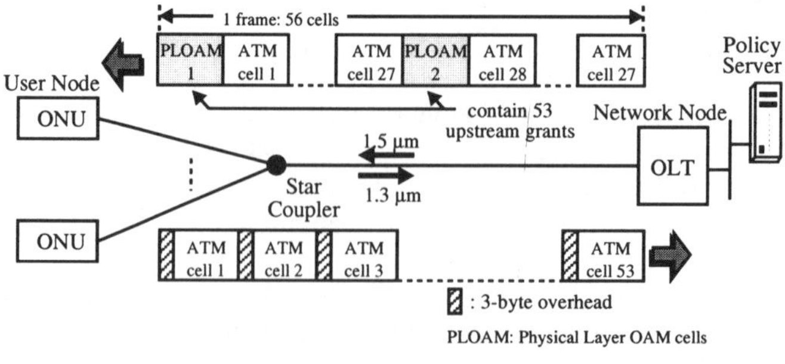

Fig. 2 ATM-PDS based system

has 53 slots of 56 bytes. The time slots are shared by all users connected in the star coupler. Up to now, research has mainly focused on how to control the bandwidth so that all active users can access the resources fairly. Therefore, while a minimum rate is guaranteed, a user cannot be sure of getting any greater bandwidth. We have studied system architectures that can set communication paths with the bandwidths required for computer communications on demand; user information pre-registered in the policy server is used for this purpose. Virtual IP circuits similar to leased circuits will be provided to the user, and transmission bandwidth is guaranteed for a specified period.

(2) Bandwidth Control

A PLOAM (Physical Layer Operation, Administration and Management) cell that gives a user node a "grant" to send cells over the upstream slots is transmitted every 28 cells in the downstream frames, so each frame contains two PLOAM cells. The network node sets the user node identifier in the grant and the identified user node can send data in the specified time slots. The downstream bandwidth is controlled by the network node setting the destination addresses of the 54 cells (two of the 56 slots are occupied by PLOAM cells) as needed. In this way, the upstream and downstream bandwidths can be controlled independently. A reservation request can be accepted from either the sender or receiver side for convenience, whereas in RSVP-based applications the request is assumed to be sent from the receiver.

We offer policy-managed time-slot control or bandwidth assignment to users. When the policy server receives a bandwidth assignment request from a user, it checks the user information data and sets up the network node to control the PLAOM cells. If the bandwidth request is issued by the sender, the policy server checks the user information and the resource usage of the network node. If the bandwidth is available, the server subsequently sends a control signal to the network node to set the desired bandwidth. The network node sends PLOAM cells containing the grant specifying the upstream rate. It also sets the downstream bandwidth of cells. When the policy server receives notification of operation completion from the network node, it sends an acknowledgement signal of bandwidth reservation to the user node. IP data to/from user PCs and the control data are transmitted by IP-over-ATM [6].

2.3 Carrier-Shared System

(1) System Configuration

An alternative configuration to the ATM or STM configuration in the PDS systems is a carrier-shared system. Multiple carrier signals are shared by all users connected to an access line. The required bandwidth is ensured by using the spare carriers. The number of carrier signals is determined by the required transmission rate. The minimum bandwidth is secured by assigning specified carrier signals at all times. Two types of systems for carrier signal transmission are considered: a WDM-type configuration (Fig. 3(a)) where multiple wavelengths are used corresponding to the number of carriers, and an FDM-type configuration (Fig. 3(b)) where a single wavelength is used for each of upstream and downstream carriers. The latter system can ensure enough bandwidth assuming that the required bandwidth is 10-100 Mbit/s at the maximum and that the number of branches at the coupler is up to 32, while the former system can provide a larger amount of bandwidth on user demand.

(2) Bandwidth Control

Multi-link protocol [7] is applied to the data transmission over multiple carrier signals.

Multiple PPP links, each of which is set on a carrier signal, are bundled to make a single data transmission path. The bandwidth is controlled by cooperation among the policy server, user nodes, and network nodes in the same way as it is controlled in ATM-PDS systems. When the policy server receives a bandwidth assignment request from a user, it checks the user information data and sets up the network node and user nodes to assign a new carrier. If a spare carrier is available, a multi-link connection is set between the network node and user nodes. When the policy server receives notification of operation completion from the network node, it sends an acknowledgement signal of bandwidth reservation to the user node.

3. Java-based Access Network Control Protocol

3.1 Basic Design

We propose an access network control protocol (ANCP) that can rapidly set up the user and network nodes in parallel. It has been designed for application to ATM-PDS systems. The policy server receives the user request via the user node. The user node provides a user interface through which the request parameters can be set and the policy server can be informed of the user request. WWW server functions will be implemented in the user node to enable the user to use any popular WWW client application to set the parameters and send the request. The policy server manages the user information as well as the node resources to achieve unified control of the nodes.

The differences among the lower layers need to be hidden to ensure wide applicability when implementing ANCP. We propose a scheme that uses a Java virtual machine (VM) [8] in the middle layer to interconnect the IP layer and lower layers. Its principle is shown in Fig. 4.

The advantages of the Java-based scheme are as follows.

- Portability: the protocol can be executed on any kind of platform having a Java interpreter
- Security: attacks against the system are impossible because all of the operation code is interpreted
- Object-oriented: system extension by class inheritance and code reuse is possible
- Tools: development is relatively easy because various development tools are available

3.2 Java Agents and Functions

The functions of ANCP and the sequences of setting up the bandwidth between user and network nodes for the case where the reservation request is sent from the sender are shown in Fig. 5. The QoS Mapping agent in the user node provides the user interface. Negotiation

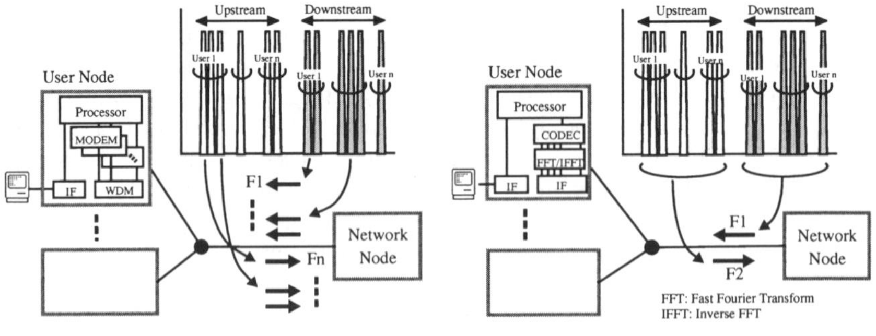

(a) WDM-Type System

(b) FDM-Type System

Fig. 3 Carrier-shared system

agents exchange the managed data information regarding resource reservation among the user nodes, network nodes, and the policy server. The Setup agent in the network node sets a Virtual Channel (VC) in the ATM layer if negotiation is successful. Management agents in the user and network nodes communicate with the Management agent of the policy server and they exchange managed data information. Monitoring agents supervise the ATM cells in the lower layers. Java Remote Method Invocation [9] is used for communication among user nodes, network nodes, and the policy server. The details of the operation sequence are given below.

Setup Phase:

When a reservation demand from a user client is received in the user node, it is sent to the QoS Mapping agent of the user node, and its format is changed to suit interagent communication.

(1)(2) The reservation demand information is transferred to a Negotiation agent in the policy server.

(3)(4) Using the reservation demand information that has been sent, the corresponding Management agent refers to the database in the policy server, judges whether it can reserve the bandwidth, and returns the result to the Negotiation agent.

(5) The Negotiation agent notifies the user and network nodes that reservation is possible.

(6) The Negotiation agent in the network node sends a VC control request to the Setup agent.

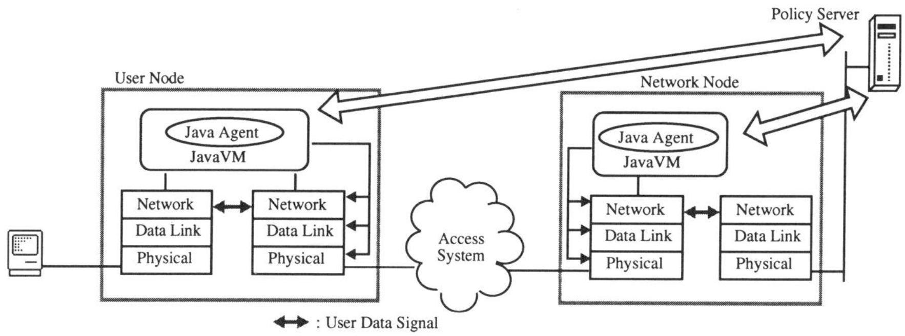

Fig. 4 Access network control protocol using Java

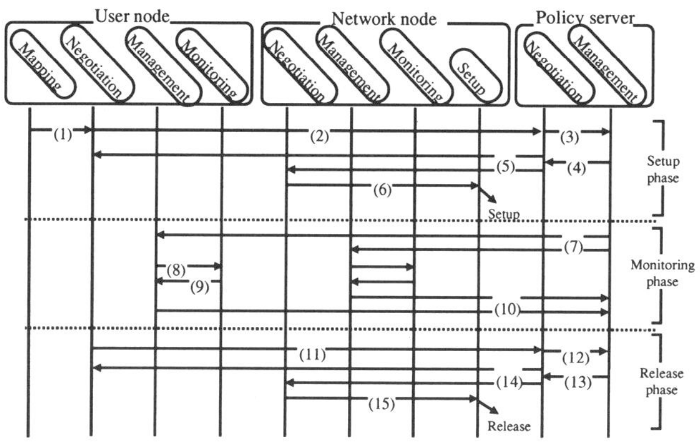

Fig. 5 Java agent and operation sequence

Monitoring Phase:

The usage of the VC must be monitored regularly in order to release the bandwidth that is not being utilized.

(7) The Management agent of the policy server sends a request for monitoring to the Management agents of the user and network nodes.

(8)(9)(10) Each Management agent communicates with each Monitoring agent and sends the monitoring information to the Management agent of the policy server.

Release Phase:

(11) If the Negotiation agent in the user node receives a release demand from the user client, it notifies the Negotiation agent in the policy server of the demand.

(12)(13) In the policy server, the Negotiation agent notifies the Management agent of the release demand and updates the database.

(14) The Negotiation agent sends the bandwidth release notice to the Negotiation agents in the user and network nodes.

(15) The Negotiation agent in the network node orders the Setup agent to drop the VC.

3.3 Implementation

We are developing an experimental system to evaluate the basic operation of the adaptive access network control protocol. It consists of two user nodes and a network node. The user and network node functions were implemented on UNIX PCs. The policy server function was implemented on a Windows-NT PC. The PC emulating the network node has multiple ATM network interface cards and an Ethernet card as shown in Fig. 6. The PC emulating the user node has an ATM network interface card and an Ethernet card for the user interface. The performance of the system is now being evaluated.

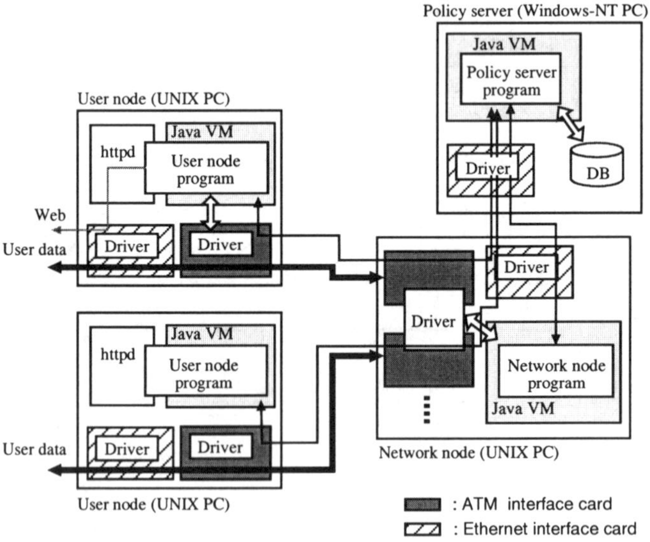

Fig. 6 Experimental system

4. Conclusion

We have designed a new policy-based access network system and a control protocol. System configurations based on ATM-PDS and carrier-shared schemes were described. We also presented a policy-managed access network control protocol that provides dynamic and unified control of the bandwidth by linking the IP layer to the lower layers. It is based on Java technologies that can hide the differences among the lower layers and supports widespread applicability. Because the system can provide virtual IP leased circuits to users upon demand, conventional applications such as FTP can be used with guaranteed rates.

Acknowledgment

The author thanks Mr. Kazuo Imai of NTT Information Sharing Platform Laboratories for his guidance. Thanks are also due to the members of the Megamedia Access Control Systems group for several fruitful discussions.

References

[1] K. Irie, et al., "Regional PC Communication Network for Residential Use Supported by Fiber-Optic Access Systems," 2nd Int. Workshop on Community Networking, June 1995.

[2] R. Braden et al., "Resource ReSerVation Protocol (RSVP)," RFC2205, September 1997.

[3] K. Kumozaki and K. Okada, "Fundamental Structure of the Fiber-Optic Passive Double Star System," IECE Trans. Comm. E-75-B, No. 9, September 1992.

[4] N. Miki, et al., "Optical Access Systems Suitable for Computer Communication," GLOBECOM'98, AN8.4, November 1998.

[5] ITU-T Rec. G.983.1, "High-speed optical access systems based on Passive Optical Network (PON) techniques," 1998.

[6] M. Laubach, "Classical IP and ARP over ATM," RFC1577, January 1994.

[7] K. Sklower, et al., "The PPP Multilink Protocol", RFC1990, August 1996.

[8] http://java.sun.com/docs/books/vmspec/html/VMSpecTOC.doc.html

[9] http://java.sun.com/products/jdk/rmi/index.html

D.W. Faulkner and A.L. Harmer (Eds.)
IOS Press, 2000 © AKM

Dynamic assignment in wavelength division multiplexed passive optical networks

M.R.Handley[1,2], M.S.Leeson[3], J.M.Senior[4], A.J.Phillips[2], P.Ball[1] and I.Wright[1]

[1]*Fujitsu (Europe) R & D Centre Ltd, 2, Longwalk road, Stockley Park, Uxbirdge, Middlesex, UB11 1AB, U.K.*
[2]*Manchester Metropolitan University, Chester street, Manchester, M1 5GD, U.K.*
[3]*Brunel University, Uxbridge, U.K.*
[4]*University of Hertfordshire, Hatfield, U.K.*

Abstract. With the expected increase in demand for broadband services, it is inevitable that wavelength division multiplexing (WDM) will migrate to the access network. Two broad strategies exist for the implementation of multiple wavelengths on passive optical networks (PON), fixed wavelength schemes and dynamic wavelength assignment. Wavelength reallocation at call setup allows a flexible response to changes in network demand. It does not, however, take full advantage of statistical multiplexing and the ATM approach. In order to achieve these benefits, two protocols have been developed which allow wavelength assignment on a frame and a cell by cell approach respectively. Simulation models of the protocols are described and their performance results in response to bursty traffic are presented.

1. Introduction

The implementation of wavelength division multiplexing (WDM) in passive optical networks (PON) can be divided into two broad approaches, namely, fixed wavelength schemes and those utilising dynamic assignment of wavelengths [1]. The former, the upstream of which is shown in Figure 1(a), are simpler and currently cheaper to implement, whilst the latter, Figure 1(b), allow more efficient use of the network resources but at the price of increased costs caused by the need for fast tunable transmitters and filters.

The benefits demonstrated for dynamic assignment in previous papers have been for systems that utilise wavelength allocation at call set-up [2,3]. Although this approach enables the network to respond to changes in the pattern of customer demand, it does not realise the full potential of wavelength reallocation.

The development of very fast tuning of components in the microsecond and even the nanosecond region [4,5] enables the development of networks which allow assignment of wavelengths on a frame by frame or even cell by cell approach. These take full advantage of statistical multiplexing and the ATM approach, further increasing the efficiency of bandwidth use in the PON.

Fixed wavelength PONs and dynamic assignment schemes on call set-up allow the use of protocols developed for single wavelength systems. From the medium access control (MAC) point of view, the network becomes in effect a number of single wavelength PONs superimposed on each other.

By contrast, allocation at frame or cell level requires the development of new MAC protocols. This requires the synthesis of ideas from protocols developed for WDM star networks and those used in single wavelength PONs.

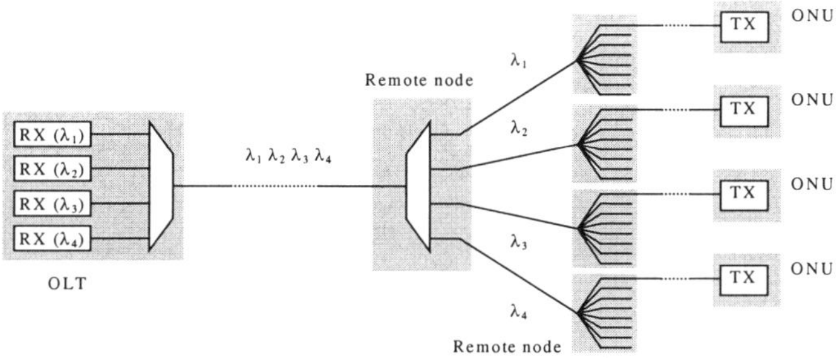

Figure 1(a): A fixed wavelength PON

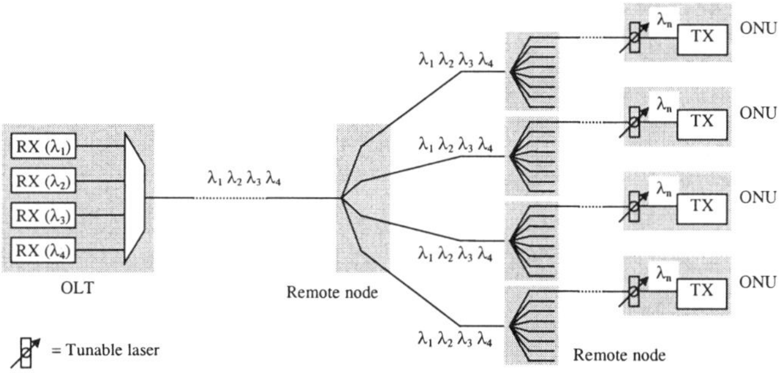

Figure 1(b) Dynamic assignment PON

Two approaches are described here. Both protocols use the frame structure outlined in the ITU-T standard for ATM over PONs [6]. The upstream, with a bit rate of 155 Mbit/s, has a frame of 53 cells. Each cell has a three byte overhead, consisting of the guardband, cell delimiter and physical layer preamble, giving a total of 56 bytes including the payload. The downstream, with no overhead, has a frame of 56 cells if the bit rate is 155 Mbit/s and 224 cells for 622 Mbit/s.

The first protocol involves the allocation of wavelengths at the beginning of each frame. This approach allows the use of components with small but non-negligible tuning times with respect to the length of a cell. Devices with tuning times from a few tens of nanoseconds [7] upto a few microseconds [4] can be used. A small amount of bandwidth is wasted during the tuning phase of the protocol.

The second approach allocates wavelengths on a cell by cell basis. Clearly, a negligible tuning time is required for such a scheme, in order not to waste an excessive amount of network capacity. Transmitters with tuning times of a few nanoseconds [8] are essential for this method.

2. Wavelength allocation at the start of the frame

Frame based protocols have been used for both single wavelength PONs [9,10] and for scheduling transmission in peer to peer WDM star networks [1,11]. All share the basic feature of a control phase in which requests are transmitted and a data phase which is for the payload. Both fixed and variable length frames have been used. The approach here follows that of previous frame-based PON protocols, with a fixed number of cells transmitted in a frame. If there are more requests than available slots in a given frame, then an arbitration mechanism allots permits to ONUs and any unsatisfied requests have to wait until the next frame.

Requests are sent in a block at the start of the frame. The request blocks consist of divided cells containing minislots [6]. Each minislot consists of 7 bytes, with three bytes used for guardband, delimiter and physical layer preamble, one byte for cyclic redundancy check (CRC) and three for request information [12]. Each time slot can thus contain upto 8 minislots. The structure of the frame is shown in Figure 2.

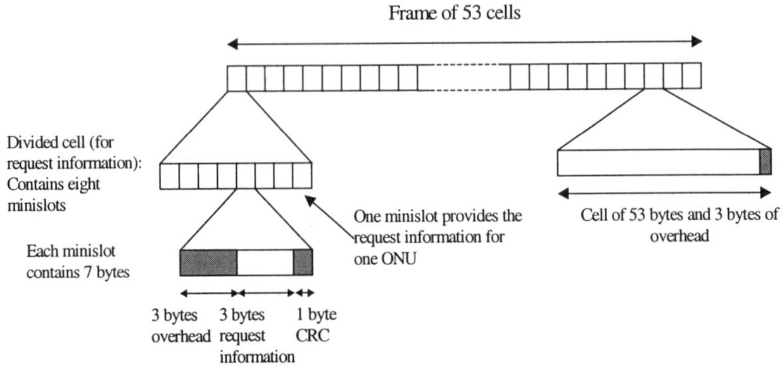

Figure2: The Frame structure for the frame protocol

When all the request blocks for a frame have arrived at the OLT, the wavelength assignment and permit allocation algorithm is implemented. This is a two stage process:

1. Wavelength allocation is performed using the LPT algorithm [13]. This assigns ONUs to wavelengths so as to balance the load as far as possible across the channels.

2. Once wavelength assignment is complete, the channel loads are checked individually for each wavelength to determine whether the number of requests is greater than the number of available time slots in the frame. If there are too many requests, then the number of permits assigned to each ONU on that channel is reduced proportionally until the channel is within capacity.

The frame protocol allows for the option of guaranteeing permits for delay sensitive constant bit rate and real time variable bit rate services. The number of guaranteed permits, determined by the connection admission control (CAC) function, is added to the number of requests for an ONU for the purpose of wavelength allocation. Once this procedure is complete, the number of guaranteed permits is known for each wavelength. The channel capacity is reduced by this number for the second stage of the protocol, as are the numbers for each ONU during the permit reduction stage.

3. The cell allocation protocol

The cell allocation protocol assigns wavelengths on a slot by slot basis. As with the frame allocation scheme, the protocol has a frame strcuture of 53 cells. Each frame is divided into 8 segments for the purpose of recording cell arrivals at the ONUs. This mechanism is aimed at reducing cell delay variation (CDV) by providing more information as to when a cell arrived within the frame. It has the same principle as the Circular Permit Processing (CPP) protocol proposed for single wavelength PONs [14]. Request information is sent to the OLT from the ONUs at a fixed point in the frame, in divided cells.

When the request information arrives at the OLT it is recorded in the Requests Matrix according to origin, frame and segment. All as yet unsatisfied requests are stored in the matrix, whether from the current frame or previous frames. The matrix is effectively circular in nature, with the first segment in the new frame being recorded in the row immediately after last segment of the previous frame. This arrangement is shown in Figure 3.

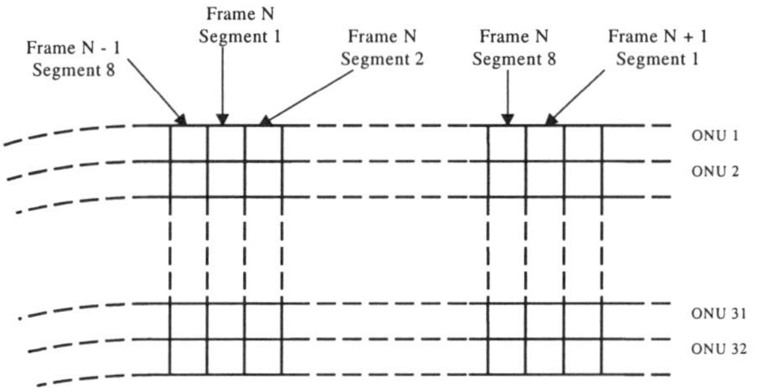

Figure 3: The Request Matrix

When all the request blocks from the frame have arrived, the construction of the transmission schedule begins. The essential mechanism for this is the round robin allocation of wavelengths, as used in the PROTON protocol proposed for WDM star networks [15]. Each segment has a pointer memory associated with it, which is initially set to a random integer between 1 and 32. At the start of the operation, the pointer is read and this indicates the ONU queue at which the permit allocation will begin. Four other memories are needed to control the operation:

The *segment number*, which indicates which segment is currently being scanned.

The *permits issued total*, which records the number of permits allocated so far during the schedule construction.

The *wavelengths indicato*r, which records the tuning of the next permit to be issued.

The *permit issued flag* for each ONU.

The sequence of events are as follows:

1. The queue length is read, giving the number of requests for the ONU in the current segment. If the number of requests is non-zero, then a permit is allocated and:

    1.1. The matrix element is decremented.

    1.2. The Wavelengths indicator is incremented.

    1.3. The Permits issued total is incremented.

2. The pointer is incremented and the next queue is read.

This process continues until all four wavelengths have been allocated for the slot or until all 32 ONU queues for the segment have been scanned. If, in the latter case, there are still wavlengths unassigned, then the pointer will be read for the next segment and this will be scanned. An ONU cannot be issued with a permit in the second or subsequent segment if its permit flag has been set. Each segment is scanned in turn until all four wavelengths are allocated or all non-empty segments have been processed.

The algorithm is repeated for each of the slots in the frame. Once the schedule is finished, the permits are sent to the permit queues whence they are transmitted to the ONUs in the physical layer operations administration and maintenance (PLOAM) cells [6]. The schedule construction for the next frame begins in the first non empty segment, so as to give priority in the next frame to unsatisfied requests from previous frames.

Like the frame protocol, the cell allocation scheme allows for permits guaranteed by the CAC. If on a given slot, a guaranteed permit is required, it is issued and only three permits are then available for the scanning process described above. This protocol also allows for the establishment of different traffic classes by using the segments within the frame for recording cell priority instead of, or as well as, cell arrivals.

4. Validation of the protocol models

Simulation models of the protocols have been developed using the Block Oriented Network Simulator (BONeS) package. In order to check the validity of the results of these simulations, the components of delay in each of the systems have been analysed and compared with the simulation results. For ease of analysis, Poisson sources at each of the ONUs have been used for this purpose. The total cell transfer delay in the PON is the sum of the following factors, the sequence of which is shown in Figure 4:

1. The wait between the arrival of the cell at the ONU queue and the sending of the request block at the end of the frame.

2. The wait of one or more frame lengths between transmission of the requests and the receipt of a permit. This period will incorporate the transmission delay of the request to the OLT, the protocol processing time and the transmission time of the permit to the ONU. The number of frames wait follows the characteristic of a batch queueing system [16] with a batch size equal to the number of cells in a frame.

3. The period between the start of the frame and the transmission of the cell. This delay is proportional to the traffic load, the arrival time of the cell at the ONU relative to the start of the frame and, in the case of the frame protocol, the randomly determined transmission start position for the ONU.

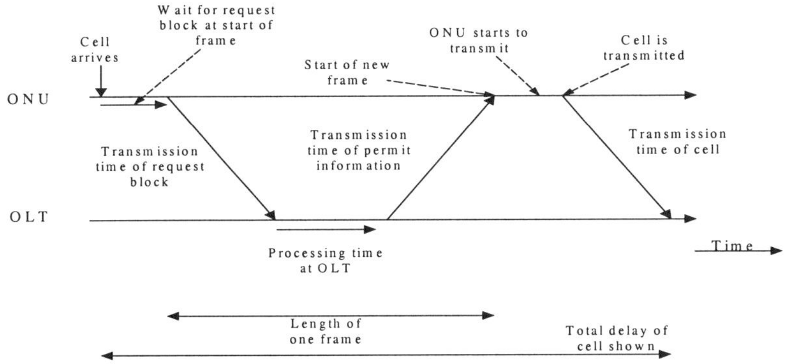

Figure 4 (a): Sequence of events in the frame protocol

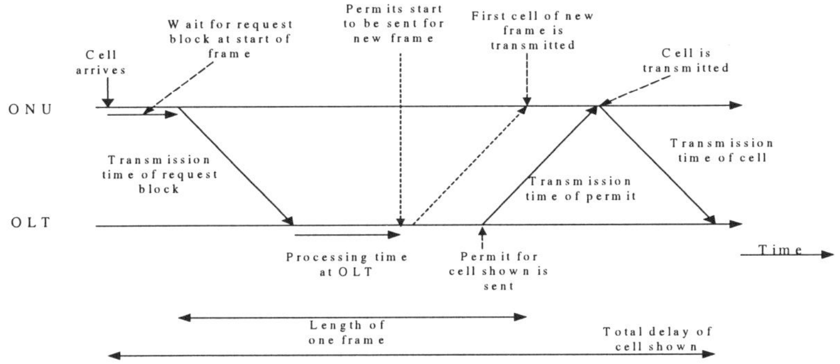

Figure 4(b): Sequence of events in the cell allocation protocol

Figures 5(a) and 5(b) show comparisons of simulation results using Poisson sources and the delay estimates based on the batch queue system and the other delays described above, for the frame protocol and the cell by cell allocation scheme respectively. A correlation between the simulation results and of 0.995 has been achieved for the frame protocol and 0.9994 for the cell allocation scheme.

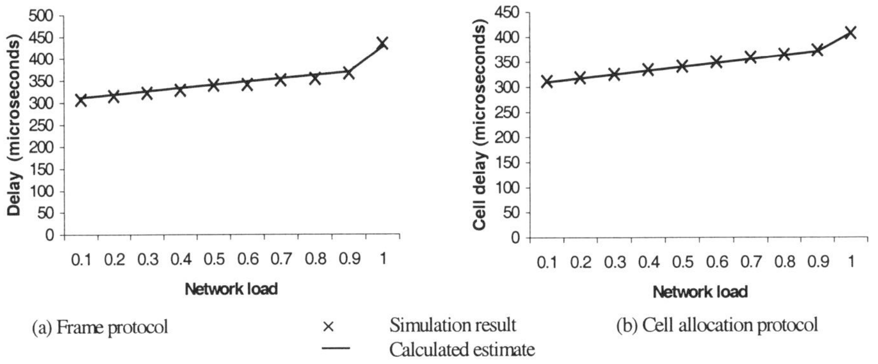

Figure 5: Validation of the protocols

5. Performance evaluation

Both protocols were simulated using bursty traffic sources, with a peak to mean ratio of 8 was used. Figure 6 shows a comparison of the cell transfer delay. This plot shows the quantiles of cell delay with a network load of 70%. The 1-point cell delay variation [17], which gives a measure of the degree of cell clumping, is shown in a similar plot in Figure 7. The results for cell transfer delay show the advantage of the cell allocation approach, with its more efficient use of bandwidth.

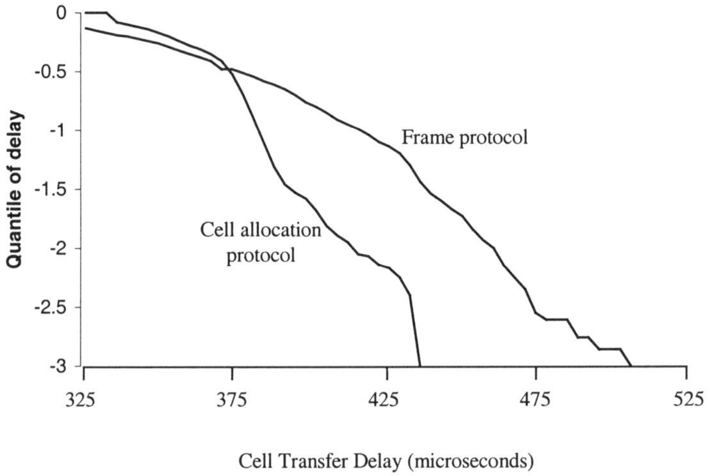

Figure 6: Comparison of quantiles of Cell Transfer Delay for the frame and cell allocation schemes at 70% load

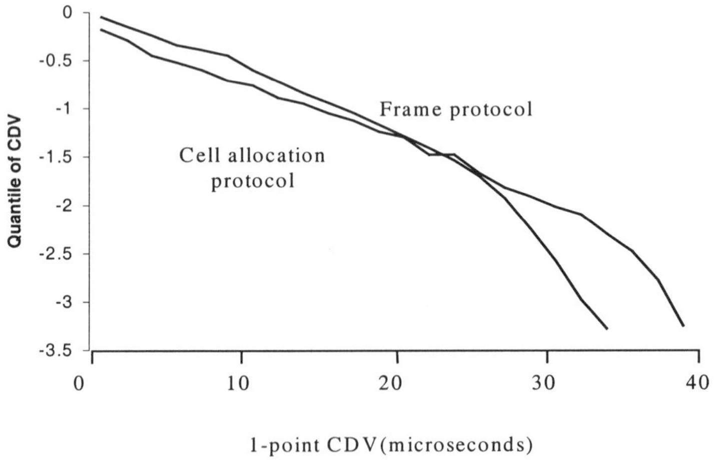

Figure 7: 1 point CDV for the Frame and Cell allocation protocols

6.Conclusions

This paper describes the development of two different approaches to the implementation of statistically multiplexed dynamic wavelength assignment for passive optical networks. Both methods are shown to work well for bursty traffic and are easily adapted to allow guaranteed permits for delay sensitive traffic. The frame protocol has the advantage of allowing non-negligible tuning times. This less stringent physical requirements may offer cost savings. The results of the simulations, on the other hand, show a better delay performance by the cell allocation scheme, the result of its more efficient permit allocation scheme. This scheme additionally offers the option of allowing the provision of different classes of service by using the segment structure to record cell priority instead of, or as well as, cell arrival time.

References

[1] J.M.Senior, M.R.Handley and M.S.Leeson, Developments in Wavelength division multiple access networking, *IEEE Communications Magazine* **36** no.12 (1998), 28-36

[2] M.R.Handley, M.S.Leeson, J.M.Senior, A.J.Phillips, P.Ball and I.Wright, "Comparison of WDM PON architectures", NOC '99, Networks and Optical Communications I, Delft, June 1999, 141-147

[3] T.Koonen,Ksteenbergen,C.van der Plaats, F.Janssen and J.Wellen, "Flexible hybrid access networks employing wavelength routing", NOC'99 Networks and Optical Communications I, Delft, June 1999, 1-8

[4] Amman and Buus, Tunable laser diodes, Artech House, ISBN 0-89-006963-8, 1998

[5] D.Mestdagh, Fundamentals of multiaccess optical fibre networks, Artech House, ISBN 0-89006-666-3, 1995

[6] ITU-T G.983.1

[7] www.Altitun.com

[8] www.nel.co.jp

[9] J.D.Angelopoulos, T.Toniatti, I.S.Veneiris and L.Verri, McFred, a frame based solution for the MAC protocol in ATM PONs, RACE Open Workshop on Broadband Access, Nimegen, Netherlands, June 1993,1-6

[10] J.D.Angelopoulos, I.S.Venieris and G.I.Stassinopoulos, A TDMA based access control scheme for APONs, J. Light. Tech. **11** no. 5/6, 1993, 1095-1103.

[11] Borella, Cancellieri and Chiaraluce, Wavelength division multiple access optical networks" ISBN 0-89-006657-4, Artech House, 1998

[12] R.Hoebeke, F.Ploumen and K.Venken, Performance improvement of ATM PONs in multi-service environments by means of dynamic MAC protocols, NOC '98, Networks and Optical Communications I, Manchester, June 1998, 81-88

[13] A.Kershenbaum, Telecommunications network design algorithms, McGraw-Hill, ISBN 0-07-112518-3, 1993

[14] J.D.Angelopoulos, S.Venieris and E.N.Protonotarios, A transparent MAC method for bandwidth sharing and CDV control at the ATM layer of passive optical networks, J. Light. Tech. **14**, no. 12, December 1996, 2625-2635

[15] D.A.Levine and I.F.Alkyidiz, PROTON: A media access control protocol for optical networks with star topology, IEEE Trans. Networking, **3**, no. 2, February 1995, 158-168

[16] T.Ryden, Waiting time distributions in buffers with batch service, IEEE Transactions on communications, **41**, no. 7, July 1993, 1027-1030

[17] ATM Forum Traffic management specification, Version 4.1, AF-TM-0121.000, March 1999

Broadband Access, WDM Metro and Network Management
D.W. Faulkner and A.L. Harmer (Eds.)
IOS Press, 2000 © AKM

Dynamically reconfigurable WDM hybrid fibre access network for IP-based services with various QoS demands

Ton Koonen, Kees Steenbergen, Jeroen Wellen
Bell Labs, Lucent Technologies Nederland B.V.
P.O. Box 18, NL 1270 AA Huizen, The Netherlands
E-mail koonen@lucent.com

Christophe Nicolas, Daniel Delprat
Corning SA
44 Av. de Valvins, 77210 Avon, France

Erik Weiss, Mirko Adamy
T-Nova Deutsche Telekom Innovationsgesellschaft mbH
Am Kavalleriesand 3, 64295 Darmstadt, Germany

Abstract

A multi-wavelength hybrid access network is proposed by the IST project HARMONICS, deploying flexible wavelength routing which enables dynamic network reconfiguration while taking different Quality-of-Service requirements of user traffic into account. Fast wavelength-switchable routers are indispensable to implement a novel integrated time-and-wavelength medium access control protocol. High network availability will be achieved with a double fibre ring feeder architecture, deploying adjustable optical power tap couplers.

Introduction

The demand for capacity in access networks is growing very fast, largely because of the steep increase of Internet usage. To keep up with this demand, in hybrid fibre access networks fibre is pushed ever closer to the customers, where it feeds the last-mile network connecting the customer. The latter network may deploy a variety of media (twisted copper pair, coaxial cable, wireless drop, …). The fibre feeder network is most economically laid out as a tree-and-branch one, with a number of splitting levels. By deploying dynamic wavelength routing in the fibre feeder network, the capacity provisioned to a particular last-mile network can be tuned to the actual traffic load of that network. Thus the system's bandwidth resources can be optimally distributed, and the network operator can maximise his revenues by minimising call blocking. This wavelength routing approach has been explored previously for fibre-coax networks in the ACTS project TOBASCO, and for fibre-wireless networks in the ACTS project PRISMA [1].

Present-day networks just provide "Best-effort" IP services, and no higher Quality of Service (QoS) is supported. In the IST project HARMONICS, "Hybrid Access Reconfigurable Multi-wavelength Optical Networks for IP-based Communication Services", a wavelength-routed hybrid fibre access network is proposed supporting higher QoS levels.

System architecture

The basic system architecture is shown in Figure 1.

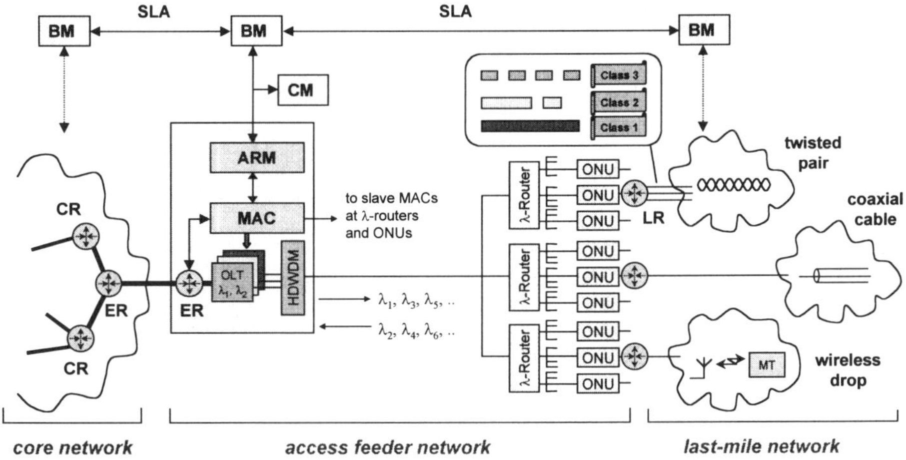

Figure 1 Wavelength-routed hybrid access network, supporting IP traffic with various Quality-of-Service levels (acronyms: see text)

By deploying a common dynamically reconfigurable fibre feeder network that supports a variety of last-mile networks, this architecture promotes the convergence of access networks. For example, a twisted pair network as designed originally for telephony, a coaxial cable network for CATV distribution, and a wireless network, may all be supported by the single fibre feeder network. In the near future, each of these last-mile networks may need to carry IP-based services with various QoS levels. Following the DiffServ QoS model, multiple service levels may be discerned, e.g. class 1 through 3. The Leaf Router (LR) acts as an intermediate between the last-mile network and the access fibre feeder network, and guards the QoS levels when handing over the traffic.

In the access fibre feeder network, multiple wavelength channels are used to communicate between the Optical Line Terminals (OLTs) and the Optical Network Units (ONUs). Flexible wavelength routers located at network splitting points steer the wavelength channels to the appropriate ONUs. The master MAC unit at the OLT site commands the slave MAC units at the wavelength routers and at the ONUs, re the wavelength channel access and the time slot access, respectively. The Adaptive Resources Manager (ARM) is supervising this master MAC, in order to optimise the allocation of the system's resources. It fulfills the Service Level Agreements (SLAs) which have been concluded upon negotiation between the Bandwidth Managers (BMs) in the respective network domains. The Configuration Manager (CM) takes care of e.g. fault and policy management. The access network interfaces to the core network via an Edge Router (ER).

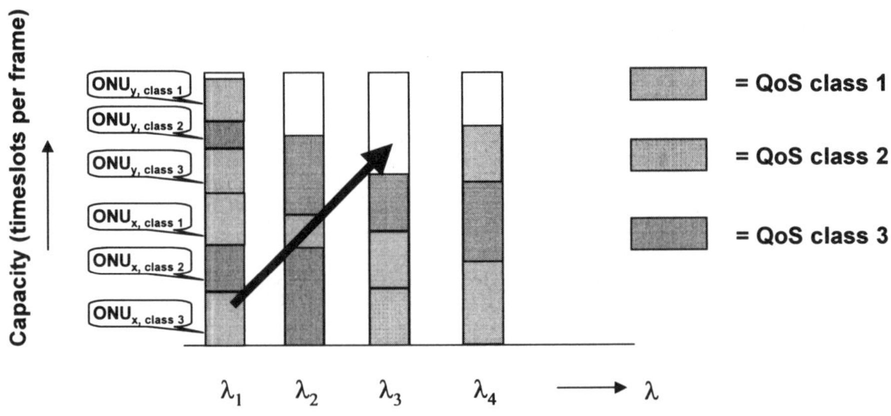

Figure 2 Reallocating wavelength channels to ONUs, taking into account QoS levels

The wavelength channels should be distributed flexibly among the ONUs in such a way that their traffic demands are met, while taking into account the QoS levels. Services at Premium level (e.g. voice, video conferences, streaming video and audio) should experience no blocking nor variable delay, and thus the wavelength distribution should be such that these services are instantly transported. Services at Best effort level (e.g. file transfer), however, may tolerate some blocking and delay, and thus may be wavelength-relocated to free up channel capacity for Premium-level services. Figure 2 shows, as an example, how several ONUs with their various service classes are hosted in the wavelength channels. When in channel λ_1 a sudden increase in traffic load of class 1 at ONU_x arises, and this channel has not enough capacity left to accommodate this increase, then the traffic of class 3 at ONU_x (and/or of class 3 at an other ONU in this channel) may be relocated to an other wavelength channel. Thus capacity is made available in channel λ_1, providing room to host the extra class 1 traffic load. Obviously, as an ONU should be capable of handling three different QoS classes, up to three wavelength channels may be allocated to an ONU.

The flexible router may be located at any of the fibre network splitting stages. When located closer to the ONUs, the granularity with which the wavelength channels can be allocated among the ONUs becomes finer. Thus the statistical gain of the flexible capacity allocation process becomes larger, resulting in a lower system blocking probability at a given traffic load on the system. However, the (relatively high) costs of the router are shared by less users. In previous studies (see [2]), an appropriate techno-economic compromise was reached when positioning the flexible wavelength router at the second splitting stage in the fibre network (as shown in Figure 1).

At the ONU, a Classify-Queue-and-Schedule operation must be performed in the Leaf Router to guarantee the specific QoS levels both for downstream and upstream traffic flows, as illustrated in Figure 3. Reflective optical modulators are used at the ONU for transmitting the upstream traffic, by remodulating the continous wave (CW) light received from wavelength-specific laser diodes located at higher split levels in the network. Several ONUs can thus share these (costly) laser diodes, reducing the per-user costs. On the other hand, this approach requires a carefully laid-out network with sufficiently low internal optical reflection levels, and is ultimately limited by Rayleigh scattering in the fibre, deteriorating the signal extinction ratio at the OLT receivers.

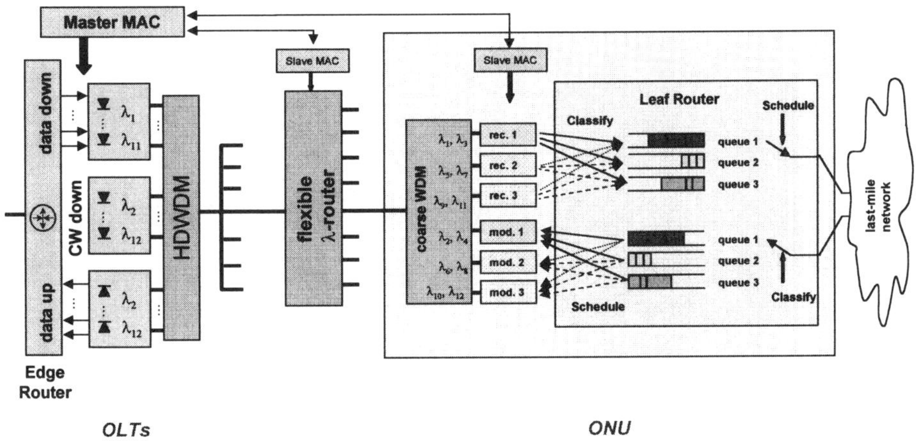

Figure 3 Two-way queueing at the ONU

Multi-dimensional MAC protocol

A new wavelength-and-time Medium Access Control (MAC) protocol is being developed for the dynamic resources allocation. This MAC protocol controls both the channel selection in the flexible wavelength router and the time-slot access in the ONU. Wavelength switching interferes with conventional TDM protocols that control a single wavelength. Therefore the MAC protocol integrates both the time and the wavelength domain. The master MAC at the Optical Line Terminals (OLTs) supervises the slave MACs at the router and the ONUs. The master MAC on its turn is controlled by the Adaptive Resources Manager, which aligns the traffic flows in the fibre feeder network with the capabilities of the core network and the last-mile network.

A fast flexible wavelength router is essential for implementation of this wavelength-and-time MAC protocol. The schematics of it are shown in Figure 4. At the OLT side of the router, a bi-directional erbium-doped fibre amplifier (EDFA) is used to compensate for the network splitting losses for both the upstream and the downstream signals. The EDFA is carefully designed in order to reduce the cross-gain modulation among the wavelength channels carrying bursty traffic. After being split among a number of wavelength (de)multiplexers, the wavelength channels to be guided to one or more ONUs are selected by fast optical gates. The gate switching time needs to be in the nanosecond range, in order to support fast packet switching. Therefore, the gating function will be performed by Semiconductor Optical Amplifiers (SOAs), which as an additional advantage offer extra amplification. Accurate timing control for switching the SOAs is needed, and is provided by the MAC. For compactness, the SOAs will be fabricated in arrays. These arrays are mounted on a silica-on-silicon platform comprising a planar waveguide wavelength (de)multiplexer or power splitter, which on its turn connects to the network's fibres. Each SOA switches only one wavelength channel, so cross-gain modulation is not an issue. To avoid electrical crosstalk, the electrical wiring of the module will be carefully designed.

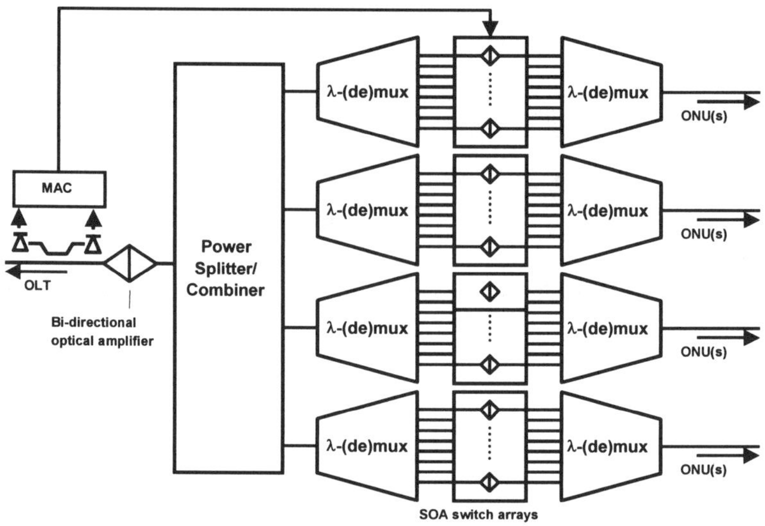

Figure 4 Fast flexible wavelength router

Network protection

Availability is a key issue in large-scale networks. In large PONs, typically a relatively long fibre feeder line is fanned out into numerous short fibre links to the ONUs. Assuming that the probability of a cable cut is proportional to its length, the feeder link is the most vulnerable part of the network. Replacing this link by a double-fibre counter-propagating ring, as shown in Figure 5, yields an efficient protection of the network. Optical power tap couplers along the ring connect to the local PONs and their various last-mile networks. In order to enable a modular design of these subnetworks, and to allow easy network extension by adding more subnetworks, it is preferable that each tap coupler extracts the same amount of optical power from the ring. This requires optical tap couplers with an adjustable power tap factor p_i for node i, assuming there are N nodes, given by the recursive relation

$$p_{i-1} = 1 - \frac{1}{1 + a_{i-1} \cdot p_i} \quad \text{with} \quad i = 2, 3, .., N \quad \text{and} \quad p_N = 1$$

When a cable break occurs, cutting both the clockwise and the counterclockwise transmitting fibre ring, the adjustable coupler with split factor p_0 close to the OLT has to be set such that the appropriate amounts of power are fed into the two ring segments. Also the tap couplers have to be readjusted, in order to feed identical amounts of power to the subnetworks. The path lengths from the OLT to the ONUs change due to this rerouting. The ranging protocol needs to accommodate these changes in order to maintain the accurate time delay control required for the upstream traffic from the ONUs.

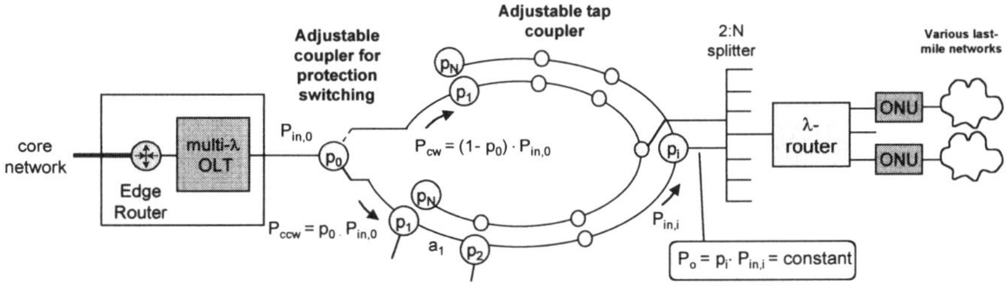

Figure 5 Network protection by double fibre ring with adjustable power tap couplers

System trial

The system will be trialed with a twisted pair VDSL user access network, and a LAN based on Polymeric Optical Fibre (POF). The field trial will involve several IP-based best-effort services like file transfer and web browsing, and higher-QoS IP-based services like voice over IP and video conferencing.

The layout of the planned field trial is shown in Figure 6. Users located on the premises of T-Nova in Berlin and in Darmstadt will be connected. These two sites will be interconnected via the WDM core network testbed. The users will be mainly technical and scientific staff members of T-Nova.

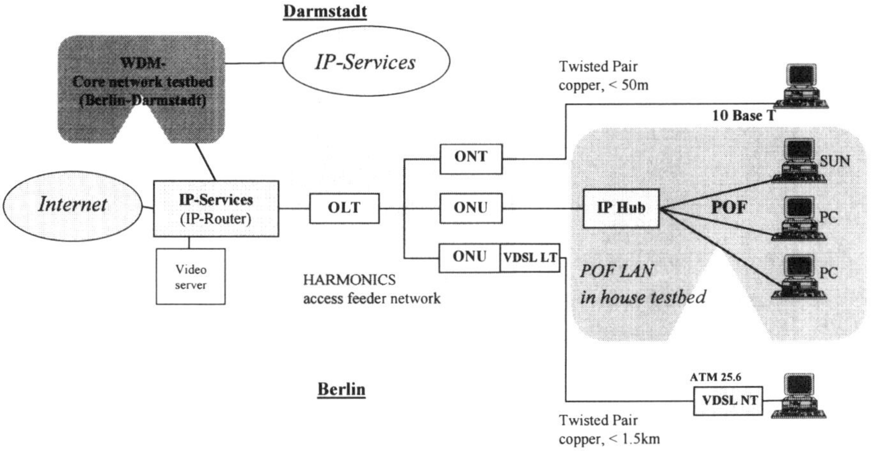

Figure 6 HARMONICS' field trial plan

Connected to the fibre feeder access network will be

- an in-house POF (Plastic Optic Fibre) LAN testbed with SUN workstations and Personal Computers with an optical interface,
- PCs using electrical Ethernet 10 Base-T interfaces in an in-house LAN (on Unshielded Twisted Pair copper cables, max. 50 metres),
- users via an ATM Forum 25.6 Mbit/s-based VDSL modem and Twisted Pair copper cables, on access lines up to 1.5 km length.

The users can deploy IP-based services like file transfer (ftp, mail), LAN emulation, web browsing, and QoS-dependent applications like Voice-over-IP, real-time video, video/audio on demand and videoconferencing.

The access to the Internet as well as to the Video server will be made by an IP Router; this will be a bursty process, since the request from the users is not predictable. Via the WDM core testbed connecting the Berlin and the Darmstadt premises, users in both locations can co-operate and communicate by means of the videoconferencing equipment.

Conclusions

A novel hybrid fibre access network architecture has been presented, which supports the transport of IP-based services to and from a large variety of last-mile customer access networks. Flexible wavelength routing in the fibre feeder part of the network allows efficient allocation of the system's resources, while taking into account the different Quality-of-Service levels of the IP traffic flows. A double-fibre ring network with adjustable optical power tap couplers improves the availability of the system.

Acknowledgement

The European Commission is gratefully acknowledged for accepting the HARMONICS project as part of the IST Programme. Thanks are also due to the other project partners for their inputs: Portugal Telecom Inovaçao, Intracom, KPN Research, Mason, IMEC – University of Gent, University of Limerick, CTIT - University of Twente,.

References

[1] Ton Koonen, Kees Steenbergen, Chris van der Plaats, Fons Janssen, Jeroen Wellen, "Flexible hybrid access networks employing wavelength routing", Proc. of NOC '99, June 22-24, Delft, The Netherlands

[2] M. Luizink, C.A.M. Steenbergen, A.M.J. Koonen, "Capacity allocation using WDM in fibre-wireless access networks", Proc. of Optical Fiber Conference (OFC'99), paper WD3-1, San Diego, Feb. 1999

D.W. Faulkner and A.L. Harmer (Eds.)
IOS Press, 2000 © AKM

Branched Single-Mode Fibre Access Network

Tarja VOLOTINEN
OFCON Optical Fiber Consultants AB, Koloniv. 12, 82450 Hudiksvall, Sweden

Abstract. Integrability, installation expenses, network capacity and easiness for an upgrading are compared for alternative FTTH-network structures. The most effective structure is a passive, along-the-route-branched two-fibre-ribbon network, which is suitable for time division multiplexed (TDM) transmission, e.g. with the dynamic transfer mode (DTM).

1. Introduction

Small business offices and private homes will be the biggest group of subscribers for new high bandwidth networks [1, 2]. These subscribers have a huge and urgent need for fast speed/high bandwidth communication. At the moment, they are obliged to use lowest bit rates of standard telephone, ISDN and mobile access lines. They pay the highest price of the market per amount of transferred bits, as a matter of fact over 100 000 times more than the drift expenses for a cost effective optical fiber access network at today's price level would be.

Field trials and first commercial installations of fibre-to-the-home/business (FTTH) and fibre-to-the-desk (FTTD) networks are starting in almost every European country. However, network structures are still under a very intensive development [1 - 8]. The most important factors for network installers and users are the cost effectiveness, integrability and upgradeability of the passive optical network (PON) structures.

In this work, the alternative single mode fibre FTTH-network structures are studied: a full-star-network, in-one-point-branched network and along–the-route-branched network. A new passive component (patent applied) further improves the integrability and cost effectiveness of the branched single-mode fibre access network.

1.1. Cost Effectiveness

Small offices and private homes really are in a huge and urgent need for the new high bandwidth communication networks. These subscribers pay at the moment the highest service rates per Gbh, (gigabit hour)[1], i.e. up to 100 000 EUR/Gbh. In most towns and villages no high speed subscriptions are available, or the service costs are extremely expensive, e.g. 2800 EUR/month for a 2 Mb/s line.

Only standard telephone lines at 64 kb/s and within 4 km from station the ISDN-lines at 64/128 kb/s are available. A one-man-office uses at least 500 h per 8 months per person of these slow speed services (ISDN: telephone/telefax at 64kb/s, internet/email at 64 - 128 kb/s). In Sweden this service costs 2200 EUR inclusive both the subscription and service expenses for 8 months. The calculated costs per Gbh[2], become as high as 100 000 EUR/Gbh.

This rate can be compared with the drift costs presented for passive FTTH-network by P. Shumate and R. Woerner [4, 5], i.e. 300 USD/year ≈300 EUR/year for a 155 Mb/s optical subscriber line, which is open all the time. This makes only 0.19 EUR/Gbh. –This kind of service would be extremely cost effective for the users.

---

[1] 1 Gbh = 1 gigabit hour = transmitted/received amount of bits corresponding to 1 hour use of a 1Gb/s communications line.

Both the expenses and the availability of communications services are very important for the economy and competing power of small office businesses. They sell services to large companies and also compete with them in the market. So their need for a low-cost high bandwidth services, already at the moment, is very huge and urgent, and furthermore rapidly increasing.

Installation expenses do not need to be close to a city buss ticket, i.e. zero, as is commonly believed by experts. Small business offices need to retain a minimum investment capital, e.g. 11000 EUR in Sweden. In consulting, software and service companies this capital is typically invested in a computer system, which looses its economic value (though not its usability) very rapidly, 30 – 50 %/year. So there is a possibility of new investments every year.

The installation of an upgradeable high bandwidth/ high bit rate access line is a valuable long-term investment for the network owner and for an owner of a small business office, private house or apartment. The installation costs shall be compared to the purchase expenses of a computer-printer-scanner-software system package. An installed single-mode fiber access line does not lose its value, usability or upgradeability like a computer system does. Estimated lifetime for such a line can be, on average, several hundred years.

– This does not mean that the homes or offices would like to pay so much as 1000 - 3000 EUR for the access of a high bandwidth communication subscription. –This is just an indication of the circumstances; there may be some potential capacity available for investments in small businesses. The estimate [4], 1500 USD (= 1300 EUR) per subscriber for the installation, cables, equipment and gateway to a high capacity system to get an all time open 155 Mb/s subscriber line, is not therefore an unrealistic price. --The problem is that these lines are not available!

1.2. Integrability and Upgradeability

The other important features of subscriber networks are integrability and upgradeability. Integrability means that new subscribers can easily be connected to the network, without extensive external cable and equipment installation work. An along-the-route-branched network (Fig. 1.) fulfils this requirement. The integration costs for later joining subscribers contain only the costs for connecting sections of two-fibre ribbon cables from the route to the subscribers and end equipment (transceiver-router), or even less, if the new subscribers are in the previously connected houses.

The upgradeability is also necessary, because the capacity need of each subscriber will rapidly increase due to the development of new high bandwidth Internet services, and people's and small businesses' own increasing interest for digital video-conferencing/videophoning and digital TV. The capacity needed per subscriber, or per one-family house containing two or three subscribers (one or two small business offices and a one private home), will rapidly exceed current estimates 10 Mb/s – 100 Mb/s per house. In five years from now, the average capacity demand per one-family house may exceed 1 Gb/s.

The adequate upgradeability can only be reached with PONs with high bandwidth fibres and components: single mode fibre, fused single mode fibre couplers, etc.

1.3. Installation

Difficulties with the installation of single mode fibre networks, and therefore also the assumed need for fibres with a thicker core, are exaggerated. People have always bought connectorized buffer cables for their electrical equipment. Professional electricians have carried out permanent cable installations inside houses. Therefore no particular reason can be found, why the house owners (other than some fibre optic engineers) would suddenly start to do these kinds of things by themselves at their homes and facilities. Market prices for connectorized single mode fibre cables are quite reasonable. Furthermore, the fibre ribbon blowing technique presented at OFC′99 [7], is faster and simpler than installation of conventional telephone or power cables.

2. An Integrable FTTH-Network Structure

The comparison of cost effectiveness, integrability and upgradeability between three alternative two-fibre PON structures (full-star, 1-point-branched and along–the-route-branched) shows that the along-the-route-branched PON structure (Figure 1) is the most suitable for FTTH-networks for small office and private home areas (Table 1).

Table 1. Comparison of PON structures

PON structure	Cost effectiveness	Integrability	Upgradeability
full star	not good	difficult	expensive
1-point- branched	rather good	rather good	rather good
along-the-route-branched	very good	extremely good	good

Figure 1. Integrable FTTH-network structure

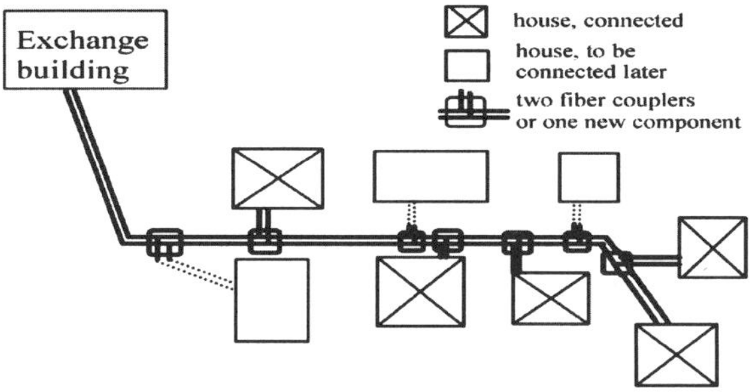

The most integrable structure for the small office and private home areas is the structure, which is branched along the route (Figure 1). The cable can be a 2-fiber ribbon duct or aerial cable. The 2-fiber pair (one fibre for upstream one for downstream) is branched by fusion splicing with two passive standard fused fibre couplers (with coupling ratio 10/90 or 20/80), or by the new passive branching component (patent pending), along the route. Each house can be reached from the branching points by a similar 2-fiber cable, which has connectors at the other end for the transceiver to be installed inside the house.

Two-fibre-ribbon is preferred because neither isolators nor different wavelengths for up and down streams are needed. When fused couplers with wide wavelength range are used there is no limitation either to use any wavelength to any direction. But it may be cost effective and easy to use similar transceiver components at both ends of the branched network, which is possible with a two-fibre solution.

In total, the installation and material costs for 8 subscribers are estimated to be slightly below 1000 EUR/subscriber (Table 2). If only 5 subscribers are to be connected from the beginning, costs would be 1200 EUR/subscriber. This includes the splice boxes and couplers for three extra branches prepared for three new subscribers to be connected later to the same route.

Table 2. Estimated installation expenses for along-the-route-branched single mode fibre networks, 8 branches

Type of expense, estimated per item, amount	**total cost**
2-fiber cable, 0.7 EUR/m, 1500 m	1050 EUR
one end connectorized cables, 120 EUR, 8	960 EUR
splice boxes, 50 EUR, 8	400 EUR
fused couplers, 50 EUR, 14	700 EUR
man days of installation work, 175 EUR, 8	1400 EUR
transceivers, 220 EUR, 9	1980 EUR
connection with a high capacity network	1000 EUR
planning + extra expenses	250 EUR
Total costs for 8 subscribers	**7740 EUR**

The work costs in Table 2 are for conventional installation techniques to existing ducts or poles. About 20 % of work costs can be saved by using blowing technique [7]. The transceiver prices in Table 2 are estimated for fully developed devices, which are expected onto the market.

Every branch of this network is further integrable for one or two new subscribers within the houses (1 or 2 small offices and a private home) using similar components and costing about 500 EUR/subscriber. New houses can be integrated along the route, by cutting the cable, splicing new 10/90 couplers into the line and connecting the new subscribers to the route. This is estimated to cost about 700 EUR per new subscriber (Table 2).

Each network route can easily be branched up for 16 subscribers, may be up for 32, without being limited by excessive power losses or bandwidth decreases for speeds 10 Mb/s – 1 Gb/s per subscriber. Distances exceeding 40 km from station can be reached with standard single mode fibres. By using high quality wide wavelength range couplers both wavelength windows 1300 and 1550 nm can be used.

This kind of network can be first installed by providing a 10 Mb/s speed to each subscriber (the estimates in Table 2). A network with a 100 Mb/s speed would not be very much more expensive either. An upgrading of speed is easily possible by replacing, or by speeding up the transceivers. Route subscribers may need to be upgraded at the same time. A single mode fibre PON with good quality broadband couplers, or the new components[2], would have in principal unlimited bandwidth for upgrades.

The planning for a field experiment on the along-the-route-branched access network has been started.

3. Safety

The safety, i.e. the reliability of the communications service and the secrecy of the contents, is also very important issue for subscribers. One possible potential safety risk is a computer driven optical node or switch, which is located in a private house cellar or similar facility close by the subscribers. These kinds of solutions are used for some full-star-access-networks with combined single-mode and multi-mode fibre cables in Fast Ethernet field trials. Computer may become stolen and users may have an access to the software and other subscriber's data in the computer. When single-mode fibres and along-the-route-branched network structure is used, the optical node and switch can be placed in a well protected telephone exchange building or in a similar facility adequately far away from the subscribers.

4. Conclusions

Integrability, installation expenses, network capacity and possibility for a further upgrading are compared for alternative FTTH-network structures. The most effective structure is a passive, along-the-route-branched 2-fiber ribbon network, which is suitable for TDM transmission, with e.g. a dynamic transfer mode (DTM).

References

[1] Penzias, The plenary session talk at IWCS'98.
[2] Short course of IWCS´98: Fiber in the Loop.
[3] Sessions TuE, TuK and TuQ at OFC´99.
[4] P. Shumate, Short course of OFC´99: Fiber to-the-X.
[5] R. Woerner, Presentation Tu1-1 at OFC`99.
[6] T. T. Volotinen, An integrable and cost effective FTTH-network structure, Techn. Digest OFC´2000, paper FA3-1, (2000) 6 - 8.
[7] Exhibition of OFC´99, EricssonCables/Amherst Inc.
[8] Private discussions with access network designers at OFC 2000 in Baltimore 4 – 10 March 2000.

Broadband Access, WDM Metro and Network Management
D.W. Faulkner and A.L. Harmer (Eds.)
IOS Press, 2000 © AKM

Traffic performance of APON and SDH rings in a FTTCab configuration

D. Goderis, I. Van de Voorde
Alcatel Corporate Research Centre,
Francis Wellesplein 1, B-2018 Antwerpen, Belgium
Tel: +32 3 240 7853, Fax: +32 3 240 9932
E-mail:danny.goderis@alcatel.be

Abstract. The article investigates the traffic performance of several broadband access technologies deployed as the feeder section for xDSL copper drop lines. The traffic performance offered by ATM cell level multiplexing on APON is compared with the performance of the resource-allocation mechanisms available on current and (future) upgrades of SDH access rings. The considered broadband services are Fast Internet Access, Video on Demand, Interactive Multimedia and High Quality Videophone. The network dimensioning methodology takes into account the multiplexing gain obtainable at the cell level, the call level and the data-burst level of a broadband service.

1. Introduction

Several studies predict an exponential growth of broadband (BB) xDSL equipment for the upcoming decade 2000-2010. According to the Lehman Brothers the number of ADSL modems in the US will grow from 503K in 2000, over 2340K in 2003 up to 5023K lines in 2007 [1]. The same study indicates a breakthrough of the broadband (BB) access technology in the US households from 1% penetration now to 30% in 2007. Also the business market experiences a continuous growth from 40% BB-penetration to 60% in the same period.

Recent technological progress makes it possible to bring high-speed multimedia to the customer on the twisted pair (e.g. ADSL and VDSL modems), the coax (cable modems) and the fibre. A mix of all these technologies will soon be present in the business market. Fibre to the Business (FTTB) solutions, in particular, have their dedicated market share based on their ability of offering very high capacity leased line services. In the residential market however, it is believed that the drop section will be dominated (the coming years) by xDSL and cable modems. Although this evolution clearly has some influence on the deployment of Fibre to the Home (FTTH) networks, an increasing demand for (cost-effective) access fibre technologies, serving as the feeder section for xDSL drop lines, can also be expected.

In this article we investigate the possibilities of the APON (ATM based Passive Optical Network) [2] and SDH-Rings in a FTTCab architecture (Fibre to the Cabinet). In such architecture, cost-effectiveness and statistical multiplexing gain capabilities of the Network Elements are tightly related. The obtainable multiplexing gain, and related traffic performance of the Access Network, is the main subject of this paper. Our results give a quantitative indication about the needed access equipment as a function of the BB-service penetration and the deployed feeder technology. This allows for making trade-offs between implementing (more) complicated traffic functionality in the Network Elements, thus generating higher costs, but getting more revenues because the access network can support higher BB-service penetrations.

The remainder of the paper is as follows. Section 2 briefly gives the basic system parameters and network dimensioning approach. Section 3 discusses the possible traffic functions in the Network Elements. Section 4 summarises the broadband service characteristics and the final section contains the results and conclusions.

2. Network Dimensioning: statement of the problem

Our basic ATM access network architecture is depicted in Figure 1.

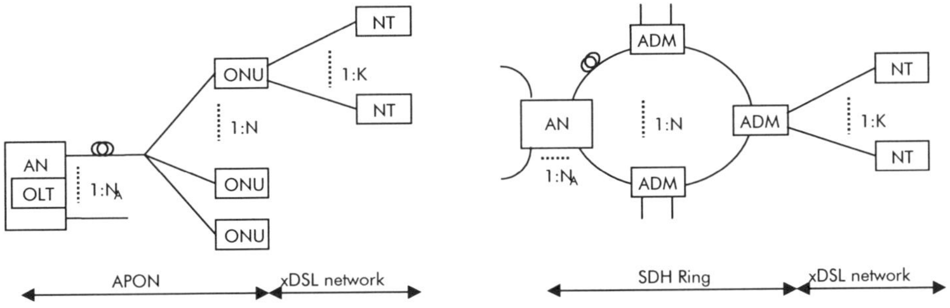

Figure 1: APON and SDH-Ring in a FTTCab architecture

ADSL or VDSL lines cover the drop section. The traffic is carried from the user NT (Network Termination) to the Optical Network Unit (ONU) or the Add Drop Multiplexer (ADM), in case of res. the APON or SDH-Ring. The drop distance depends on the type of xDSL technology. Table 1 summarises the basic traffic parameters of the copper drop lines.

xDSL Service	PBR downstream	PBR upstream	maximum drop distance
ADSL	2 Mbps (*)	0.128 Mbps	3 km
VDSL residential (A-symmetric)	< 26 Mbps	< 2 Mbps	1 km
VDSL business (Symmetric)	< 13 Mbps	< 13 Mbps	1 km

(*) Mbps = Megabit per second

Table 1: ADSL and VDSL Peak Bit Rates (PBR) and drop distances

The access technology of the feeder section is the APON or SDH-Ring with a symmetric up- and downstream capacity of 155.52 Mbps, corresponding with an SDH VC-4 container. The feeder distance R is supposed to be smaller than 10 km and the total service area $A_{service}$ is bounded by 300 km² ($A_{service} \leq \pi.R^2$).

Network dimensioning consists in calculating the (access) equipment needed for covering the customer needs in a certain service area. This complicated problem may be quantified by three dimensioning parameters:

- K, the (maximum) number of drop lines per ONU or ADM.
- N, the (maximum) number of ONUs per APON or ADMs per SDH-Ring.
- N_A, the number of Access systems needed for covering the service area. N_A measures the required bandwidth or capacity, i.e. N_A x 155.52 Mbps.

Calculating these parameters is very complicated and dependent on user-behaviour, service-characteristics, offered quality of service, and – last but not least – the available technology and traffic functionality in the Network Elements. Quantitative results can only be obtained by making further assumptions about technology (section 3), traffic patterns (section 4) and user-distribution. The latter is supposed to have the following properties:

- The combination user density – service penetration, is not too high such that a drop area can certainly be served by a capacity of 155.52 Mbps. Therefore the total number of ONUs or ADMs (≡ NxN_A) is determined by the service area $A_{service}$ and the drop area A_{drop}. It is obtained by filling $A_{service}$ by the smaller areas A_{drop}. The latter depends on the xDSL technology and is obtained from the drop distance. The filling should also take into account appropriate scaling factors related with hexadecimal filling and street patterns.

- The number of drop lines K per ONU (ADM) is determined by the required customer lines in the drop area, i.e. the combined user density – service penetration. Thus, we don't take into account any equipment practice constraints due to rack restrictions or others. Furthermore only homogeneous user- and service distributions are considered.

Thus, the number of subscribers and topological constraints determine the parameters K and N, resulting in a (simplified) dimensioning problem with only one parameter, i.e. N_A. In case of very small BB-service penetrations, the user demands can be covered by only one Access System ($N_A \equiv 1$) because $A_{service} \cong \pi.R^2$. N_A reaches his upper bound if the full 155 Mbps is needed to cover a single drop area, i.e. all access systems are reduced to Point to Point links (N=1).

3. Traffic Functions in the Access Network

The required capacity N_A is clearly related with the traffic performance of the Network. Especially in the Access Network, performance and multiplexing gain are extremely important because it allows for cost sharing of the equipment amongst several (xDSL) users. The multiplexing gain is essentially determined by the multiplexing method and the traffic functions in the diverse Network Elements, i.e. the ONU, ADM and Access Node (AN). In the sequel we distinguish the following functions:

- **The multiplexing method**, i.e. ATM cell or SDH Virtual Container (VC) multiplexing.
- **The Connection Admission Control (CAC)** algorithm, i.e. Peak Bit Rate (PBR) or Effective Bandwidth (BW_{eff}) allocation for a service (see also next section).
- **The Medium Access Control (MAC)** protocol. A static or dynamic MAC protocol only matters in case of the APON, which is indeed a shared medium.
- **The signalling functionality**, i.e. Permanent or Switched Connections.

APON System

Concerning the APON, the main distinction is made according to the functionality of the ***MAC protocol*** [3]. The distinction is only relevant for the *upstream* traffic, i.e. from ONUs to OLT. Indeed, all *downstream* ATM cells are simply multiplexed at the OLT and broadcast from OLT to the ONUs. CAC for downstream traffic can be based on BW_{eff} -allocation.

In case of a *static* MAC protocol, the upstream bandwidth is statically dedicated to the ONUs through the Network Management System (NMS) of the operator. At ATM cell level, the bandwidth must be allocated at the Peak Bit Rate (PBR) of the connection. The Access Concentrator function is situated at the ONU.

In case of a *dynamic* MAC protocol, full cell multiplexing for upstream traffic is possible because upstream capacity is allocated *on demand*. The *dynamic* APON acts as an ATM distributed queuing system, where the bandwidth allocation (or scheduling) algorithm is situated at the OLT and the (distributed) queues are situated at the ONUs. Therefore the allocation of capacity (at the OLT) to the ONUs can be realised full dynamically according to the bandwidth needs of the ATM connections. BW_{eff} –CAC implementation is possible and the Access Concentrator function is situated at the OLT.

Broadband upgrades of SDH-Rings

The main point for SDH-Rings is the ***multiplexing method***. SDH networks have been designed with synchronous services in mind. Resource allocation is done by multiples of the SDH VC-12 capacity (2.048 Mbps), which themselves are integer multiples of the basic 64Kbps channel. ATM traffic can nevertheless be carried directly over an SDH network. The ATM cells are then transported on SDH End-to-End paths and routing is based on SDH-VC multiplexing. The major drawback of this method is the inefficiency of SDH in carrying non-standard rates. An ATM connection of e.g. 6 Mbps has to be carried in a SDH VC-3 container of 50 Mbps, i.e. the smallest SDH container larger than the 6 Mbps of the connection. Clearly this is mostly an unacceptable waste of bandwidth. There are essentially two techniques to overcome this basic problem. The first one is called *Inverse Multiplexing for ATM* (IMA) [4]. The second strategy is based *on Integrated SDH-ATM* (ISA) overlay networking [5].

IMA enables the transport of a high bitrate ATM cell stream onto a number of lower-rate physical links, e.g. E1 lines, or SDH VC-12s. This is accomplished by grouping these lower-rate links into a single *logical* channel. The Bandwidth granularity becomes 2 Mbps, which solves to a large extent the basic problem described above. However, multiple nodes on the ring can not share the SDH-VC's carrying broadband traffic. The reason being that SDH-VC's are statically dedicated to ADM nodes on the ring by the Network Management System. The Access Concentrator function is therefore situated at the ADM node.

The ***ISA*** solution is more flexible and is based on the introduction of ATM-switching functionality in the SDH access equipment (Figure 2).

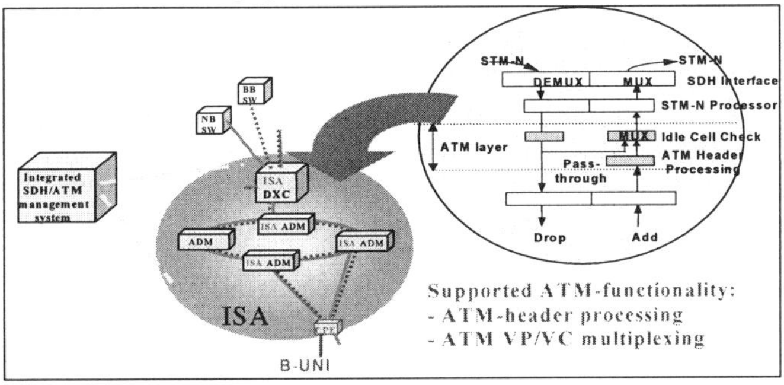

Figure 2: Integrated SDH-ATM overlay networking

A basic ISA-requirement is that the SDH-nodes are able to add, drop, pass through and discard on a cell-level basis, rather than on a STM time-slot basis. This enables a more efficient consolidation of ATM (non-standard rate) traffic into the SDH containers. The ISA-solution allows for simultaneous transport of STM and ATM traffic whereby the STM and ATM traffic must be carried in separate SDH VC's. ATM traffic is carried in separate SDH VC's (normally SDH VC-3's) which are terminated at each ring node equipped with ISA-functionality. Therefore the multiple nodes can share the same SDH-VC and, by consequence, the Access Concentrator function can be located at the Access Node itself!

4. Broadband service characteristics and multiplexing gain

The characterisation of (symmetric) Narrowband services is straightforward. At the cell (or circuit) level the required capacity of the service ($C_{service}$) is usually a multiple of the basic channel (DS0=64Kbps). At the call level, the user statistical behaviour is modelled as a Poisson arrival process of calls with exponentially distributed call duration. The statistical parameters are the Busy Hour Call Attempt (BHCA) and Mean Call Duration (MCD).

The description of Broadband services is more complicated and, in general, three levels are needed to characterise the service: the cell, the file or application and the call level. Table 2 summarises the characteristics of some Broadband services, which served as input for our simulations. Naturally these values are only approximations and, as such, they indicate rather orders of magnitude than exact values.

Broadband Service	Fast Internet Access	Video on Demand	Interactive Multi-Media	High Quality Videophone
Cell Level				
PBR downstream	2 Mbps	6 Mbps	1 Mbps	1 Mbps
BW_{eff} downstream	1.3 Mbps	5.8 Mbps	0.65 Mbps	0.96 Mbps
PBR upstream	0.128 Mbps	0.128 Mbps	<symmetric>	<symmetric>
File Level				
BHCA (data-bursts/hour)	30	N/A	120	N/A
MCD (seconds)	4	N/A	1	N/A
Call Level				
BHCA (calls/hour)	0.2	0.05	0.2	1
MCD (seconds)	3600	7200	3600	120

Table 2: Characteristics of Broadband services

The ***cell level*** description is the most common approach in packet-based networks. The Peak Cell Rate (PCR), Sustainable Cell Rate (SCR), Minimum Cell Rate (MCR) and Mean Burst Size (MBS) describe bursty data-services supported by an ATM-network. CAC-algorithms are sometimes based on only one parameter, the *effective bandwidth* BW_{eff} of a service. The ratio PCR/BW_{eff} measures the burstiness and the CAC algorithm allocates a capacity BW_{eff} instead of the (larger) PCR. The effective bandwidth can be calculated from analytical models, such as the fluid ON/OFF approximation or more complicated Markovian sources [7]. Sometimes BW_{eff} is obtained from "real-time" network measurements.

The ***call level*** describes the user-statistical behaviour of setting up a connection, such as picking up the phone or logging on for surfing on the Internet. As for synchronous NB-services the calls are modelled as a Poisson process with BHCA and MCD as statistical parameters.

The ***file level*** is meant for describing the statistical behaviour of data-transport in a time-scale between the cell- and call-level. A user surfing on the Web for example only retrieves information from time to time by clicking an URL. Clearly neither the cell- nor the call level is appropriate for describing the statistical distribution of the "clicks". The IP session (the "call") may take hours, the cell time scale is in the order of microseconds while the mean "click"-arrival rate might be something like every two minutes. The file level is also relevant if PCR and SCR differ some orders of magnitude. Indeed, the instantaneous peak bit rate obtainable by an ADSL modem might be 2 Mbps, while the mean rate of the entire IP session is rather in the order of 30 to 40 Kbps.

Statistical Multiplexing Gain

Network dimensioning consists of having a good estimation of the multiplexing gain as a function of the network and service characteristics. For synchronous Narrowband services, only a gain at the call level is possible and its calculation is usually based on the Erlang-B formula. For broadband-services however, a statistical gain is possible at all three levels.

The ***cell level multiplexing gain*** depends on the bursty behaviour of the service on cell or packet level. In case of e.g. Constant Bit Rate services no gain is possible. The *possible* gain can be expressed by the ratio PCR/BW_{eff}. However this gain can not always be *realised* by the network; advanced traffic functions are sometimes required. For example the APON should be equipped with a dynamic MAC protocol, otherwise PCR-allocation for upstream traffic is required. In case of a dynamic MAC protocol the CAC algorithm only has to allocate a capacity of BW_{eff}.

The ***call level multiplexing gain*** is determined by the ON/OFF behaviour of the service on call level. The Leased Line service for example is a permanent service and no gain is possible. A gain can only be realised in case of dynamic signalling (switched connections) and is usually calculated by the Erlang-B formula.

The ***file level multiplexing gain*** is determined by the file- or data-burst behaviour of the service. There exists no unified approach for calculating this gain. In [6] the authors base the dimensioning on the **Erlang-B** formula by taking the *file arrivals* (BHCA) and the *file download time* (MCD) as the statistical parameters. Another possibility is the **Burst Scale Loss Model**, which is a typical broadband model. The dimensioning is based on the Cell Loss Probability caused by burst scale congestion [7]. Some results can be seen in Figure 3, which gives the multiplexing gain (G) in function of α, the probability that the source is ON. Figure 3 (and other simulations) show that both models give the same *qualitative* behaviour, i.e. G behaves exponentially as a function of α.

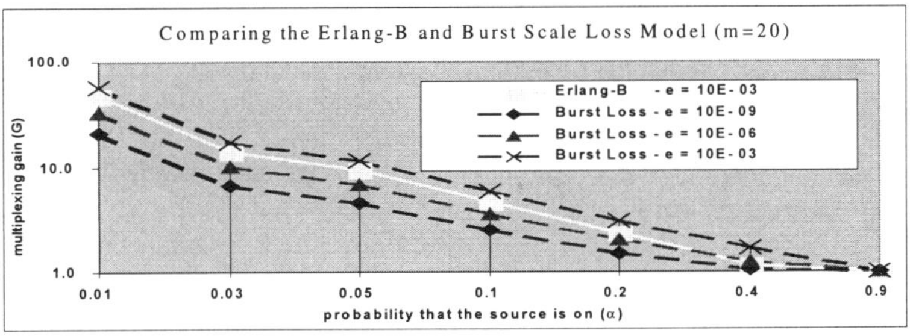

Figure 3: Calculations of the multiplexing gain

G is calculated for a single Node Multiplexer with simple input FIFO queuing and an output link with capacity C_{link} bps (Figure 4). The node has K (input) drop lines, which all carry the same service with a bitrate of $C_{service}$ bps.

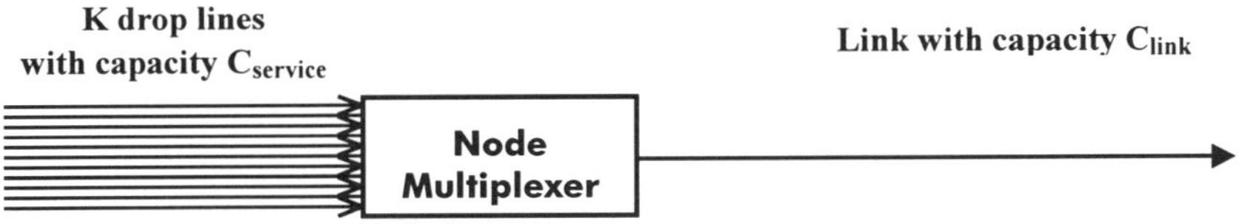

Figure 4: Node Dimensioning

Roughly speaking, the node can *simultaneously* handle *m* active sources, where m is the ratio $m \equiv C_{link} / C_{service}$. The node can however carry more lines because the probability that all sources are simultaneously active is very small. The (statistical) multiplexing gain G_ε is therefore defined as the ratio K/m. ε denotes the Quality of Service dependency of G, which may e.g. be measured by the call blocking or cell loss probability.

The results of Figure 3 can be derived from the following formulas.

Erlang-B formula:
$$P_{blocking}(K,\alpha,m) = \frac{(K \cdot \alpha)^m / m!}{\sum_{i=0}^{m} (K \cdot \alpha)^i / i!} \le \varepsilon_{blocking}$$

Burst Scale loss formula:
$$P_{loss}(K,\alpha,m) = \frac{1}{K\alpha} \sum_{i=m+1}^{K} b(i,K,\alpha) \cdot (i-m) \le \varepsilon_{loss}$$

$$b(i,K,\alpha) = \frac{K!}{i!(K-i)!} \alpha^i \ (1-\alpha)^{K-i} \text{ (Binomials)}; \ \alpha = \frac{BHCA \ \times MCD}{3600} \quad m = \text{int}(\frac{C_{link}}{C_{service}})$$

5. Case Studies and Conclusions

The traffic performance of the APON and SDH-Ring is calculated for Fast Internet Access, Video on Demand, Interactive Multimedia and High Quality Videophone. *Fast Internet Access* can be carried over ADSL while the other services require VDSL. Table 3 gives some details about the simulation scenario, which differs slightly for ADSL and VDSL.

30.000 POTS users	ADSL access	VDSL access
Service Area	300 km^2	100 km^2
Drop Area (scaling correction included)	12 km^2	1.5 km^2
User Density	100/km^2	300/km^2
Total number of ONUs or ADMs (NxN_A)	25	73

Table 3: user density and service area

Figure 5 shows the results for each (homogeneous) scenario, i.e. one particular BB-service is combined with 100% POTS penetration. The required capacity (N_A x155Mbs) is calculated for each value of the BB-service penetration, which varies from 0% up to 40% Calculations take into account the multiplexing gain at the cell-, file- and call-level, as explained in previous sections.

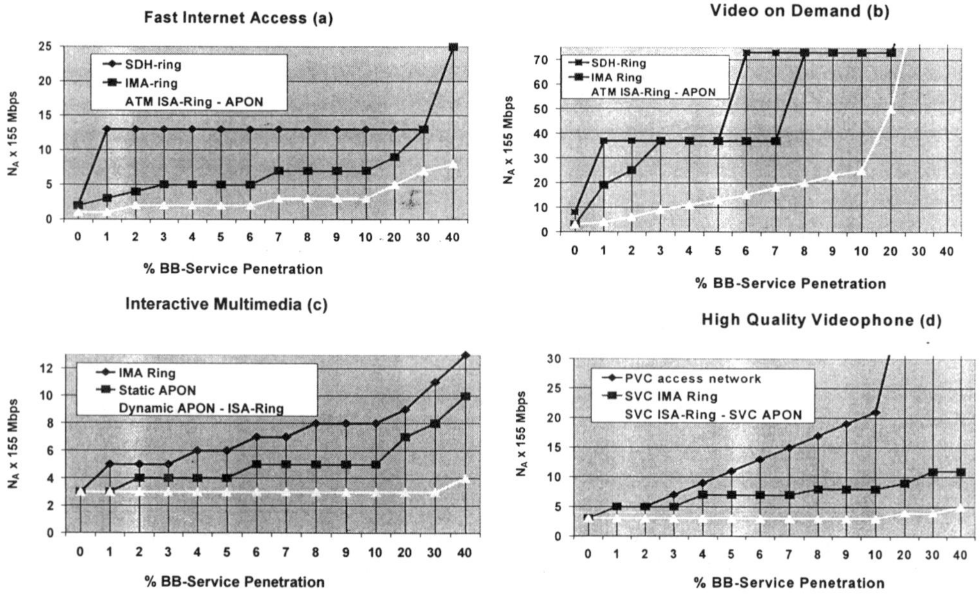

Figure 5: Network dimensioning results

- The native SDH–Ring always performs badly due to the inefficiency of SDH in carrying non-standard rates. For e.g. Fast Internet Access, a BB-penetration of only 1-% already requires a VC-3 per drop area for carrying the BB traffic. Therefore a Ring can have at most two ADMs because another VC-3 is reserved for the POTS traffic.
- The difference between the IMA and the ISA Ring in Figure 5a is due to the *location* of the access concentrator. In case of the IMA Ring the concentrator is located at the (individual) ADMs, while for the ISA Ring the Access Node itself is the concentrator. The concentrator determines the available bandwidth for (statistical) multiplexing the services. The larger the available bandwidth, the larger the multiplexing gain. For the

ISA-Ring, this is 155 Mbps, while for the IMA Ring, the bandwidth is 155 Mbps divided by the number of ADMs on the Ring.

- For asymmetric services, the bandwidth bottleneck is in the downstream direction. Therefore a dynamic MAC protocol has no influence on the APON performance for Fast Internet Access and Video on Demand. In case of Interactive Multimedia however, a dynamic MAC protocol has a major influence.

Complexity versus traffic performance

In all cases, it is important to determine the level where the main ON/OFF behaviour of a service occurs. This level determines to a large extent the realisable multiplexing gain. In case of e.g. Fast Internet Access, this is the *File* level, while for High Quality Videophone it is the *call* level. *Realising* the multiplexing gain requires advanced traffic functions. For example statistical multiplexing gain at the *call* level requires dynamic signalling, which explains the difference in performance between Permanent and Switched Virtual Connection networks. Statistical gain at the file level for upstream APON-traffic is only possible with a dynamic MAC protocol. Thus the operator should make a trade-off between implementing more complexity and the (statistical) gain that can be realised by the higher system. This largely depends on the type of service and the underlying technology!

Conclusions concerning the Fibre Access Technologies

- Support of BB-services for an **SDH-Ring** requires undoubtedly extra (ATM) functionality. Even for the smallest BB-service penetrations, a native SDH-Ring is almost equivalent with the Point to Point solution. Moreover, the ISA-upgrade technology is for sure much more future safe than the IMA solution. IMA offers *in the best case* a BB support for service penetrations less than 10%. ISA allows a smooth transition from a hybrid architecture, i.e. separate SDH-VC pipes for Narrowband and Broadband traffic, to a full ATM overlay ring.
- For asymmetric services, the static **APON** performs as well as the ATM-ISA Ring. For symmetric services, the static APON is also satisfactory for service less than 10%. A dynamic MAC protocol is necessary for higher service penetrations. The general conclusion is that the static APON is the appropriate feeder technology if ADSL lines cover the drop area (assuming the lines carry asymmetric services), while the dynamic APON is more advisable if the APON has to serve VDSL lines.
- The choice for the **APON versus SDH-Ring** technology depends mainly on the installed base of the operator. Both technologies can be made equally efficient from a pure traffic point of view. It can be noted that the APON has the intrinsic advantage that it supports cell multiplexing, while an SDH-Ring must be ATM-upgraded. In addition, other parameters such as cost, topology, reliability should be taken into account when one decides for a certain technology.

References

[1] ADSL versus Cable Modems. The Lehman brothers, Telecommunication services. June 2, 1999

[2] ITU-T, "Broadband Optical Access Systems based on Passive Optical Networks (PON)", ITU-T Recommendation G.983.1, 10/1998

[3] Performance improvement of ATM PONs in Multi-service environments by means of dynamic MAC protocols, R. Hoebeke, F. Ploumen, K. Venken, Proceedings NOC'98, Manchester, UK

[4] Inverse Multiplexing for ATM (IMA) ATM Forum Committee specification, version 1.0 (AF-PHY-0086.00), July 1997

[5] ATM over SDH: Bandwidth Management. M. Huterer, F. Berrouet, Proceedings of ISS'97, Toronto September 1997

[6] Dimensioning methodologies for the BroadbandLoop Access Network, S. Sargento, R. Valadas, A. de Sousa, Proceedings NOC'99

[7] Broadband Network Teletraffic – Final Report of Action COST 242 J. Roberts, U. Mocci, J. Virtamo (ISBN 3-540-61815 Springer-Verlag - 1996)

Broadband Access, WDM Metro and Network Management
D.W. Faulkner and A.L. Harmer (Eds.)
IOS Press, 2000 © AKM

APON versus SDH for Multiple E1 services

Vince BARKER
Fujitsu Europe Telecom R&D Centre, 2 Longwalk Road, Stockley Park, Uxbridge, Middx., UB11 1AB, UK

Abstract. ATM PON shows great potential as a future multi-service platform for the Access Network. The initial introduction of ATM PON technology, however, needs to be justified to a great extent on the basis of existing services. This paper presents a relative techno-economic analysis that compares the provision of multiple E1 services over an APON and over SDH point-to-point optical customer links. In performing this analysis, account is taken not only of the Access Network equipment, but also the equipment and optical fibre required for transporting the traffic up to the Core Network connection point. A variety of deployment architectures for both the APON and SDH equipment are considered. A key feature of this analysis is that it assumes a data-centric core network. The results of the study indicate that APON architectures can provide a cost-competitive for a limited number of E1s.

1. Introduction

APON has been gaining considerable interest worldwide, especially with development of international standards by the Full Services Access Network (FSAN) group which have been adopted by the ITU-T, see [1]. A key factor in the introduction of new technology into a telecommunications network is the ability to justify it on the basis of providing existing services. Demand for these is generally felt to be better understood and often given greater weight than forecasts of future new services.

It is apparent that there is a growing migration of the core network from circuit based towards a data oriented Multi-Service Platform (MSP), which is currently expected to be based on IP and/or ATM. This is driven in great part by the "Data Wave" whereby circuit traffic has been overtaken by data traffic. It is still necessary however for access technologies to support circuit based services. Another trend that is apparent is the migration of SDH from the core network down into the access network, often referred to as Access SDH (ASDH).

This paper analyses the relative merits of six network architectures, four APON based and two ring-based SDH with optical point-to-point customer drops. Configurations are analysed in which an APON can provide a cost competitive solution for the provision of multiple E1 services once the core network has migrated to a more data oriented MSP. All models include equivalent assumptions for the network and customer traffic levels. Results of a techno-economic analysis are presented which identify the relative merits of the various architectures determined by varying a range of system parameters.

2. Architectures

This study considers four APON network architecture variants and two SDH ring based optical point-to-point customer drop architecture variants. All architectures assume that services are carried through a Multi-Service Platform (MSP), and as such all assume traffic is carried back up to a Tier 2 node for connection to an ATM edge switch, IP Router or equivalent. All customer access connections are assumed to be fed from equipment located at a Tier 3 node, which here is always assumed to be located away from the Tier 2 node.

Previous Access Network costing studies, see for example [2], have shown that it is important to consider costs associated with the core network when considering the merits of different access architectures. This study takes into account the outer core infrastructure required, both in terms of equipment, including interfaces to the MSP, and also the costs of outer core fibre.

2.1 APON OLT at Tier 2

In this architecture, depicted in Figure 1, the OLT is located at Tier 2, and feeds PONs at one or more Tier 3 nodes. For brevity this architecture is generally referred to as OLT at T2.

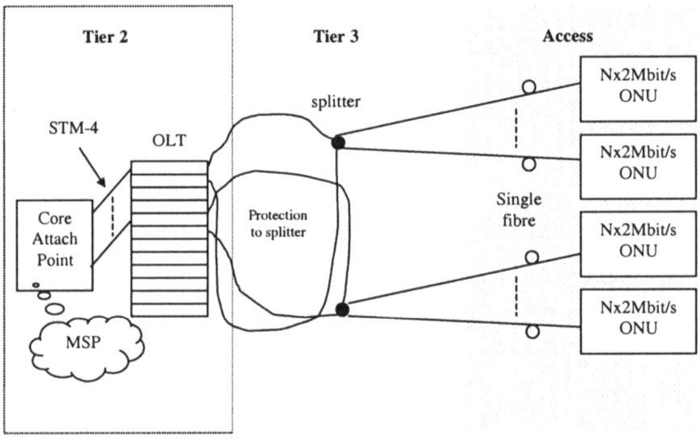

Figure 1: APON OLT at Tier 2

The following assumptions are made:

- The splitter is located at Tier 3, with no further splits in the Access (each PON serves only one Tier 3 node).
- Protection is provided to the first split, with the protection route set to be 1.8 times the distance of the working route.
- The ONU provides the Circuit Emulation function.
- Connections to the ATM Edge switch are STM-4, unprotected.

2.2 APON OLT at tier 3

In this architecture, depicted in Figure 2, the OLT is located at Tier 3, and is connected back to the MSP via one or more SDH rings provided by SDH ADMs. Two variants of this model are considered, one based on a STM-4 ring, the other on a STM-16 ring. For brevity these architectures are generally referred to as OLT 4 at T3 and OLT 16 at T3.

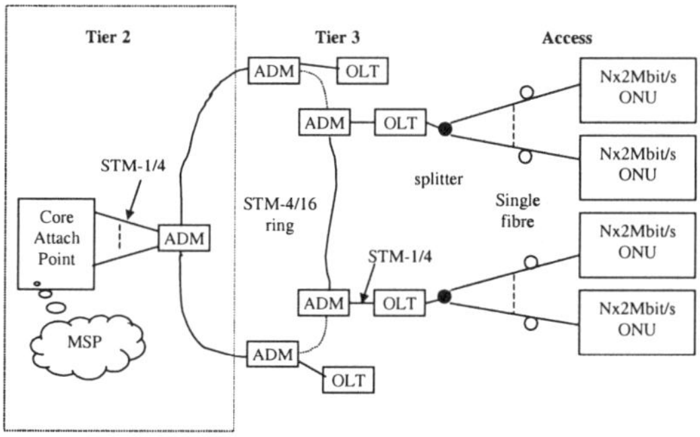

Figure 2: APON OLT at Tier 3

The following assumptions are made:

- The splitter is located at Tier 3, with no further splits in the Access.
- No protection is provided to the first split.
- The ONU provides the Circuit Emulation function.
- Connections to the ATM Edge switch are STM-1 for the STM-4 ring and STM-4 for the STM-16 ring, unprotected.
- Connections between the ADMs and the OLTs are STM-1 for the STM-4 ring and STM-4 for the STM-16 ring, unprotected.

2.3 OLT in STM-16 Ring Mode

In this architecture, depicted in Figure 3, the APON OLTs are deployed in ring mode at STM-16. For brevity this architecture is generally referred to as OLT 16 ring.

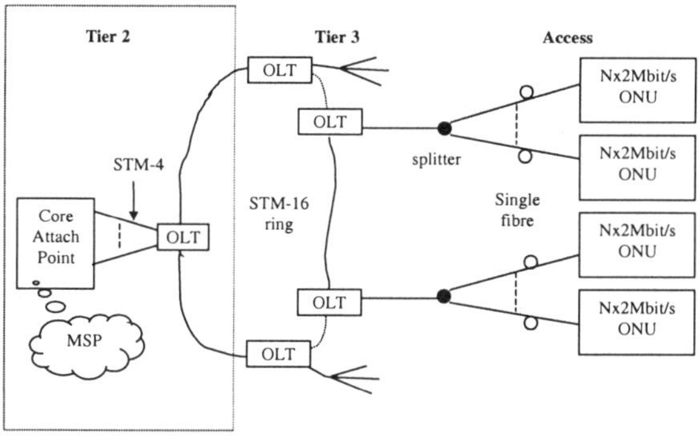

Figure 3: APON OLT in STM-16 Ring Mode

The following assumptions are made:

- The splitter is located at Tier 3, with no further splits in the Access.
- No protection is provided to the first split, protection to Tier 3 being provided by the SDH ring.
- The ONU provides the Circuit Emulation function.
- Connections to the ATM Edge switch are STM-4.

2.4 SDH based Optical Point-to-Point

In this architecture, depicted in Figure 4, the optical point-to-point customer links are provided by an SDH ADM located at Tier 3, which is connected back to the MSP via a

SDH ring. Two variants of this model have been implemented; one based on a STM-4 ring, the other on a STM-16 ring. For brevity these architectures are generally referred to as SDH 4 and SDH 16.

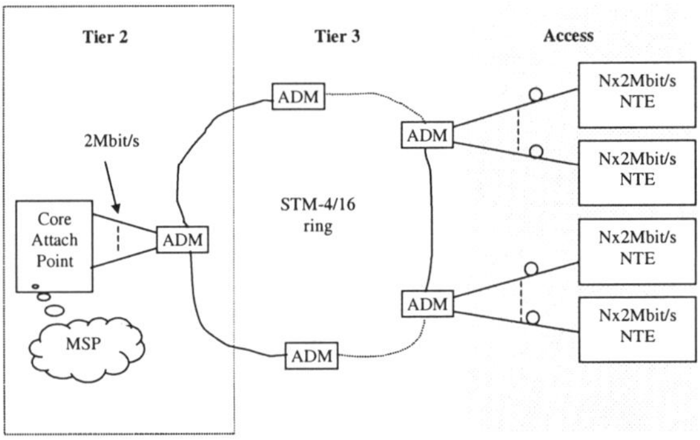

Figure 4: SDH based Optical Point-to-Point

The following assumptions are made:

- The Core attachment equipment (ATM Switch/IP Router) provides the Circuit Emulation function on its input ports.
- Connections to the Core attachment equipment are 2Mbit/s electrical, unprotected.

3. Techno-economic modelling

Techno-economic models of the preceding architectures have been built. All of the models are based on homogeneous averaged traffic over both customers and nodes. All models include equivalent assumptions for the network, in particular:

- The same number of Tier 3 nodes.
- The same distance between adjacent network nodes (whether Tier 2 or Tier 3), with a nominal value of 10km generally being used. The cost of this fibre, referred to here as the "Core Fibre Cost" (CFC), is based on two fibres, and quoted as cost per km.
- SDH rings are dual fibre.
- The architecture for the STM-16 ADM is assumed to have 16 tributary slots and be capable of dropping its full aggregate at STM-1.
- The architecture for the STM-4 ADM is assumed to have the same number of tributary slots as the STM-16 ADM.
- It is assumed that the Optical Line Terminal has 32 slots available for either APON or network interface cards. This assumption means that the OLT is equivalent to the STM-16 ADM, providing capacity for 16 PON interface cards and 16 STM-1 network interface cards. In addition it is further assumed that the OLT STM-16 ring interface cards occupy further additional interface slots.
- Optical customer interface cards on the ASDH ADMs are assumed to be capable of providing 3 single fibre optical customer links.
- 2Mbit/s interface cards on the SDH ADMs and on the MSP interface equipment (ATM Edge Switch, IP Router etc.) are assumed to be to be capable of providing 21 connections.
- Prices used for the equipment have been set to values that are considered to be reasonable for large-scale deployment of equipment over the next few years.
- All traffic is taken up to the MSP for onward routing to its destination.

4. Results

The techno-economic models of the architectures have been used to calculate system costs for a range of bandwidths. Included in the calculated system costs are: all equipment prices from ONU/NTEs up to and including the interfaces to the "Core Attach Point"; together with the costs of the fibre connections for the Tier 2 to Tier 3 and for the Tier 3 to Tier 3 links.

As noted earlier the cost of the outer core fibre (CFC) is included in the model. This cost may vary greatly, either between operators, or in different parts of a network. This variation has been allowed for by considering a range of values, with the effect of not considering it shown by generating results at zero cost. Other parameters varied in the study are the number of Tier 3 nodes per Tier 2 and the number of E1s per customer.

As this study is concerned with the relative merits of the different architectures, the results are expressed relative to the SDH 4 architecture value. The SDH 4 values are always expressed as 100%, with figures showing 10% contour bands and white highlighting the band around 100%. Higher percentages indicate higher costs.

Results for a fixed pair of values of CFC and number of E1s per customer for a range of Tier 3 nodes per Tier 2 have been plotted, and are shown in Figure 5 - Figure 16. These have been calculated at a fixed total bandwidth per Tier 3 node of 960Mbit/s, with the number of customers per Tier 3 being 480 for 1xE1, 120 for 4xE1 and 60 for 8xE1. Similar results have also been obtained for a lower bandwidth value of 640Mbit/s, and also for higher bandwidth values of 1820Mbit/s and 3200Mbit/s.

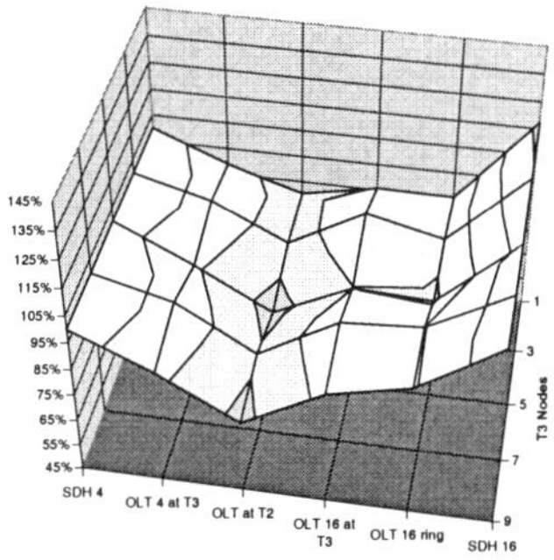

Figure 5: £0 CFC and 1xE1

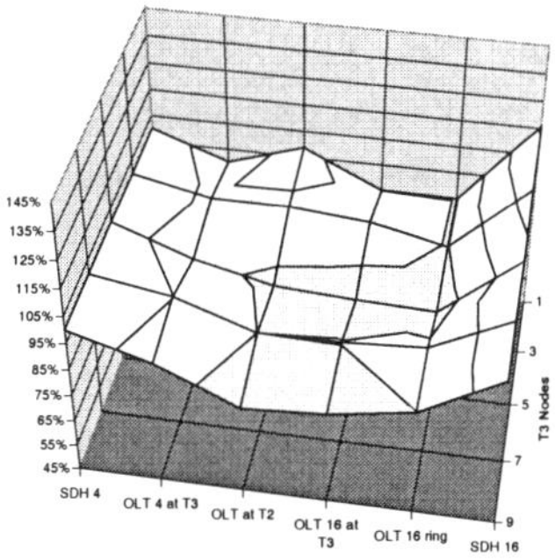

Figure 6: £500 CFC and 1xE1

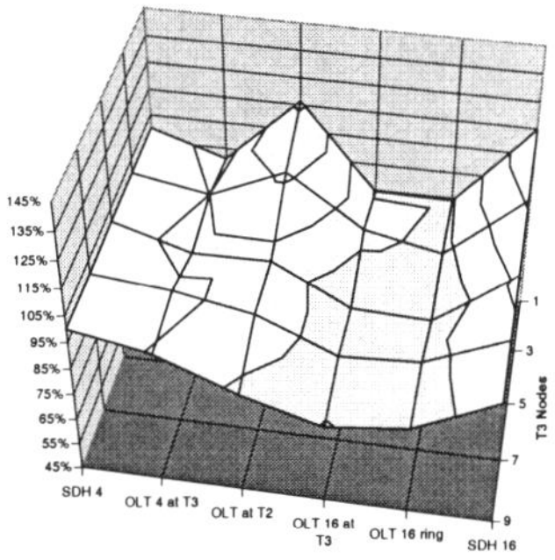

Figure 7: £1000 CFC and 1xE1

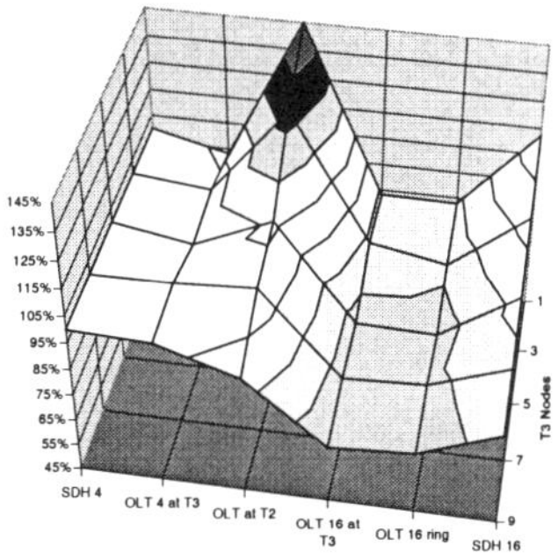

Figure 8: £2000 CFC and 1xE1

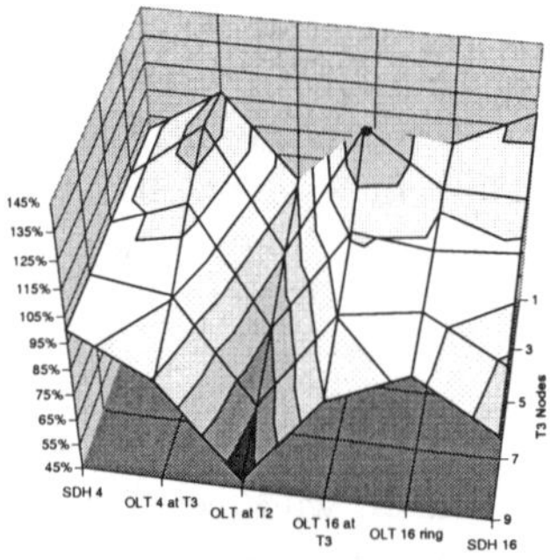

Figure 9: £0 CFC and 4xE1

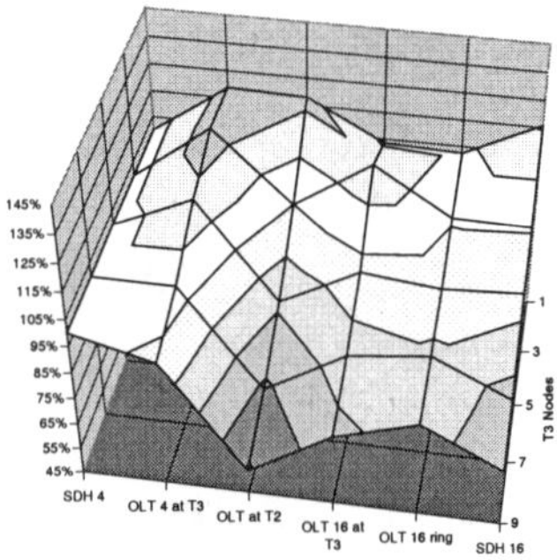

Figure 10: £500 CFC and 4xE1

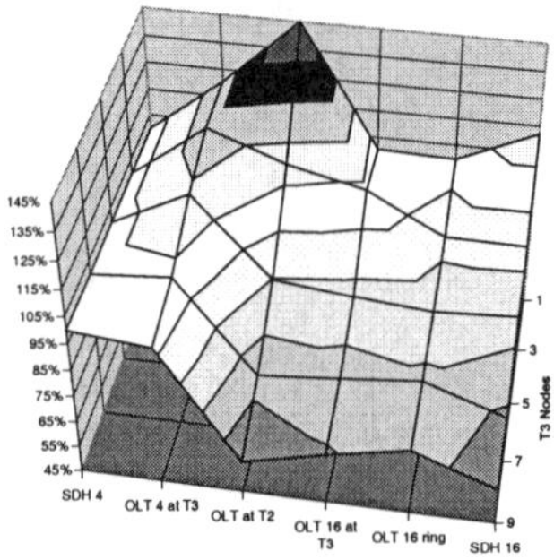

Figure 11: £1000 CFC and 4xE1

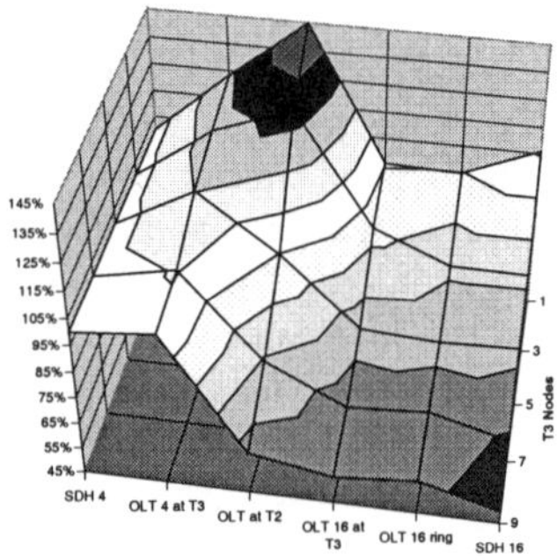

Figure 12: £2000 CFC and 4xE1

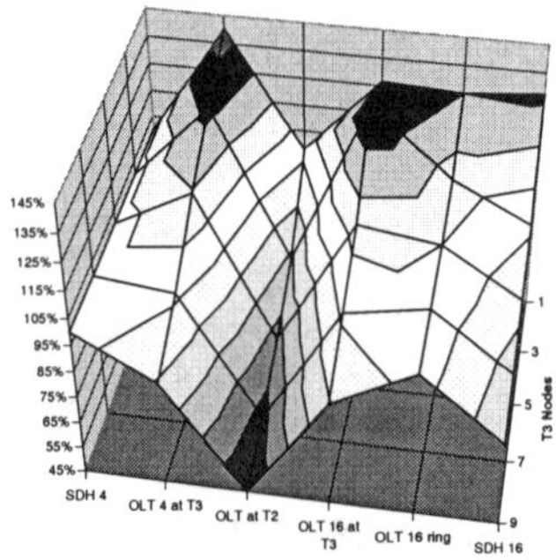

Figure 13: £0 CFC and 8xE1

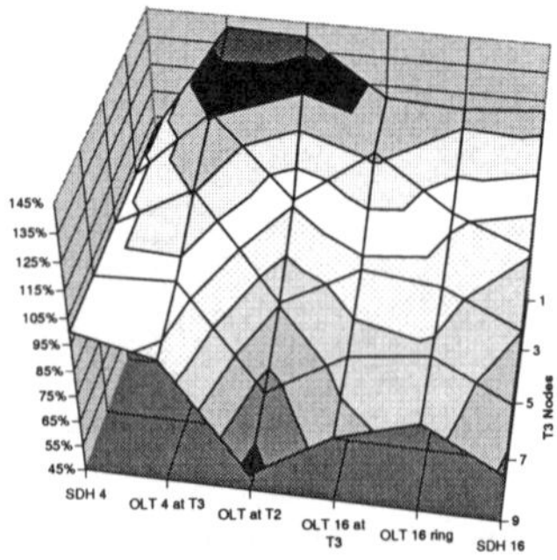

Figure 14: £500 CFC and 8xE1

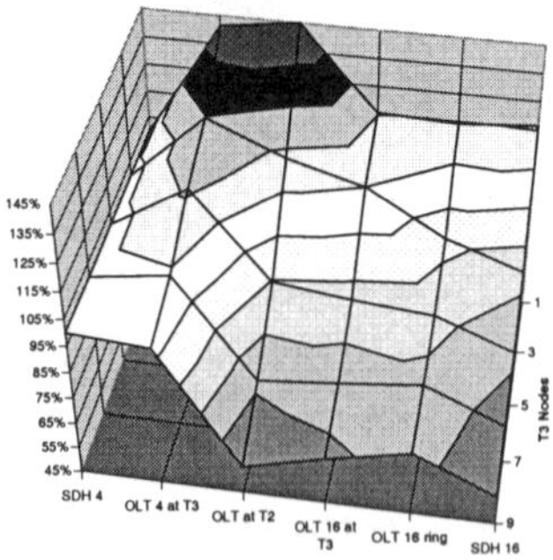

Figure 15: £1000 CFC and 8xE1

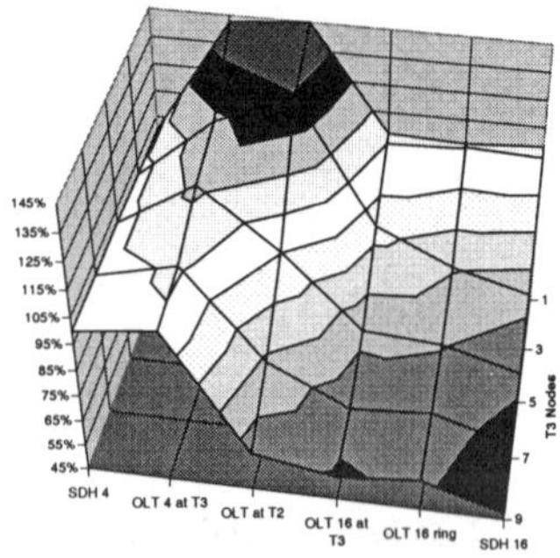

Figure 16: £2000 CFC and 8xE1

5. Observations

5.1 Network Finance Core Fibre Cost (CFC)

The figures clearly indicate that the value of CFC has a significant effect on the relative merits of the differing architectures.

The most striking effect of CFC values is on the OLT at T2 architecture. This is due to the larger number of fibres required, as each PON requires two fibres from Tier 2 to Tier 3. Thus at low numbers of Tier 3 nodes the number of fibres required is relatively high. The number of fibres per Tier 3 does not increase as the number of Tier 3 nodes per Tier 2 increase, however, which is the case for the ring architectures. Thus this accounts for the high relative cost at low number of Tier 3 nodes which decreases as the number of Tier 3 nodes increases.

As expected, the results generally indicate that the STM-16 architectures improve relative to the STM-4 as the value of CFC increases. This is due to the extra rings, and hence fibre, required to be deployed for the STM-4 architectures at the bandwidths considered here.

5.2 Number of Tier 3 nodes per Tier 2

Generally the results indicate that all architectures improve relative to the SDH 4 architecture as the number of Tier 3 nodes increase. The reasons for this though are dependant on the architecture and are discussed below.

5.3 OLT 4 at Tier 3

This architecture shows rough parity with that of SDH 4 at 1xE1. At other numbers of E1, there is a very noticeable trend from relatively high, down to parity, as the number of Tier 3 nodes increases. This effect is based on the requirement to deploy both an OLT and an ADM at Tier 3 for this architecture. But as the number of Tier 3 nodes increases, the number of ADMs per Tier 3 increases less for this architecture, while the number of OLTs remains constant. This effect is due to the assumptions in the models that the architectures have evolved as traffic has grown and that optimum repacking has not occurred as the bandwidth grew.

5.4 OLT at Tier 2

Results for the OLT at T2 architecture are the most striking. These indicate that this is clearly the best solution at zero CFC, and generally a competitive solution up to 8xE1s for higher values of CFC providing the number of Tier 3 nodes is above about 5. But the results also dramatically show that this architecture is generally the least competitive as the CFC value increases and the number of Tier 3 nodes remains low. This effect at low numbers of Tier 3 nodes is due to the efficient fill of the OLT being overcome by the extra cost of the fibre to the Tier 3 node.

It should be noted that the fibre cost plus all other non customer interface related equipment costs per Tier 3 node remains constant as the number of Tier 3 nodes increases. For the other architectures, however, these costs rise. This latter effect is the reason that the OLT at T2 architecture becomes increasingly cost competitive as the number of Tier 3 nodes increases.

5.5 OLT 16 at Tier 3 and OLT 16 Ring

These two APON STM-16 ring based architectures are both reasonably similar. They show competitiveness with the SDH 16 architecture up to about 8xE1 per customer. As for the SDH 16 architecture they show increasing cost competitiveness as the value of CFC rises, indicating that they make more efficient use of fibre than the STM-4 architectures.

In addition they both also show similar trends as the number of Tier 3 nodes increases. This arises for the same fibre use efficiency reasons.

5.6 SDH 16

Generally this architecture is relatively more cost competitive than the SDH 4 architecture. The exception to this is at low numbers of Tier 3 nodes for 1xE1. For this latter case, the number of customers mainly determines the number of ADMs deployed leading to similar numbers for a small number of Tier 3 nodes. As the STM-16 ADMs are more expensive than the STM-4, this accounts for this result.

6. Conclusions

The results from this study clearly show that where there are a low number of E1s per customer and where the core network is a MSP, that STM-16 based APON architectures offer a cost competitive solution to the SDH based architectures. The OLT at Tier 2 architecture may also offer a cost competitive solution when the number of Tier 3 nodes per Tier 2 is greater than about 5. This study therefore is indicative that, when services are carried through a multi-service platform, APON solutions offer a good candidate for multi-service access, as they can competitively support existing E1 services, but also offer the ability to support data services.

The analysis so far assumes that optical splitters are located only within the Tier 3 site, thus equalising the access fibre costs with optical point-to-point delivery. Allowing splitters into the access network could offer significant savings, which could increase the range over which an APON based architecture was competitive. But as these saving are highly dependent on the exact deployment architecture they have not been considered as part of this study.

7. References

[1] ITU-T recommendation G.983.1, Broadband optical access systems based on Passive Optical Networks (PON).

[2] V.R.Barker et al, Access SDH: Techno-Economic Analysis as services migrate from Circuit to Cell, Broadband Access and Technology, NOC'99, IOS press, Amsterdam, 1999, pp 94-102.

A Study of Data Channel Reservation Protocol in a WDM Single-hop Network

Tomofumi TAMURA, Mikio TAKAHARA and Masanori HANAWA
Yamanashi University, 4-3-11 Takeda, Kofu-shi, Yamanashi-ken 400-8511, Japan

Abstract. In a WDM single-hop star network, a dynamic scheduling protocol which can effciently transmit variable-sized messages without any global information has been proposed. In this paper an improved protocol though the dynnamic scheduling protocol is a simple one, the throughput characteristics are improvable a little. Then we propose a improved protocol and compare the performance of the proposed protocol with that of the original one by computer simulations. As the results, the performance based on the proposed protocol shows better than the original one.

1 Introduction

In a WDM single-hop network, a number of data channel reservation protocols using a control channel have been proposed[1]. Among them, the dynamic shceduling protocol is a simple and efficient protocol to transmit variable-sized messages without any global information[2]. Thus, any new node can join the network without any network initialization.

The protocol, however, has two points to improve the throughput characteristics; one is the vacant minislot in the control channel at each finishing point of transmitting data packets, and the other one is the retransmission of control packets failed to reserve. The former especially influences the throughput characteristics in case of the transmission consisting of short packets mainly. Moreover, transceivers need to have a wide tunable range and good tuning stability.

In this paper, we propose an improved protocol to solve two problems mentioned above. The improved performance based on the new protocol by computer simulation is shown.

2 Proposed protocol

Fig.1 shows the WDM single-hop star network architecture. There are M nodes in the network, and each node is connected by a passive star coupler. Also a wavelength λ_0 is used for the control channel, and wavelengths $\lambda_1, \lambda_2, \cdots, \lambda_N$ are used for data channels. Thus, each node has one fixed-wavelength transceiver to access to the control channel, and one tunable wavelength transceiver to access to any one of data channels. All signals sent from those transceivers are transmitted to nodes of the whole network.

Fig.2 shows an example of the reservation procedure of the proposed protocol. Each message consists of some fixed data packets. All channels including the control channel are divided into time slots of T_d, as shown in Fig.2. One time slot of the control channel is further divided into the end-to-end propagation delay τ and p minislots of T_c. In the

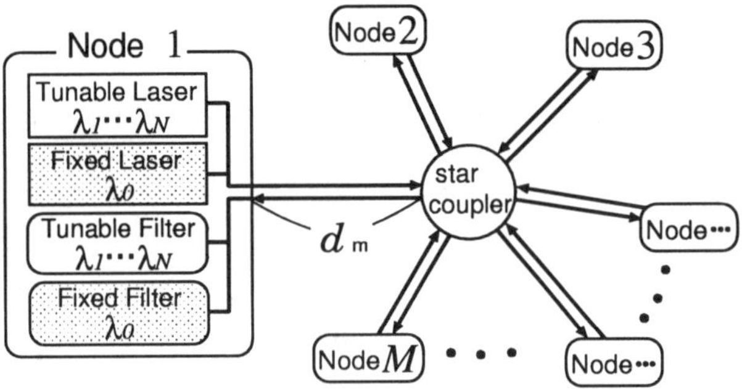

Fig. 1: Network architecture with a passive star topology

proposed protocol, a control packet has the information for the number of the remaining data packets in addition to the destination address different from the original protocol having the destination address only.

Now, the reservation procedure of the proposed protocol is explained as the following using Fig.2. When a message arrives at a node, the node transmits its control packet to the control channel according to the slotted ALOHA protocol. Then the reservation of a data channel is accomplished if and only if all the following three conditions are satisfied.

1. the control packet does not collide with others in the time slot(condition 1).
2. the control packet does not have the same destination address of other control packets having been reserved successfully(condition 2).
3. the number of previous successful reservations exceeds the number of the data channels N(condition 3).

If a node has failed to reserve the data channel, the control packet is retransmitted at the next time slot. Here, the proposed protocol adds the following procedure to original one. The control packet unsatisfied condition 3 is transmitted to the minislot in which the control packet does not collide with others at the next time slot. In the time slot 2 of Fig.2, four control packets are transmitted successfully to the control channel. The packet destined for the node 6, however, fails to transmit the data packets because three sendable data channels have been reserved by other three packets. Thus, the control packet destined for the node 6 is allocated the 4th transmittable minislot without collision at the time slot 3.

When all the conditions are satisfied, the node reserves the data channel corresponding to the minislot transmitted the control packet and transmits the data packets through the same data channel till the end of the transmitting message. In the proposed protocol, all nodes can know the transmitting situation by the number of the remaining data packets in the minislot. In 1st minislot of the time slot 1, the control packet destined for the node 1 is transmitted. All nodes know the transmitting message finishes at the time slot 1, then any node can try to catch the 1st minislot in the time slot 2. Thus, the proposed protocol does not have the vacant minislot showing the finishing point of the transmitting message.

It is expected the proposed protocol can improve the throughput characteristics of original one by above procedures.

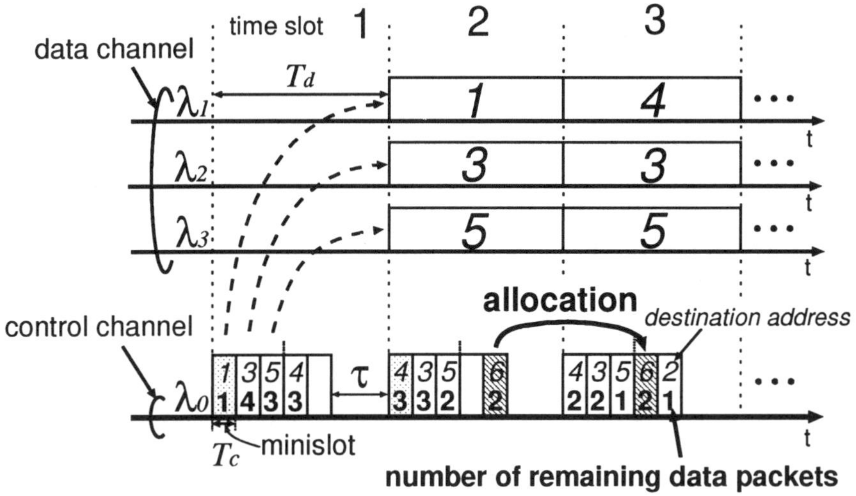

Fig. 2: The reservation procedure of the proposed protocol.

3 Computer simulation results

In this section, the performance of the proposed protocol is compared with that of the original one in Figs.3, 4. The simulation parameters used are shown in Table 1. In the computer simulation, the tuning times of all transceivers are assumed to be zero, and the distance between all nodes and star coupler is assumed to be 200 m.

Table 1: Simulation parameters

number of nodes	M	50
number of data channels	N	16
number of minislots	p	24
data packet size		656 bytes
bit rate per a channel		2.5 Gbit/s
network size	d	200 m

Fig.3 shows the network throughput versus message arrival rate characteristics for various average message lengths. In the high load region over the message arrival rate $\sigma = 0.1$, the throughput characteristics are better than that of the original ones. Collisions between/among control packets are the main reason degrading the throughput characteristics in the original protocol system. The proposed protocol makes the degradation decrease by decreasing collisions between/among control packets. Moreover the shorter message case such as $L = 1312$ bytes is improved better than the longer message case such as $L = 6560$ bytes. Because the shorter message has the finishing point of the message more than the longer message. Therefore it is considered that the throughput characteristics can be improved by the proposed protocol.

Fig.4 shows the average delay time versus message arrival rate characteristics for various average message lengths. The characteristics are improved similar to the throughput characteristics. The reason is the natural due to the decreasing of collisions.

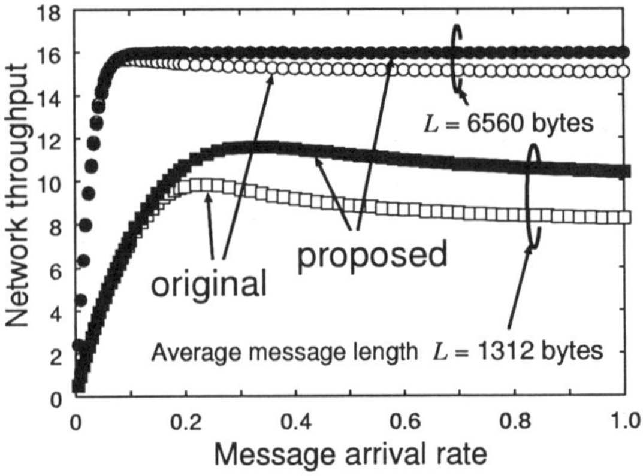

Fig. 3: Network throughput versus message arrival rate for various average message lengths

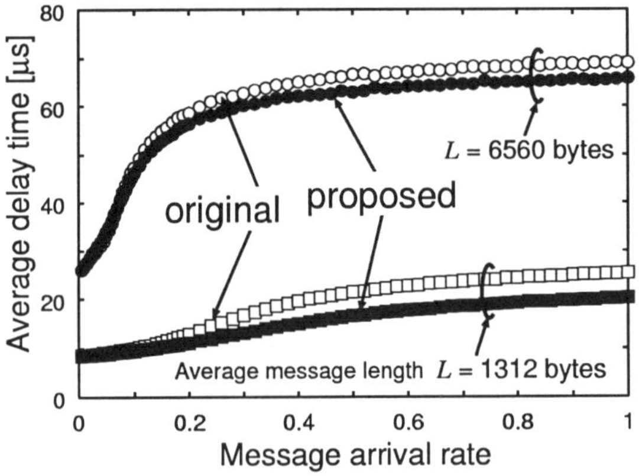

Fig. 4: Average delay time versus message arrival rate for various average message lengths **(proposed : proposed protocol, original : Dynamic scheduling protocol)**

4 Conclusions

In this paper, an improved dynamic scheduling protocol has been proposed. The performance of the proposed protocol is compared with that of the original one by computer simulation. As the result, since the vacant minislot and collisions between/among control packets in the control channel are decreased, the performance of the original one can be improve by 15% at the message arrival rate of 0.4. by the proposed protocol Especially, the proposed protocol has the advantage in the short message dominant network.

References

[1] H.B. Jeon and C.-K. Un, "Contention-Based Reservation Protocols in Multiwavelength Optical Networks with a Passive Star Topology," IEEE Trans. Commun., vol.43, No.11, pp.2794–2802, November 1995.

[2] J.H. Lee and C.-K. Un, "Dynamic Scheduling Protocol for Variable-Sized Messages in a WDM-Based Local Network," IEEE J. Lightwave Technol., vol.14, No.7, pp.1595–1600, July 1996.

Broadband Access, WDM Metro and Network Management
D.W. Faulkner and A.L. Harmer (Eds.)
IOS Press, 2000 © AKM

Design Issues Related to Grooming Functionality for SNCP Ring Interconnection

Didier Colle, Piet Demeester
Department of Information Technology, The Ghent University – IMEC
Sint-Pietersnieuwstraat 41, B-9000 Gent, Belgium
Email: didier.colle@intec.rug.ac.be

Kris Struyve
Engineering, Network R&D and Integration, GTS Network Services
Terhulpsesteenweg 6A, B-1560 Hoeilaart, Belgium
Email: kris.struyve@gtsgroup.com

Abstract: Traffic is growing at a very fast pace, resulting in the requirement for a high availability of the network. A popular solution to this problem is applying self-healing rings inside the network.
A network typically contains a set of interconnected rings. These rings are interconnected in nodes called gateway nodes. Gateway nodes are often chosen to perform grooming. This implies that the content of the higher bitrate signals is reordered.
Our study examines how this grooming functionality can be provided in the gateways interconnecting SNCP-rings at the VC-4 layer and what the impact is on network planning. One can choose between installing a cross-connect in the gateway, or distributing the grooming functionality across the ADMs on the rings.

1 Introduction

Optical fiber technology has been adopted to carry the exponential growing traffic demand. However, huge amounts of data can get lost due to a single fiber cut. High reliability requirements can be met by introducing redundancy in the network.

Self-healing rings [1] are widely used as a popular recovery mechanism. The reason is that a (self-healing) ring is rather simple from an equipment and operational point of view. Rings are often equipped with so called add-drop multiplexers (ADMs): an ADM is able to cross-connect traffic from(/to) its tributaries to(/from) its aggregates.

Although ring networks may have interesting properties, a typical network consists of more than just one ring. A common way of working is to interconnect several rings in order to span the complete network. Thus, a ring can be seen as a sub-network. Previous studies [2] have shown that transporting high bitrate signals through the network, on an end-to-end basis, is only cost-efficient for networks limited in size (in terms of nodes and not of distance) or for large demands. Since one (or more) rings form a sub-network, reordering of the information inside the higher bitrate signals is performed at the edge of those sub-networks (or in other words: inside the gateway nodes, performing ring interconnection). This reordering is referred to as the grooming functionality. This paper focus on STM-16 SNCP-rings [1] at the VC-4 level, interconnected in gateway nodes performing VC-12 grooming. We consider HO ADMs (ADM 4/4), containing a VC-4 matrix, and HO/LO ADMs (ADM 4/4/1), containing a VC-4 *and* VC-12 matrix. We also

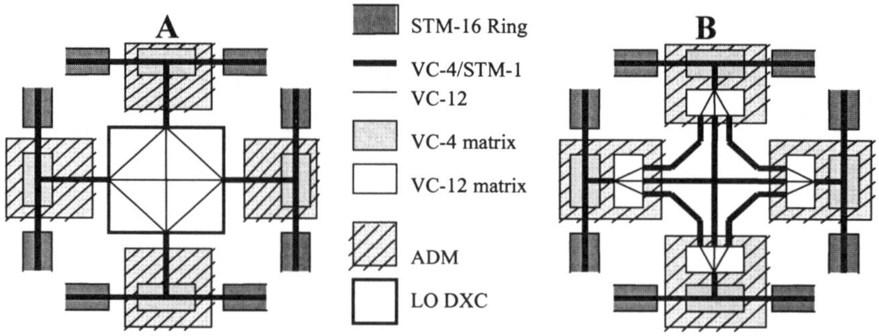

Figure 1: Providing grooming functionality in a gateway node

assume that both types only host STM-1 interface cards. In the near future a similar situation will arise, where optical self-healing rings are interconnected in gateways, performing VC-4 grooming.

2 Problem Description

As discussed in the previous section, gateway nodes are a good place to perform grooming. However, the problem is which architecture to choose for these gateways: for example, HO ADMs are not able to perform VC-12 grooming.

A first option simply connects all HO ADMs to a LO DXC (DXC 4/1, containing a VC-12 matrix) performing VC-12 grooming. The disadvantage is of course the installation of an expensive cross-connect.

In a second option, HO/LO ADMs are manually interconnected with STM-1 trib-to-trib interconnections. The grooming functionality is hence distributed over all interconnected HO/LO ADMs inside a gateway site. The consequence of replacing an expensive LO DXC and HO ADMs by HO/LO ADMs, is the corresponding design complexity: the number of tributary interfaces, required on the HO/LO ADM, is not easily dictated, by the number of tributaries accessed on the ring. Theoretically, it would be possible to have each VC-12 on one ring to be routed to another ring (assuming an unrealistic worst case amount of rings) and thus a HO/LO ADM placed on an STM-16 ring, has to be capable to host 16*63=1008 (STM-1) interface cards. It is obvious that most ADMs don't have that much interface slots.

An example comparing both architectures is shown in Figure 1. Four STM-16 rings are interconnected with each other and a single VC-12 is exchanged between each pair of rings (thus, on each ring a single VC-4 with 3 VC-12s as content is terminated). The figure shows that the architecture, with the central cross-connect (Figure 1A), only requires a single STM-1 interface card per HO ADM and that the other scenario (Figure 1B) needs three STM-1 trib-to-trib interfaces on each HO/LO ADM. Note that it is also possible to obtain less trib-to-trib interfaces per HO/LO ADM, as clearly shown in Figure 2. Although both rings contain two VC-4s (towards different destination on the ring) each containing a single VC-12, only a single STM-1 trib-to-trib interconnection is required in the gateway.

Figure 3 shows how to solve the problem that too many interfaces are required per HO/LO ADM. A possible way of working is to install ADMs in tandem and to distribute the interface requirements over these ADMs. However, this may result in less efficient

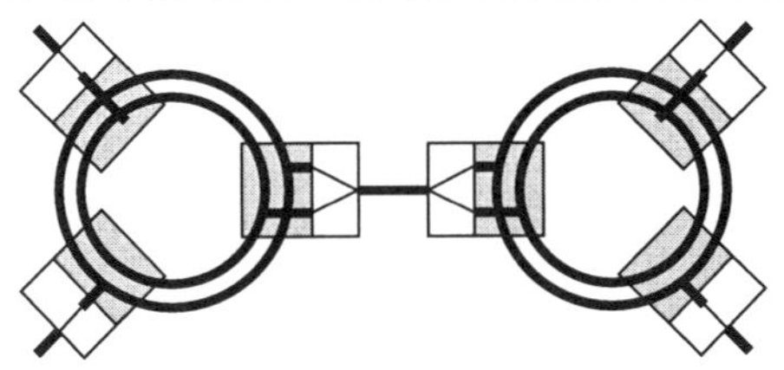

Figure 2: less trib-to-trib interfaces

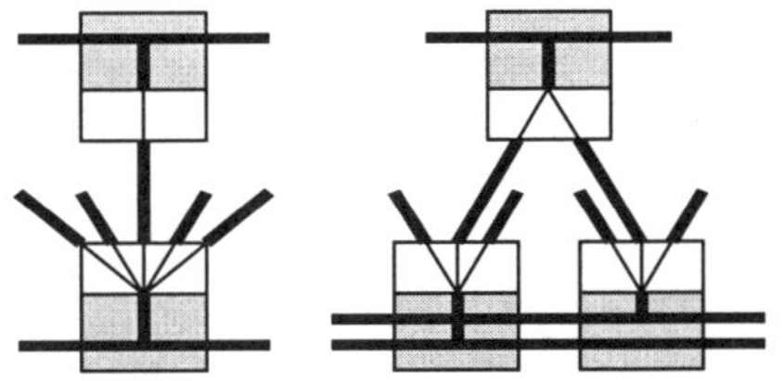

Figure 3: reducing #interfaces/ADM

packing of the VC-12s in the VC-4s on the ring, since only traffic towards a single ADM can be groomed into a single VC-4. Consequently, a larger number of VC-4s may result in an additionally stacked STM-16 and in an increased number of required interfaces on the interconnected rings (Figure 3). Similar problems may occur on these interconnected rings, due to the increased interface requirements on those rings. In this way, reducing the interface requirements of a single HO/LO ADM may propagate to a network wide effect. These problems do not exist in the case with a central cross-connect, since the HO ADMs don't break up the VC-4s and thus a ring-per-ring design remains possible.

For operational purposes it may be desirable as well to put ADMs in tandem: i.e. dedicated ADMs for access and transit traffic in each node. This means that for each node(/site) pair, a single VC-4s is replaced by at most four VC-4s (A→A, A→T, T→A and T→T). In other words this means that the required capacity may increase in the worst case to 400 %!

3 Case Study

We have performed calculations on a particular case study for the GTS network [3] (21 rings, 39 nodes and 16 gateways). We started from a network dimensioning/design obtained with a commercial tool based on simulated annealing. First a design of the VC-12 layer is done: roughly speaking each ring in the VC-4 layer provides a full mesh of links in the VC-12 layer. Secondly these links of the VC-12 layer are routed as VC-4 demands over the underlying rings.

Starting from that design, we perform first a non-optimized timeslot allocation, on each ring. Out of this, the amount of VC-12s flowing from one STM-16 ring to another can easily be calculated.

From this point on, we can calculate the required amount of trib-to-trib interfaces per ring. In the case that a central cross-connect is applied the required amount of STM-1 interfaces is the amount of dropped VC-4s. In the other case, where the grooming functionality is distributed over the HO/LO ADMs, the number of trib-to-trib interfaces is dicated by the amount of VC-12s flowing between each pair of STM-16 rings. Note that this implies that we assume HO/LO ADMs, able to contain an unlimited amount of interface cards (in order to avoid placing ADMs in tandem as discussed in Figure 3) and thus we don't care about the problem of network wide effects. However, we have provided the functionality to install dedicated ADMs for access and transit traffic.

Figure 4 confirms our theory that placing ADMs in tandem (here due to separation of access and transit ADMs) requires more capacity per ring (average: 37%), resulting in a lower link filling. Even more, although a single STM-16 has to be sufficient, the separation of access and transit ADMs requires another stacked STM-16 ring in 9 of the 21 rings.

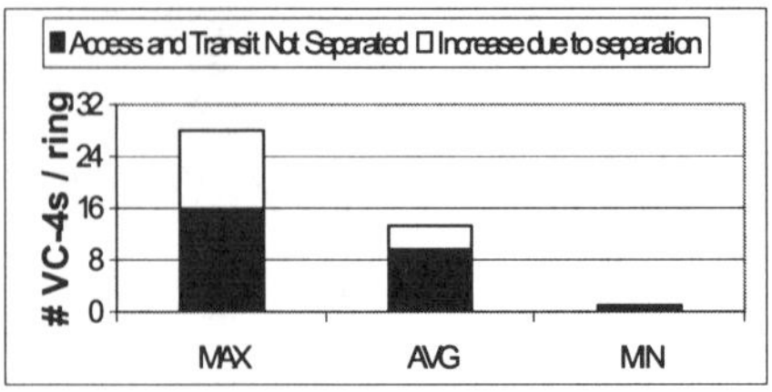

Figure 4: capacity requirements

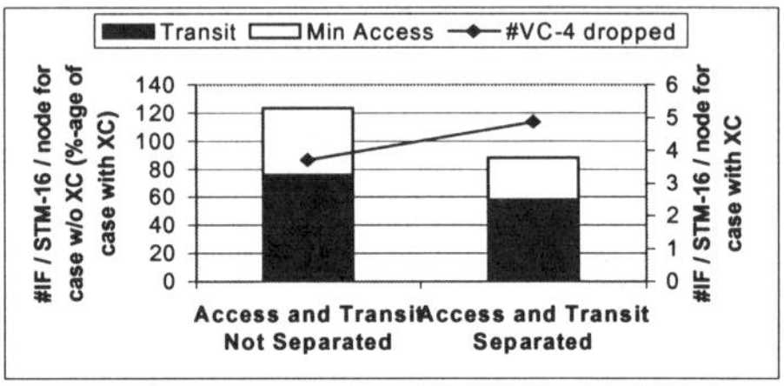

Figure 5: interface (IF) requirements

Figure 5 shows the interface requirements on the ADMs, taking into account the assumption that access traffic (towards the customers) can be packed as efficient as possible into the STM-1 interfaces (therefore the picture shows the absolute minimal number of interfaces required for access traffic). The line represents the absolute amount of dropped VC-4s (or thus the number of interfaces in the case with a central cross-connect). The bars represent the case without cross-connects and give the required number of interfaces relative to the amount of dropped VC-4s (represented by the line). As we expect, distributing the grooming functionality over the ADMs may result in a higher amount of interfaces required on the ADMs, compared to the case where a central cross-connect is provided. However, Figure 5 also confirms that a lower link filling typically results in a higher possibility to merge traffic from multiple VC-4s on the ring into a single STM-1 trib-to-trib interconnection (see also Figure 2): on the average the content of 4.88 (instead of 3.72) dropped VC-4s is merged into 4.31 (instead of 4.6) STM-1 trib-to-trib interconnections.

4 Conclusions

This paper presents two main options to provide grooming functionality in gateway nodes interconnecting (self-healing) rings. The first scenario uses in each gateway node a central cross-connect performing the grooming functionality and in the second scenario this functionality is distributed over all ADMs. The latter option may save the cost of an expensive DXC, but result in huge interface requirements on the ADMs: in order to reduce these requirements per ADM, one can distribute these requirements over multiple ADMs (placed in tandem). However, this may affect the network design on a network wide basis instead of on a ring wide basis. Placing ADMs in tandem reduces the link filling, but makes the use of HO/LO ADMs more attractive, compared to HO ADMs in combination with a LO DXC.

References

[1] ITU-T Recommendation G.841 (1995), "Types and Characteristics of SDH Network Protection Architectures"

[2] P. Arijs, D. Colle, M. Gryseels, M. Pickavet, H. Opstaele, P. Demeester, N. Wauters, M. Groisman, P. Falcao, "SDH protection in long-distance networks: a practical case study", proc. DRCN'98, May 17th-20th 1998, Brugge Belgium

[3] K. Struyve, N. Wauters, P. Falcao, P. Arijs, D. Colle, P. Demeester, P. Lagasse, "Application, Design, and Evolution of WDM in GTS's Pan European Transport Network", IEEE Commun. Mag., March 2000, vol. 38, no. 3, pp. 114-121

Broadband Access, WDM Metro and Network Management
D.W. Faulkner and A.L. Harmer (Eds.)
IOS Press, 2000 © AKM

An Algorithm to Design Logical Topologies in Wavelength-Routed Optical Networks

Fernando GONZÁLEZ, Ignacio de MIGUEL, Juan Carlos AGUADO,
Patricia FERNÁNDEZ, Rubén M. LORENZO, Evaristo J. ABRIL, Miguel LÓPEZ
Dpt. of Signal Theory and Communications and Telematics Engineering
E.T.S.I. de Telecomunicación, University of Valladolid
Campus "Miguel Delibes", Cno. del Cementerio s/n, 47011 Valladolid (Spain)
Tel: +34 983 423665, Fax: +34 983 423667, E-mail: {fernan, ignmig}@pesquera.tel.uva.es

Abstract. We propose an algorithm to design logical topologies in optical networks. It attempts to minimise the congestion in the network by establishing lightpaths in a manner that a maximum amount of traffic can be routed in at most two hops. A branch-exchange process is then performed to improve the results. Due to the computational hardness of this process, we also propose another method that considerably reduces the computing time at the expense of slightly increasing the congestion.

1. Introduction

Wavelength-Routed Optical Networks (WRONs) have become the preferred option for applying Wavelength Division Multiplexing (WDM) to Wide Area Networks (WANs). In WRONs, transparent optical channels, called *lightpaths*, are established between pairs of nodes that are not necessary adjacent. Each lightpath is assigned a wavelength on every link that it traverses and, obviously, if two lightpaths span the same link, they must be assigned different wavelengths. In the absence of wavelength converters, a lightpath must occupy the same wavelength on every link it traverses. This wavelength is used at intermediate nodes to properly route the lightpath towards its destination [1–4].

Lightpath establishment allows a *logical* (or *virtual*) *topology* to be embedded into the physical one [5–7]. Designing an optimal logical topology consist of [5]:

i) Determining a good logical interconnection pattern, that is, determining the source and destination nodes of the lightpaths.
ii) Routing the lightpaths over the physical topology of the network
iii) Assigning wavelengths to the lightpaths.
iv) Routing the offered traffic on the logical topology.

In this paper we will centre in subproblems i) and iv). We will search for a logical topology that minimises the network congestion [6–7], defined as the maximum offered load to a logical link [6]. We suppose that the externally offered traffic can be partitioned into different routes from its source to its destination. As it is unlikely that lightpaths will be established between every pair of nodes with external traffic between them, an amount of external traffic should flow over different lightpaths to reach its destination.

2. Minimum Congestion Logical Topology Design

By minimising the congestion, the capacity of the network to carry growing traffic is maximised. Then, our aim is to determine the optimal logical interconnectivity pattern (subproblem i), and to route the offered traffic through the logical links so that the congestion is minimised (subproblem iv).

In [6] and [7], Mixed Integer Linear Programming (MILP) formulations for solving both problems in joint are proposed. These formulations are computationally extensive, so separated (suboptimal) formulations for each subproblem are desirable.

A heuristic to select a good logical topology is to establish logical links (lightpaths) between those nodes with the maximum traffic between them. Following this principle, we propose the algorithm CP2, which attempts to route the maximum amount of traffic from its source to its destination node in at most two hops. The algorithm is shown below.

Algorithm CP2

<u>Given:</u> N, number of nodes of the network
T, logical nodal degree (number of transmitters and receivers per node).
$\boldsymbol{\Lambda} = \{\lambda^{sd}\}$, offered traffic matrix. λ^{sd} is the offered traffic from node s to node d.

<u>Output:</u> $\boldsymbol{Z} = \{Z_{ij}\}$, logical connectivity matrix. $Z_{ij} = 1$ if there is a lightpath from node i to node j, otherwise $Z_{ij} = 0$.

<u>Data:</u> Tx_i, number of logical links already established with node i as origin.
Rx_i, number of logical links already established with node i as destination.
L, total number of logical links already established.
x, intermediate node for two-hop routing.

<u>Process:</u>
```
Tx_i = Rx_i = 0, i = 1, 2, ..., N
L = 0
While (L < N·T)
        Search for (i, j) such as λ^ij = max {λ^ij : Z_ij = 0}
        If (Tx_i < T) and (Rx_i < T) /* A direct logical link is possible */
                Z_ij = 1
                Increment L, Tx_i, Rx_j
        Else-If (Tx_i < T) /* There are not free receivers at destination node */
                If (∃ x : Z_xj = 1, Rx_x < T)
                        Z_ix = 1
                        Increment L, Tx_i, Rx_x
                End-If
        Else-If (Rx_i < T) /* There are not free transmitters at source node */
                If (∃ x : Z_ix = 1, Tx_x < T)
                        Z_xj = 1
                        Increment L, Tx_x, Rx_j
        End-If
End-While
```

Note 1: If several intermediate nodes are candidates for establishing a two-hop route, it is chosen the one that maximises the single-hop traffic that can be carried through the new link.

Note 2: If the number of iterations is greater than N^2, it means the algorithm cannot finish while maintaining the logical degree. In this case, we break the 'while' loop and complete the connectivity pattern by means of branch-exchange operations.

In [7], an Integer Linear Program (ILP) formulation, called CP, is proposed for subproblem i). It maximises the amount of external traffic that can be routed in one hop, subject to the degree constraints. After obtaining an initial topology with CP, a local-search process using branch-exchanges is performed. A branch-exchange replaces two existing logical links (i, j)

and (m, n) for other two new links (i, n) and (m, j). The local-search process in [7] applies branch-exchanges to the least congested links. This method can also be used after CP2. We denote the whole process (initial solution and branch-exchange) by CPB and CP2B respectively.

Local-search takes much more time than solving CP or CP2, so executing both CP and CP2 instead of only one does not increase much the resolution time. Hence, we propose the following heuristic (MISB: Multiple Initial Solution and Branch-exchange):

1. Find an initial solution using CP and another one using CP2.
2. Select the one with the lowest congestion.
3. Perform the branch-exchange process using that solution.

For computing the network congestion we use the Linear Programming (LP) formulation proposed in [6] and [7], called RP.

When the number of nodes in the network is high, solving the whole problem (initial solution and branch-exchange) becomes intractable as resolution time increases exponentially with the number of nodes. Hence, a solution is to use only the two first steps of MISB. We denote with MIS to this approach.

3. Results and Conclusions

We have carried out several experiments with randomly generated traffic patterns of different models (quasi-uniform, quasi-ring, random and servers). LP problems have been solved using *lp_solve* [8].

In Figure 1, we plot the percentage reduction on congestion obtained by MISB when compared to CPB for 150 experiments with networks from 8 to 10 nodes. MISB results in better congestion values in the 76.7% of the experiments, compared to the 66% of CPB and the 34% of CP2B. In the 32% of cases, MISB gets different results than CPB, with an average congestion reduction of 2.27%.

Other 120 experiments have been carried out with networks from 12 to 16 nodes, to evaluate the goodness of MIS. The results are showed in Figure 2, where congestion reduction provided by MIS when compared to CP is plotted. MIS gets the same results than CP in 59.2% of cases and in the other 40.8% of cases MIS provides an average congestion reduction of 10.4%.

Note that both MISB and MIS present the worst results for servers traffic, where no gain from CPB (respectively CP) is achieved. It agrees with [9], where it is shown that scenarios with servers traffic are the most difficult to tackle.

Figure 3 shows mean values and 95% confidence interval of solving times (Pentium II 400 MHz, 128 MB RAM, Windows NT 4.0) for both CP and CP2 with different traffic patterns. CP2 proves to be less time-expensive than CP, but both CP and CP2 resolution times are despicable when compared to the branch-exchange process (which increments the initial-phase time in 2100% for 8-node networks and more than 4100% for 12-node networks). Due to the computational hardness of the branch-exchange process, it would be interesting to quantify its benefits. As shown in the histogram in Figure 4, CPB improves the congestion in 52% of cases, and in the rest of cases MIS, which does not execute any local-search, gives equal or even better results than CPB. Therefore, for networks with medium or large number of nodes, the use of MIS is recommendable, as it provides an acceptable value on congestion with a little computing time.

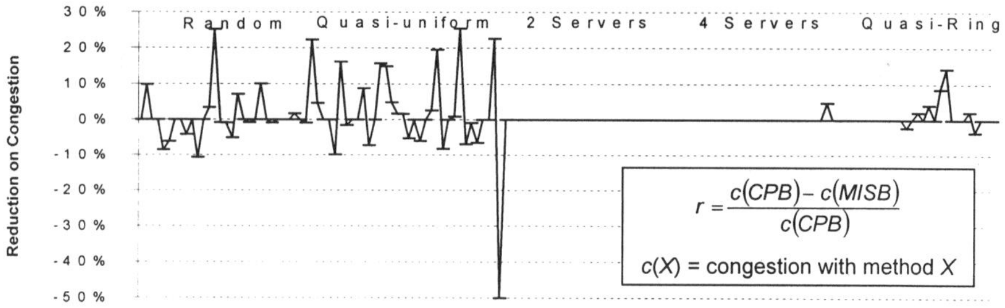

Figure 1: Congestion reduction *r (%)* provided by MISB when compared to CPB.

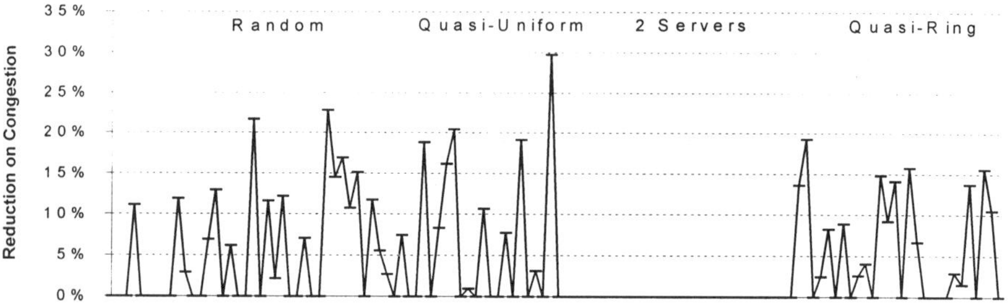

Figure 2: Congestion reduction provided by MIS when compared to CP.

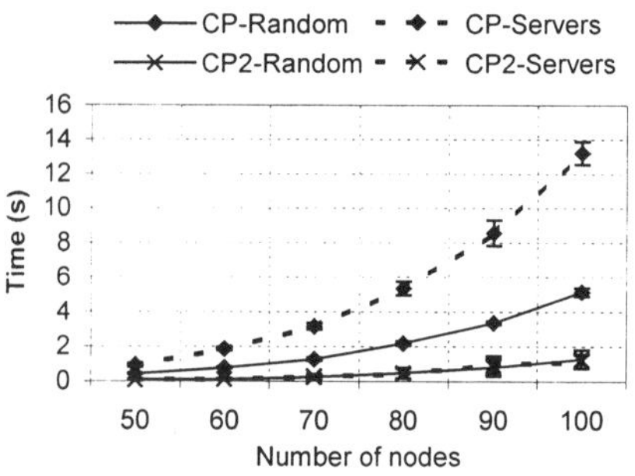

Figure 3: Solving times of CP and CP2

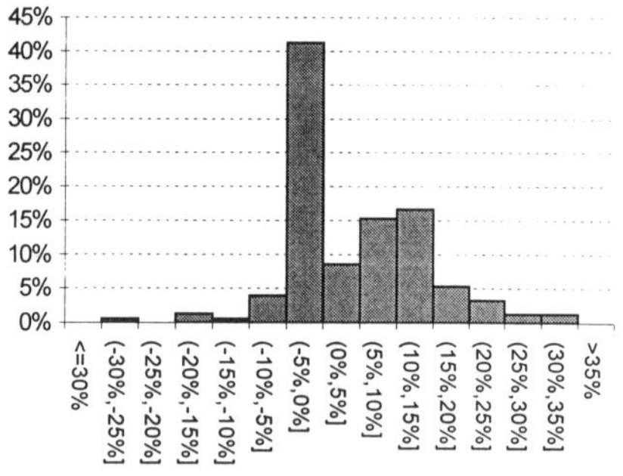

Figure 4: Congestion reduction of CPB vs. MIS

References

[1] R. Ramaswami and K.N. Sivarajan. *Optical Networks: A practical Perspective.* The Morgan Kaufmann Series in Networking. Morgan Kaufmann Publishers, Inc. San Francisco, 1998.

[2] I. Chlamtac, A. Ganz, G. Karmi, Lightpath Communications: An Approach to High Bandwidth Optical WAN's, *IEEE Transactions on Communications*, **40** (1992) 1171–1182

[3] R. Ramaswami, Multiwavelength Lightwave Networks for Computer Communication, *IEEE Communications Magazine* **31** (Feb. 1993) 78–88

[4] O. Gerstel, On the Future of Wavelength Routing Networks, *IEEE Network.* **10** (Nov./Dec. 1996) 14–20

[5] B. Mukherjee, D. Banerjee, S. Ramamurthy and A. Mukherjee, Some Principles for Designing a Wide-Area WDM Optical Network, *IEEE/ACM Transactions on Networking.* **4** (1996) 684–696

[6] R. Ramaswami and K.N. Sivarajan, Design of Logical Topologies for Wavelength-Routed Optical Networks, *IEEE J. on Selected Areas in Communications,* **14** (1996) 840–851

[7] J.-F.P. Labourdette and A.S. Acampora, Logically Rearrangeable Multihop Lightwave Networks, *IEEE Transactions on Communications.* **39** (1991) 1223–1230

[8] M. Berkelaar and J. Dirks. *Lp_solve: The mixed integer linear program solver.* Accessed at Feb. 2000 from ftp://ftp.es.ele.tue.nl/pub/lp_solve/

[9] E. Leonardi, M. Mellia, and M. Ajmone Marsan, Algorithms for the Logical Topology Design in WDM All-Optical Networks, *Optical Networks Magazine*, **1** (Jan. 2000) 35-46

Broadband Access, WDM Metro and Network Management
D.W. Faulkner and A.L. Harmer (Eds.)
IOS Press, 2000 © AKM

Analysis of the "Dispersive" Transmission of Chirped Digital Optical Signals

S. Betti (*), E. Bravi (**), M. Giaconi (***)

(*) Università degli Studi de L'Aquila - Dipartimento di Ingegneria Elettrica - Poggio di Roio, 67040 L'Aquila - ITALY

(**) Fondazione Ugo Bordoni, viale Europa 190, 00144 Rome - ITALY

(***) Scuola Superiore G. Reiss Romoli, via G. Falcone 25, 67100 L'Aquila - ITALY

Abstract. In optical transmission systems, the propagation of high-speed digital signals is affected by the joint action of the AM-FM conversion due to the nonlinear chirping effect in semiconductor laser and the chromatic dispersion of the optical fiber. The optical signal turns to be changed in shape in such a way that the performance can be severely constrained. In this paper, the "dispersive" propagation of chirped digital signals in optical fiber is evaluated by an analytical model, which overcomes asymptotic approaches of previous models and, unlike simulative approaches, permits us to maintain sensitiveness to the physical parameters governing propagation and to follow the evolution of the propagating optical pulse.

1. Introduction

The transmission of high speed digital signals in a "dispersive" optical fiber [1], jointly to AM/FM conversion due to semiconductor laser chirping effect [2] gives rise to distortions, which constrain the performance. Optical fiber propagation of chirped pulses has been so far analyzed assuming Gaussian shaped pulses [3, 5]. However the assumption of linear frequency chirp imposed on the pulse is limited to high values of the laser extinction ratio. In actual high bit-rate optical communication systems, where the extinction ratio is low [4], classical models [3] can lead to an erroneous evaluation of the pulse distortion. The proposed model takes into account both the finite value of the extinction ratio and the nonlinear frequency chirp: closed expressions are derived for the evaluation of the "dispersive" propagation of chirped optical pulses. Unlike approaches based on numerical simulations, it permits us to maintain direct sensitiveness to the physical parameters governing pulse propagation and to realize the physical effects giving rise to pulse deformation.

2. Propagation in "Dispersive" Optical Fibre of Optical Pulses Affected by Semiconductor Laser Chirping

The propagation of Gaussian shaped pulses is a reference model in high-speed, digital optical transmission systems [3, 5] therefore, in the case of direct modulation, the signal emitted by a semiconductor laser can be expressed as

$$v(t)=\sqrt{S(t)}\cos[\omega_0 t+\phi(t)]=\sqrt{S_m \exp[-(t/T)^2]+S_b}\,\cos[\omega_0 t+\phi(t)] \qquad (1)$$

where ω_0 is the central angular frequency, $\phi(t)$ the phase modulation due to chirping, S_m the peak power of the optical pulse, S_b the optical power corresponding to the bias current, and T the pulse time constant. If the signal bandwidth is lower than the relaxation-oscillation frequency of the semiconductor laser, the angular frequency shift due to chirping is given by [5]

$$\delta\omega(t)=\frac{d\phi(t)}{dt}\approx\frac{\alpha}{2}[\frac{1}{S(t)}\frac{dS(t)}{dt}] \qquad (2)$$

where α is the laser linewidth enhancement factor. From (2) the following nonlinear expression can be derived for $\delta\omega(t)$

$$\delta\omega(t)=\frac{d\phi(t)}{dt}\approx-\frac{\alpha t}{T^2}[\frac{1}{1+\varepsilon\exp(t/T)^2}] \qquad (3)$$

with $\varepsilon=S_b/S_m$ (0.1-0.15) [4], which is reported in Fig.1, where the Gaussian pulse S(t) is shown, as well. It is worth noting that $\delta\omega(t)$ shows a linear behaviour in the central portion of the pulse but clearly not in correspondence of the pulse tails. This leads to changes of the group velocity that are no more proportional to time. Therefore, unlike the model reported in [3], it is to be expected that only the central portion of the pulse is affected by uniform deformation while this does not hold any more for rise and fall edges, which are affected by nonuniform deformation. Such effect arises from the phase modulation $\phi(t)$ due to chirping, expressed by

$$\phi(t)=\frac{\alpha}{2}\ln[1+\frac{1}{\varepsilon}\exp[-(t/T)^2] \qquad (4),$$

which is not tractable easily. It is possible to consider an expression that can be analytically managed in simpler way. As a matter of fact, the following Gaussian shaped function

$$\phi(t)\approx\frac{\alpha}{2}\,k\cdot\exp[-(t/T)^2/k] \qquad (5)$$

with $k=\ln(1+1/\varepsilon)$, is an approximation for $\phi(t)$, which is identical to the exact espression (4) in the pulse central interval and close to (4) in correspondence of its tails.

Therefore, the baseband equivalent signal propagating in optical fiber is angularly modulated by chirping and results in

$$v(t)=\sqrt{S_m\exp[-(t/T)^2]+S_b}\cdot\exp[j\frac{\alpha}{2}k^{e[-(t/T)^2/k]}]\approx$$
$$\approx\sqrt{S_m}\{A(\varepsilon)\cdot\exp[-(t/T)^2/2B(\varepsilon)]+\sqrt{\varepsilon}\}\cdot\exp[j\frac{\alpha}{2}k\cdot e^{[-(t/T)^2/k]}] \qquad (6)$$

where $A(\varepsilon)=[(1+\varepsilon)^{1/2}-\varepsilon^{1/2}]$, $B(\varepsilon)=(1+\varepsilon)^{1/2}[(1+\varepsilon)^{1/2}-\varepsilon^{1/2}]$ and the finite value of $\varepsilon=S_b/S_m$ has been considered to overcome the excessively simplified assumption εÅ0 [3]. From (6) the following expression can be derived

$$v(t) \approx \sqrt{S_m}\ \{A(\varepsilon)\cdot \exp[-(t/T)^2/2B(\varepsilon)] + \sqrt{\varepsilon}\}\cdot \sum_{n=0}^{\infty}\left(\frac{j\alpha k}{2}\right)^n \frac{\exp[-\frac{n}{k}(t/T)^2]}{n!} \tag{7}$$

in which phase and quadrature terms can be envisaged according to the value of j^n. It is worth noting that such terms are not singolarly phase modulated by chirping, whose effect arises from their interaction. The previous relationship can be rewritten as follows

$$v(t) \approx \sqrt{S_m}\ A(\varepsilon)\sum_{n=0}^{\infty}\left(\frac{j\alpha k}{2}\right)^n \frac{\exp[-\frac{1}{2}(t/T)^2(\frac{2n}{k}+\frac{1}{B(\varepsilon)})]}{n!} + \\ +\sqrt{S_m}\ \varepsilon\cdot \sum_{n=0}^{\infty}\left(\frac{j\alpha k}{2}\right)^n \frac{\exp[-\frac{1}{2}\cdot\frac{2n}{k}(t/T)^2]}{n!} \tag{8}$$

which shows how also the term corresponding to bias contributes to the overall distortion due to chirping. Such effect is not pointed out in a simplified analysis, which assumes $\varepsilon=0$. Propagation along an optical fiber of length z is characterized by a quadratic phase response, that is [4]

$$H(\omega) = A\exp[\frac{j}{2}\beta_2\omega^2 z] \tag{9}$$

where $A=10^{-az/10}$, with a the optical fiber attenuation, $\beta_2=-\lambda^2 D/2\pi c$, D being the chromatic dispersion coefficient (ps/nm·km), λ the optical carrier wavelength and c the light velocity. Assuming $x=t/T$ and introducing the adimensional parameter $\mu=\beta_2 z/T^2$, the following expression can be derived for the received optical signal w(t) [3]

$$w(x,\mu) = \sqrt{S_m}\ A(\varepsilon)\sum_{n=0}^{\infty}(\frac{j\alpha k}{2})^n\{\frac{1}{[1-j\mu\cdot(2n/k+1/B(\varepsilon))]^{1/2}}\}\exp\{-\frac{x^2\cdot[2n/k+1/B(\varepsilon)]}{2[1-j\mu\cdot(2n/k+1/B(\varepsilon))]}\}\frac{1}{n!} + \\ +\sqrt{S_m}\varepsilon\sum_{n=0}^{\infty}(\frac{j\alpha k}{2})^n\{\frac{1}{[1-j\mu\cdot(2n/k)]^{1/2}}\}\exp\{-\frac{x^2\cdot(2n/k)}{2[1-j\mu\cdot(2n/k)]}\}\frac{1}{n!} \tag{10}$$

The series (10) can be summed giving rise to a function that can be developed with respect to μ, resulting in

$$w(x,\mu) = w(x,0) + \frac{d}{d\mu}w(x,\mu)\Big|_{\mu=0}\mu + \frac{d^2}{d\mu^2}w(x,\mu)\Big|_{\mu=0}\frac{\mu^2}{2!} + \ldots \tag{11}$$

where the terms can be expressed by closed analytical relations. The sign of β_2 has obviously repercussions on μ giving rise, for $\beta_2<0$, to alternate sign functions in (11).

In Fig.2 the signal power $|w(t)|^2$, normalized to S_b, resulting from (10), is shown for different values of the fiber length z, and compared with a classical numerical approach based on FFT. With increasing z, the pulse changes its shape and becomes no more Gaussian. Because of nonlinear chirping, for a sufficiently high value of z, the pulse is not only Gaussian any more, but it turns out severely distorted. The splitting-like effect shown for z=60 km can not be explained by classical models assuming linear chirping [3] and such distortion effects could represent an ultimate limit to very high bit-rate optical fiber transmission systems.

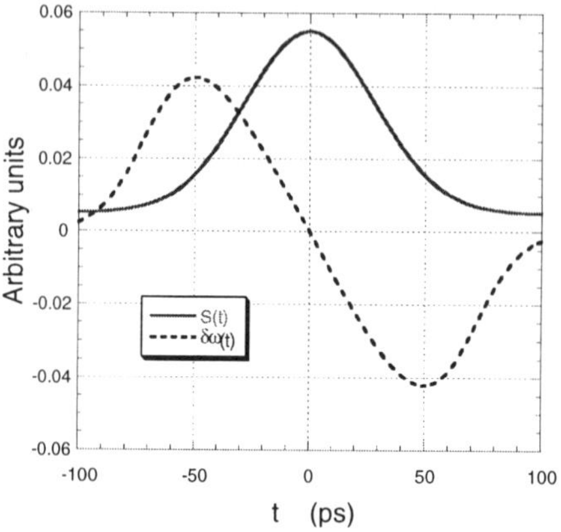

Fig.1 Gaussian pulse S(t) and nonlinear chirping (ε=0.1; T=40 ps; α=2).

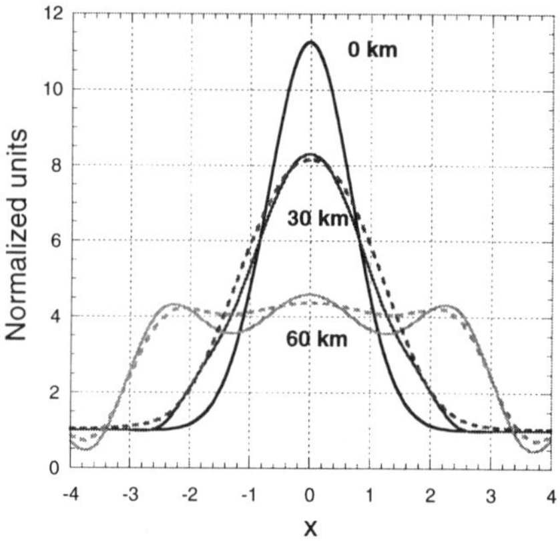

Fig.2 Normalized signal power for different fiber lengths: proposed model (dashed curve); numerical solution based on FFT (solid curves) (ε=0.1; β_2=20 ps^2/km).

3. Conclusions

So far, asymptotic approximations have been used to model the "dispersive" propagation of chirped, digital pulses in optical fiber. Previous models assumed linear chirping and high values of the semiconductor laser extinction ratio. In high bit-rate systems, such parameter is instead low and this imposes more accurate analysis. In this paper, a model is presented that takes into account the real phase modulation due to nonlinear chirping and analyzes its effect on the evolution of the pulse along the fiber. In particular, the analysis shows that, for realistic values of the laser extinction ratio, the propagating pulse changes from its initial Gaussian shape and turns out severely distorted.

References

[1] G.J. Meslener, "Chromatic dispersion induced distortion of modulated monochromatic light employing direct detection", IEEE-J. Quantum Electron., vol. QE-20, 1984, pp.1208-1216.

[2] T.L. Koch, J.E. Bowers, "Nature of wavelength chirping in directly modulated semiconductor lasers", Electr. Lett., vol. 20, 1984, pp.1038-1039.

[3] G.P. Agrawal, *Fiber-Optic Communication Systems*, New York, John Wiley & Sons, Inc. 1992.

[4] ITU-T Recommendation G.957 (07/95).

[5] G.P. Agrawal, N.K. Dutta, *Long-Wavelength Semiconductor Lasers*, New York, Van Nostrand Reinhold Company Inc., 1986.

Broadband Access, WDM Metro and Network Management
D.W. Faulkner and A.L. Harmer (Eds.)
IOS Press, 2000 © AKM

Experimental Analysis of Distortion Effects in Analog Externally Modulated Lightwave Systems

E. Bravi (*), F. Curti, Member IEEE (*), S. Betti (**), M. Giaconi (***)

(*) Fondazione Ugo Bordoni, viale Europa 190, 00144 Rome (Italy)
(**) Università degli Studi di L'Aquila, Dipartimento di Ingegneria Elettrica, Poggio di Roio, 67040 L'Aquila (Italy)
(***) Scuola Superiore G. Reiss Romoli, via G. Falcone 25, 67100 L'Aquila (Italy)

Abstract. A generalized analysis for the evaluation of intermodulation distortions due to nonlinear response of a Mach-Zehnder external intensity modulator has been proposed. The analysis holds for every modulating signal that can be assumed as a Gaussian process (e.g. SCM, CDMA). In this paper an experimental confirmation of the theoretical results and a proposal for standard definition are presented.

1. Introduction

Hybrid fiber-CDMA (Code Division Multiple Access) cellular radio networks and SCM (SubCarrier Multiplexing) systems are promising to meet the increasing demand for mobile radio communications and for optical access networks, respectively. In order to overcome some of the problems arising in these systems when using directly modulated semiconductor laser [1, 2], electro-optic external intensity modulators [3, 4] can be adopted. This device is robust against distortion effects since it can be biased in such a way that second order distortion components are nulled out.

In this letter, a recently proposed theoretical model for the performance evaluation of systems based on external intensity modulators [4] is used to explain experimental results obtained in the case that the modulating signal can be assumed as a Gaussian process. This represents a good approximation for CDMA and SCM signals. As a matter of fact, starting from the central limit theorem, it can be demonstrated that the more the channels number is high, the more the resulting signal is Gaussian. Typically, for N>10 the drive signal can be modelled as a Gaussian random process. This approach has been already adopted in the performance evaluation of SCM lightwave systems [1, 2] and hybrid fiber-CDMA radio systems [5].

2. Distortion Effects due to External Modulator Nonlinearities

In an optical link using external intensity modulator, an important source of nonlinearities is due to the optical power-voltage response of this device (e.g., integrated optic $LiNbO_3$ Mach-Zehnder modulator), which can be expressed as [4]

$$w(t) = w_0\{1+\sin[x(t)]\} \tag{1}$$

where w_0 is the average optical output power, $x(t)=\pi v(t)/v_\pi$ with v(t) the applied RF voltage and v_π the half-wave voltage, that is the voltage required to achieve a π optical phase shift.

Since x(t) can be assumed as a Gaussian process, the autocorrelation function $R_y(\tau)$ of the process $y(t)=w(t)-w_0$ can be derived by Price theorem, resulting in

$$R_y(\tau) = w_0^2 e^{-\sigma_x^2} \sinh[R_x(\tau)] = w_0^2 e^{-\sigma_x^2} \sum_{n=0}^{\infty} \frac{R_x^{2n+1}}{(2n+1)!} \tag{2}$$

with $\sigma_x^2=\pi^2\sigma_v^2/v_\pi$, σ_v^2 being the variance of the Gaussian process v(t). From (2), the power spectral density $\Phi_y(\omega)$ can be determined, resulting in

$$\Phi_y(\omega) = w_0^2 e^{-\sigma_x^2} \sum_{n=0}^{\infty} \frac{[\phi_x(\omega)\otimes\phi_x(\omega)\otimes..]_{(2n+1)\text{times}}}{(2n+1)!} \tag{3}$$

where $\otimes$ indicates the convolution operation.

The series expansion (2) for $R_y(\tau)$ and (3) for $\Phi_y(\omega)$ take into account the useful signal and all the distortion components due to device nonlinearity.

3. Experimental Set-up

Distortions due to the nonlinear optical power-voltage response depend on the modulating signal power. The experimental set-up is shown in Fig.1: it has been used to analyze the evolution of the power spectral density of the overall output signal. The white Gaussian modulating signal is limited within the band from 350 MHz to 650 MHz. In order to measure in-band distortions, a narrow hole (10 MHz) is generated by a tunable notch filter. This avoids that spurious second order distortion components, due to bias voltage shifting, affect the measurement frequency. The modulating signal is generated by a white noise generator, a pass-band filter, an amplifier, a variable attenuator and a notch filter. This signal modulates the optical carrier generated by a laser diode by means of an external Mach-Zehnder modulator. The optical signal is then detected by a standard optical receiver.

The measurements have been performed with different values of the modulating signal power. Particular care is required to optimize the bias of the external modulator in order to avoid second-order distortions. Notch filter is centered at 620 MHz. The modulating signal power is selected in the range from 2 dBm/MHz to 16 dBm/MHz. A comparison between theoretical model and experimental data is reported in Fig.2, where the ratio signal/noise+intermodulation against modulating signal power is shown. Because of the nonlinear response, with increasing the modulating signal power, the overall distortion increases.

4. Proposal for Standard Definition

The use of Gaussian distributed modulating signals leads to a simple mathematical model for the performance analysis of systems that can be dealt with as analogue. The measurement instrumentation required when using Gaussian distributed signals as test signal is not complex and provides reliable and repeatable results. Moreover, a Gaussian signal, which is white in the useful band, leads to test the system behaviour uniformly within the frequency interval of interest. From an experimental point of view, instrumentation for system testing based on sinusoidal carriers as test signal, is more complex and expensive, and this technique does not simply lead to reliable and repeatable values of the system parameters that are required by a standardization procedure. Moreover, the approach based on Gaussian distributed signal has been already adopted for FDM (Frequency Division Multiplexing) radio systems, similar to SCM systems, after that several attempts based on sinusoidal carriers failed.

In conclusion, even in this case the use of Gaussian distributed signals as test loads seems a good and feasible method for standard definition.

5. Conclusions

Distortions due to the nonlinear response of external intensity modulators is analyzed in the case of signals that can be dealt with as analouge (e.g., SCM, CDMA systems). An analytical model,

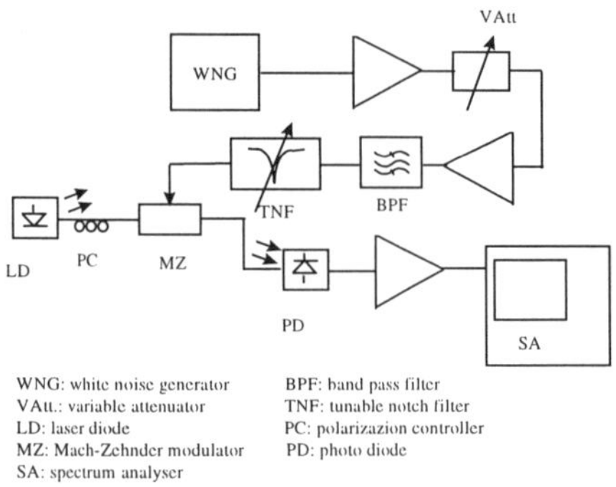

Fig.1 Experimental set-up.

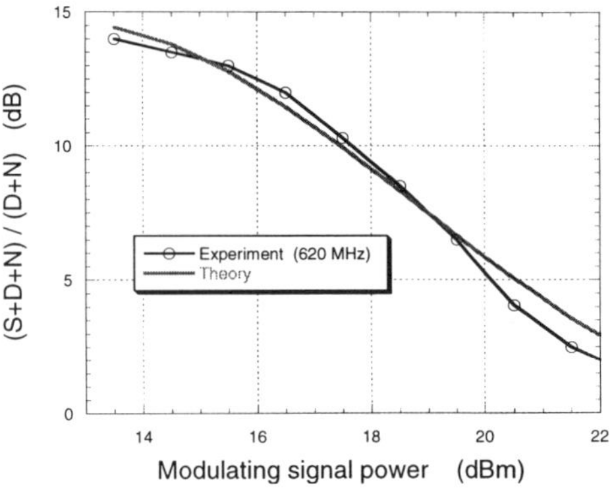

Fig.2 (S+D+N) / (D+N) versus modulating signal power at the frequency of 620 MHz

obtained assuming the modulating signal as a Gaussian process, is used to explain experimental results. The model leads to propose the use of Gaussian distributed signals as test loads for standard definition.

References

[1] S. Betti, E. Bravi, M. Giaconi, “Nonlinear Distortion due to the “Dispersive” Transmission of SCM Optical Signals in the presence of Chirping Effect: an accurate Analysis”, *IEEE Photonics Technology Letters*, vol. 9, n° 12, pp 1640-1642, December 1997.

[2] S. Betti, E. Bravi, M. Giaconi, "Effects of Intermodulation Distortions due to the Joint Action of Dynamic Chirping and "Dispersive" Transmission of SCM Optical Signals", *IEEE Photonics Technology Letters*, vol. 11, n° 6, pp 680-682, June 1999.

[3] R. Olshansky, "Optimal design of subcarrier multiplexed lightwave systems employing linearized external modulators", *IEEE J. Lightwave Technology*, vol. 10, n° 3, pp 378-382, March 1992.

[4] S. Betti, E. Bravi, M. Giaconi, "Analysis of Distortion Effects in Subcarrier-Multiplexed (SCM) Externally Modulated Lightwave Systems: A Generalized Approach", *IEEE Photonics Technology Letters*, vol. 9, n° 1, pp 118-120, January 1997.

[5] S. Betti, A. Bizzotto, E. Bravi, M. Giaconi, “Effect of Semiconductor Laser Intrinsic Dynamic Distortions in Hybrid Fiber-CDMA Cellular Radio Networks”, Microwave and Optical Techn. Letters, vol.20, n°3, pp.207-211, February 1999.

Managing Convergent Networks

Rajesh Thomas
WIPRO Technologies, K-312, 5th Block Koramangala, Bangalore – 560095, India

Satish Singla
WIPRO Technologies, K-312, 5th Block Koramangala, Bangalore – 560095, India

Surbhi Sinha
WIPRO Technologies, K-312, 5th Block Koramangala, Bangalore –5600 95, India

Abstract. The boom in Internet technology is changing the way today's networks are evolving. The drive towards a data centric network has given birth to technologies like ATM, ATM over SONET, IP over SONET etc. This has made the management of these kind of networks extremely complex. Bandwidth crunches has now led to the evolution of the DWDM technology adding an optical layer over the existing SONET/SDH layers. This has added an additional level of complexity managing the wavelength provisioning in this layer. This paper analyses the challenges that the current network management is facing and how the network management would evolve as we have more complex networks in future.

The drive today is towards a completely convergent network where voice and data traffics are merging. The future is now in the DWDM and Optical Networking technologies to quench the thirst for speed and bandwidth. The various technologies enabling this are ATM over SONET, IP over SONET, IP over Fiber etc. This has been made possible by the advances in the devices that are now commercially available in the market. Equipment vendors have come up with native ethernet interfaces to Optical Networks. NORTEL's interWAN and CISCO's DPT technology are in this direction. Ciena also offers its DirectConnect technology, which enables service providers to transparently transfer all kinds of network traffic – voice, video/data. The future of the optical networking lies in the MEMS and OXC techlogies. Tellium is doing extensive work in the area of OXCs. Corvis claims that it's suite of products can transport 400 Gbps per fiber (160 channels) upto 3200 kms without regeneration. Avici is one of the start-ups aiming at creating very high-speed IP routers, the Terabit Switch Router (TSR) would be capable of handling 400 or more OC-48 ports and/or lesser number of OC-192 ports. So the drive for a convergent network is not a dream for the future and is very much a reality today.

The networks are largely dependant on the network operators and their interests greatly influence the future of these networks. Due to the recent mergers & acquisitions the business model of the service providers is also changing. Vertical service integration of IXCs, ISPs & IECs is causing new network management challenges. Traditionally, IXCs & ISPs had their own network management system. Due to the mergers it would be rational to have unified network manager which would in turn cause efficient usage of services and lead to cost effectiveness. The acquisitions of Sprint by MCI-Worldcom and Cerent by CISCO are all in this direction.

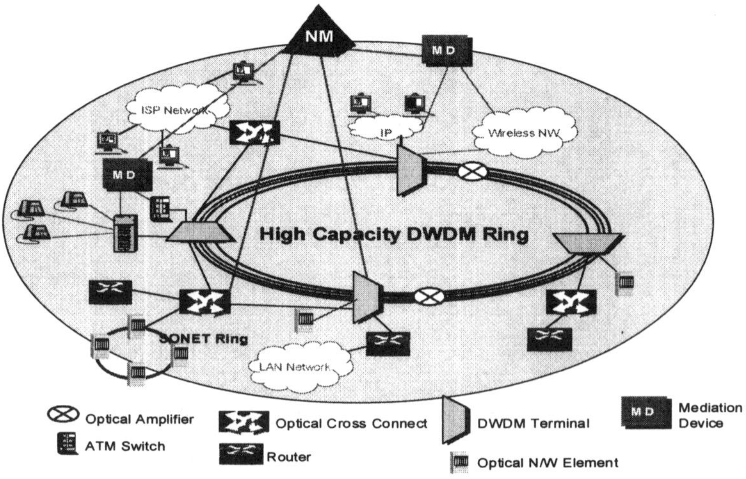

Figure 1 : Next Generation Convergent Network

Talking about Netwok management, we need to address each of the key issues i.e. the FCAPS. The mergers of the service providers have led to the need for integrated OAM&P on these networks. Let us analyse each of these areas in detail and see what kind of solutions would eventually evolve in each of these area.

Fault Handling - The key issues addressed here are fault detection, correlation and isolation. Most of the current DWDM products are promising the less than 50 ms fault detection using SONET. Evolving wavelength ADMs would be able to do the restoration at the optical level without using SONET. The Integrated Network Management system should have necessary intelligence to handle interface issues between different kind of devices like the ATM & SONET interface, IP router & optical devices interface etc.

Commisioning / Configuration – This is going to be a real challenge for the Network Manager(NM) with DWDM and optical layer getting added. A futuristic network as depicted in Figure 1, shows the variety of interfaces that this NM needs to handle. A simple drag and drop should be able to provision a trail through the complete network. E.g. the operator should be able to provision a OC12 connection starting from access network, getting added as a wavelength in DWDM ring and dropped onto another OC48 ring. With OXCs and WADMs evolving, there will be an obvious need for wavelength management capability. As in SONET timeslot interchange(TSI), timeslot assignment(TSA) and ADD/ DROP is possible, a similar function in the optical layer would be required.

Administration – This function as it mainly involves with billing, could continue to remain distributed. A convergence in this might be too involved. The billing would still be a stand alone component.

Performance Monitoring -. Performance monitoring at the network level could be used as a strong tool to identify network problems before the fault actually occurs. Such a set-up would enable the service providers to be more proactive in administering their networks and also providing the QoS for the various diverse services being offered.

Security – The security issues with regards this is well understood and is handled better at the individual ISP levels.

Having analysed all this, the possible approaches for a Convergent Network management as shown in Figure 1 could be

1. Mediation Devices – This is the simplest possible and the most widely deployed approach where we would have mediation devices to take in the various traffic. This could have interfaces with the various service providers. So these mediation devices would require
   - One network manager with generic interfaces
   - Network manager should be highly modular. This would enable use of different components for each type of network devices.
   - NM would be very bulky with lot of interface ports.
2. Intelligent Network Managers – Intelligence could be built into the Network Manager that efficiently handles the OAM&P issues.
   - The main issues that we face here are intelligent wavelength management. This is something that needs to be worked on. Real time wavelength provisioning is still to be a reality.
   - If in future a need based bandwidth allotment becomes possible in SONET/SDH based systems, the network manager would be required to handle this also. Existing components need to be enhanced to handle this.
   - A complete Optical Cross Connect would enable the provisioning of any kind of configuration whereby all the Network elements in the metro Network could be terminated on this one OXC and any configurations can be provisioned in real time from the NM.
   - Standalone software components could be developed which can interface with Network Management System (NMS) for enhanced fault correlation and isolation, collection and interpretation of PM data.

Broadband Access, WDM Metro and Network Management
D.W. Faulkner and A.L. Harmer (Eds.)
IOS Press, 2000 © AKM

Improving the cost/performance ratio in multicast communications

Andrea Borella, Giovanni Cancellieri
Dipartimento di Elettronica ed Automatica, Università di Ancona, 60131 Ancona, Italy

Abstract. The problem of including QoS requirements in the multicast routing protocols is not yet resolved. In this paper a new algorithm, called CDSAMA, is proposed to match either performance or cost requirements. Its behavior is compared with that of alternative existing techniques.

1. The multicast problem

QoS considerations for a multicast application include tolerance to jitter, delay, and lost packets. In order to satisfy an application QoS requirements, a protocol must be able to reserve and control network resources. This implies that finding the shortest path to a destination when constructing a delivery tree is no longer sufficient. In fact, the route must satisfy other QoS constraints as well. This adds considerable complexity to the routing task. A common metric used by routing algorithms is a count of the number of hops in the path from the source to the destination. In order to satisfy QoS requirements when setting up a route, parameters such as delay, bandwidth, loss probability, cost, and jitter must be considered as well. Since it is difficult to satisfy multiple constraints simultaneously, a possible approach would be to define an order in which the various parameters are processed and to evaluate the second parameter, for example, only for those paths considered best with respect to the first parameter. In essence, three quality parameters can be considered in setting a multicast connection. More precisely it is advisable to minimize [1]:

1. the maximum delay from source to any destination,
2. the maximum difference in the delays from source to destinations,
3. the connection cost.

Having assumed the cost as the ratio between the amount of bandwidth resources utilized to transmit information and the available bandwidth, we can define the maximum source-destinations delay Δ_{max} as the age of obsolescence of the packet's information, which is reached at time Δ_{max} after the transmission of the packet itself. Moreover, the maximum spread of the source-destination delay δ_{max}, also called differential delay, is the maximum difference of the end-to-end delay for every couple of destinations belonging to the same multicast group.

Inside a network N=N(V,A), where V is the set of nodes and A is the set of links, let us consider a routing tree T, with cost C(T), corresponding to a multicast group G(T) with source s. For every node $u \in G(T)$ a route P(s,u) exists in T. Saying r a generic branch of T with cost c(r) and correspondent average delay D(r), the above conditions may be analytically written as [2]:

$$C(T) = \sum_{r \in T} c(r) \le C_{max} \qquad \text{(cost constraint)} \qquad (1)$$

$$\sum_{r \in P(s,u)} D(r) \le \Delta_{max} \qquad \text{(maximum delay constraint)} \qquad (2)$$

$$\max_{u,v \in G(T)} \left| \sum_{r \in P(s,u)} D(r) - \sum_{q \in P(s,v)} D(q) \right| \le \delta_{max} \qquad \text{(differential delay constraint)} \qquad (3)$$

where C_{max} is the maximum acceptable cost of a multicast connection. According to the previous definitions, the values of C(T) and C_{max} are upper bounded by the number of the links of the network. Starting from the knowledge of the three parameters Δ, δ and C_{max}, the heuristic algorithms that we have developed work to find a sub-optimal multicast routing tree. In this context, they offer a possible way of obtaining reasonably good performance considering either the service provider necessities (which needs to increase the resource utilization at the most) or the user point of view (that asks for high QoS levels).

2. Multicast routing by CDSAMA

The multicast routing algorithm that we propose here is called CDSAMA (Cost and Delay Sensitive Applications – Multicast Algorithm) [3].

Having defined the source of a multicast communication session, the multicast group (including the destinations of the considered transmission) and the constraints for the parameter indicated at points 1., 2. and 3., CDSAMA chooses a minimum cost tree T by means of the Dijkstra algorithm. However in this case Dijkstra operates in this context considering the cost of the network links, instead of their proper delays. After that, among all the destinations, CDSAMA selects the one (node w) characterized by the minimum delay on the path that connects itself with the given source along T. Then the algorithm looks for other alternative s-w paths, not belonging to T but having relatively low costs. In any case, their own delays must be lower than the relative constraint. Such paths are found in a sub-network generated by erasing in N every link present on the s-w path in T, once per time.

For every path selected as illustrated above, and working on a subset of the network obtained by the original one by erasing the links and the nodes already belonging to the current tree, CDSAMA finds the best link connecting a new destination to the tree itself. The choice is taken according to the following rules:

a) if only one path is characterized by a maximum delay Δ with $\Delta \le \Delta_{max}$, the latter is selected;
b) if multiple paths exist, having a maximum delay lower than, or equal to, Δ_{max}, the new path is selected as the one that is the best with respect to the maximum differential delay δ, in comparison with all the other destinations already reached by the tree;
c) if multiple paths are characterized by either maximum delay $\Delta \le \Delta_{max}$ or maximum difference of delay $\delta \le \delta_{max}$, the path having the minimum cost is selected.

In this way, node after node, repeating the same procedure illustrated above, CDSAMA provides a multicast tree that connects all the destinations with the given source. In practice, when possible, the algorithm identifies a tree that is compatible with the constraints on the maximum and maximum differential delays. Among all the trees that satisfy such limits, it chooses the cheaper. When the algorithm fails, in any case, it suggests the choice of the closest solution to the optimal one

A possible modification of CDSAMA, here called CDSAMA2, is obtained when the first multicast tree T is generated by the Dijkstra algorithm considering the link delays, instead of their corresponding costs. It is evident that, privileging the delay requirements, CDSAMA2 accepts higher costs as a counterpart of a better performance. In other words, this approach reduce the network efficiency (that, in some way, is correlated to the tree cost) but causes a further improvement on the performance. In fact, the delay is considered as a primary quality parameter since the beginning of the procedure. In spite of this, simulations shows that the resulting costs are not too far from that offered by alternative procedures.

3. Numerical results

In this analysis the following routing algorithms have been compared: DVMA [4], CCDVMA[5], CDSAMA and CDSAMA 2. Each of them has been applied on the same set of randomly generated meshed WANs (composed by 30 nodes and with a bit rate on the links equal to 622 Mbit/s.), where a source addresses packets towards a multicast group at 0.5 Mbit/s. This multipoint connection is superposed on a background of uniformly distributed unicast traffic, generated by all the nodes (multicast source excepted) with a packet generation probability per slot that varies between 0.1 and 0.8 (step 0.1). The dimension of the multicast group has been considered equal to the 6 and 12, varying their distribution within the network randomly. The position of the multicast source is randomly selected as well.

The graphs shown in the sequel represent Δ (Fig. 1), δ (Fig. 2) and C (Fig. 3), as a function the unicast traffic (represented in terms of packet probability at the nodes). The parameters Δ and δ are measured in tens of time slots, whereas C is reported in centesimal units. In every figure two graphs are reported, relative to multicast group size equal to 6 and 12 respectively. Comparing the three subsequent pairs of plots, we can understand the difference among CDSAMA and CDSAMA2.

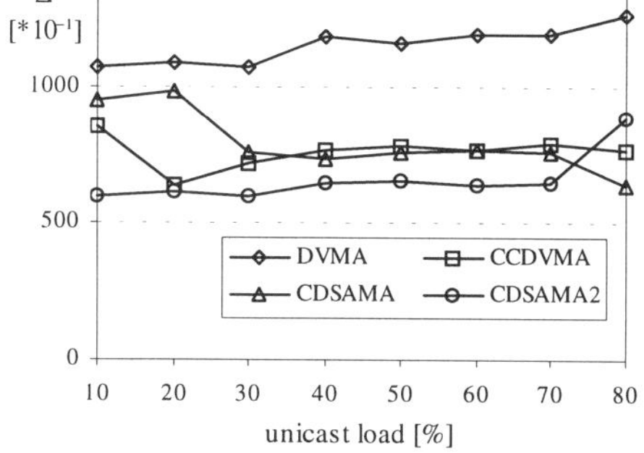

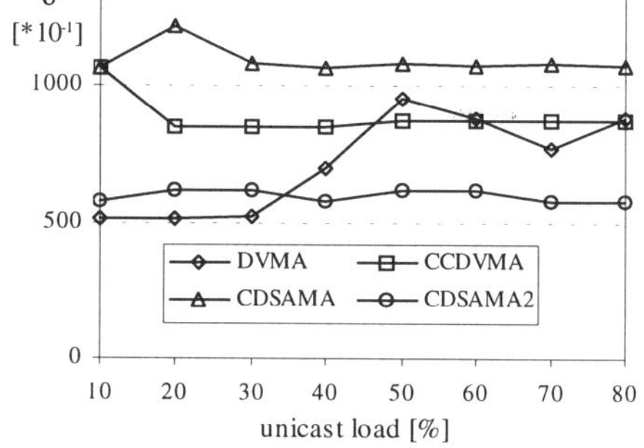

Fig. 1 - Maximum delay for 6 destinations (left) and 12 destinations (right).

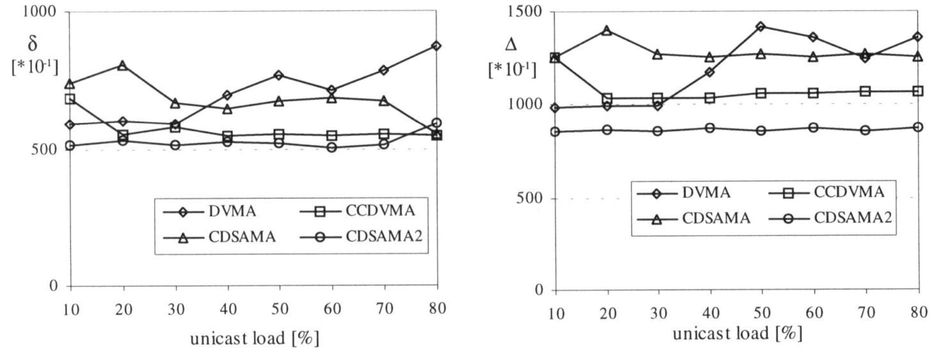

Fig. 2 - Maximum differential delay for 6 destinations (left) and 12 destinations(right).

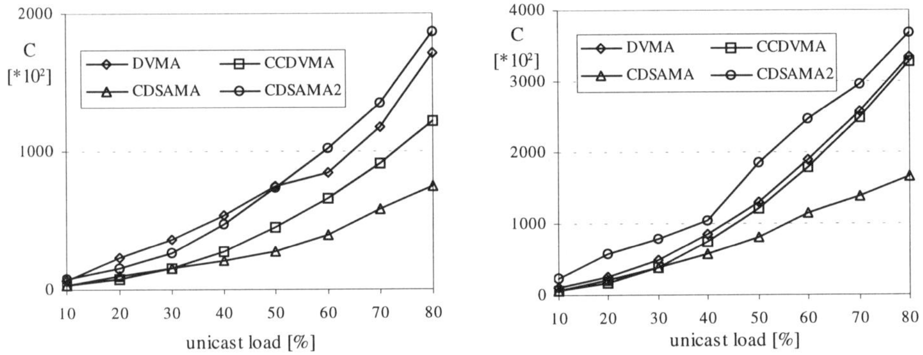

Fig. 3 - Connection cost for 6 destinations (left) and 12 destinations(right).

In fact CDSAMA2, beginning its tree construction with a Dijkstra procedure oriented to delay minimization, finds better multicast paths as regards Δ and δ. The opposite occurs to CDSAMA, prevalently influenced by cost constraints. Moreover, it is interesting the compared analysis with DVMA, which is an algorithm that does not take into account the cost parameter, and with CCDVMA that, instead, is cost constrained too. In fact despite its characteristics DVMA is not the best solution, even confining the comparison to the delay performance. CCDVMA, on the other hand, exhibit in general an intermediate behavior between our new routing strategies. Similar results, not reported here for the sake of brevity, can be obtained by simulations applied on a more extended network casuistry. They demonstrate the good performance of the proposed technique.

References

[1] A.Amorini, A.Borella, G.Cancellieri, "DVMA-CC algorithm for multicast connections in optical networks" (in Italian), Proc. of Fotonica 99, Trento, June 2-4, 1999, pp. 433-436.

[2] A.Borella, "A cost-delay constrained multicast algorithm for interactive multimedia applications", Proc. of NOC '99, Delft, June 22-24, 1999, pp. 278-284.

[3] A.Borella, G.Cancellieri, "Remote time-alignment of interactive services through efficient multicast algorithms", submitted for the presentation to 1st IEEE ECUMN'2000, Colmar, October 16-18, 2000.

[4] G.N.Rouskas, I.Baldine, "Multicast routing with end-to-end delay and delay variation constraints", IEEE Journal on Selected Areas in Communications, Vol.15, No.3, April 1997, pp.346-356.

[5] B. Haberman, G.N. Rouskas, "Cost, Delay, and Delay Variation Conscious Multicast Routing", Technical Report TR-97-03, North Carolina State University, 1997.

Design Considerations for a Slotted OTDM Ring-LAN

K. Bengi, S. Aleksic
Vienna University of Technology,
Institute of Communication Networks
Favoritenstrasse 9/388, A-1040 Vienna, Austria

Abstract. In this paper, two different node architectures for a high-speed packet-switched ring LAN based on the optical time-division multiplexing technique are proposed. Due to the incorporated low-speed packet headers the presented ring can be viewed as an optically (bit rate) transparent network. In both cases, the node can be implemented by using the current opto-electronic technology. Moreover, access protocols based on these proposed architectures are introduced employing the Slotted Ring scheme with destination release. Thereby, the specific processing delay requirements of the nodes are taken into account. Additionally, simulation results are shown for both configurations evaluating throughput and access delays as main performance measures. The paper reveals whether the deployment of additional (and expectably costly) OTDM multiplexers/demultiplexers at each ring node is really justified considering the system performance.

1. Introduction

New bandwidth-intensive applications such as supercomputer interconnections, interactive TV and multimedia banks will place severe demands on the network bandwidth available from a single-mode optical fiber in local and metropolitan area networks. Therefore, future photonic LANs are supposed to provide transmission speeds in the upper Gbit/s range on instantaneous demand with high medium efficiency and minimal delay. A very promising technique in this field is Optical Time Division Multiplexing (OTDM) which yields aggregate transmission rates beyond the reach of high-speed electronics. Used within LANs, the OTDM technique guarantees flexible and fast access to the medium.

However, at present, it is not possible in a medium access node located at the end-user to keep an optical signal completely in optical form. It has to be converted into the electrical domain, processed, saved and interpreted by electronics. For point-to-point connections between end-users within high-speed optical LANs, the payload must be converted to lower data rates for processing at the nodes. Therefore, by employing the optical packet switching technique implying that optical packets of constant duration consist of variable bit rate payloads and headers with fixed data rates, an optical transparent LAN may be developed [3].

In this paper, we present two different node architectures for such a high-speed packet-switched ring-LAN employing optical packet compression and decompression (bit-rate conversion) and also using OTDM multiplexer/demultiplexer units. Additionally, in a first step, we introduce static Medium Access Control (MAC) protocols based on these architectures. These fixed-assignment access protocols are based on the Slotted Ring scheme with destination release using several virtual channels and are appropriately taking the processing overhead of the considered node configurations into account.

2. Network Architecture

The proposed ultra high-speed OTDM ring-LAN is based on the slotted single-ring architecture. For saving costs, a single ring is considered in this paper, in contrast to the previous works [3,4], where counter-rotating double rings were analyzed.

In Fig. 1, the considered network architecture is demonstrated. The ring network operates at a medium rate of up to 100 Gbit/s and all attached nodes are supposed to be capable of attaining this data rate for payload transmission (differing from [4], where a heterogeneous network has been considered) by means of approaches given in section 3 below.

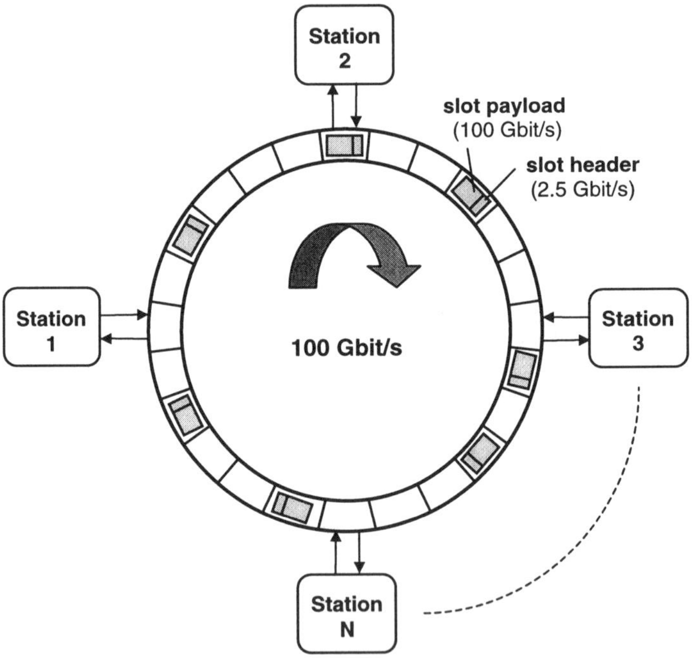

Figure 1: Network architecture of the proposed high-speed ring LAN

Fixed-period slots with duration T_s separated by guard bands circulate permanently on the ring and may carry data packets. Each slot consists of a *header* and a *payload* section, as depicted in Fig. 2. The slot header consists of an empty/full flag, destination address, source address, and additional protocol control flags. It may also contain information about the format, bit-rate, and type of the payload. The payload itself is bit-rate variable and may be transmitted at any data-rate up to 100 Gbit/s.

Transmission rate transparency is achieved by implementing a fixed lower bit-rate packet header. In this specific case, it is assumed to be 2.5 Gbit/s, which makes it possible to realize fully electronical header processing without need for data rate conversion. This feature makes a node cheaper and easier to realize. Moreover, the nodes without data rate conversion facilities can also read the packet header. Consequently, the packets can pass through the network transparently and can be routed not only by all-optical switches in a packet-switched all-optical network but also by electronic switches with optical interfaces [5].

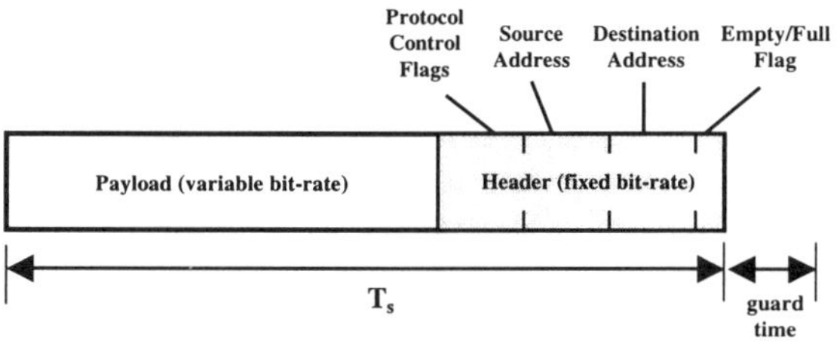

Figure 2: Slot format for the proposed high-speed ring LAN

In this paper, two design strategies for a high-speed (100 Gbit/s) node deploying optical packet payload compression and decompression for a high-speed packet-switched ring-LAN based on the OTDM technique are compared. In addition, we design and analyze a MAC (medium access control) protocol for the proposed ring network, in which the specific processing delay requirements given by the node architectures are considered.

3. Node Design

A slotted system with empty/full slot markers allows a design of a very simple high performance MAC protocol. It permits concurrent access to the media by multiple, geographically separated nodes. This principle advantage of slotted systems can be achieved only by medium access at the full media rate, but current limitations by electronics make a processing of a 100 Gbit/s bit-stream infeasible.

However, the existence of optical buffers and optical rate conversion systems enable the access nodes to receive a packet of a data stream at the media rate and convert it down to the acceptable rate for electronically processing and interpreting. As a consequence, the down conversion delay must also be taken into account. Two possible node architectures for a slotted high-speed ring LAN with 100 Gbit/s medium access are shown in Fig. 3.

In Fig. 3(a), nodes providing only one service type at a data rate of 2.5 Gbit/s and require optical packet compression (and decompression) in order to achieve a target transmission speed of 100 Gbit/s (and convert back to 2.5 Gbit/s) for an one-channel signal is presented. Indeed, it is possible to generate and process a 2.5 Gbit/s bit-stream electronically. There are customary components available for that (e.g., a synchronous FIFO memory with an access time less than 8 ns [6]). For serial/parallel and parallel/serial conversion, a 2.5 Gbit/s 20 bit transceiver [7] can be used.

Several methods of optical rate conversion (compression and decompression) have been proposed to date [8,9,11]. Most of these schemes are based on an optical buffer (optical storage loop) and a sampling technique. However, there are some restrictions concerning bit-rate and packet size. In [10], the storage of 9-kbit packets at 80 Gbit/s for over 30 seconds is reported along with the prediction that by using optical modulation, the operating rates of these fiber loop memories may be extended to 100 Gbit/s.

The optical packet compression scheme in [10] is based on the idea of rational harmonic mode-locking. The relation between the fundamental frequency of the storage ring f_0, the sampling clock frequency f_1, and the target high bit-rate f_2 is as follows:

$$f_2 = m \cdot f_1 = (m \cdot h - 1) f_0 \ ,$$

where m represents the required number of round-trips in the loop to obtain the desired bit rate and h denotes the number of bits entering the storage loop in each round-trip yielding the packet size mh - 1. In our case, we assume $h \leq 213$ and $m = 40$, so that the maximal slot size can be set to $l_{s,max} = (mh - 1) = (213 \cdot 40 - 1) = 8519$ bits (corresponding to 20 ATM cells encapsulated in one slot plus 39 bits reserved for additional control information exchange or a error-control method, e.g. CRC32).

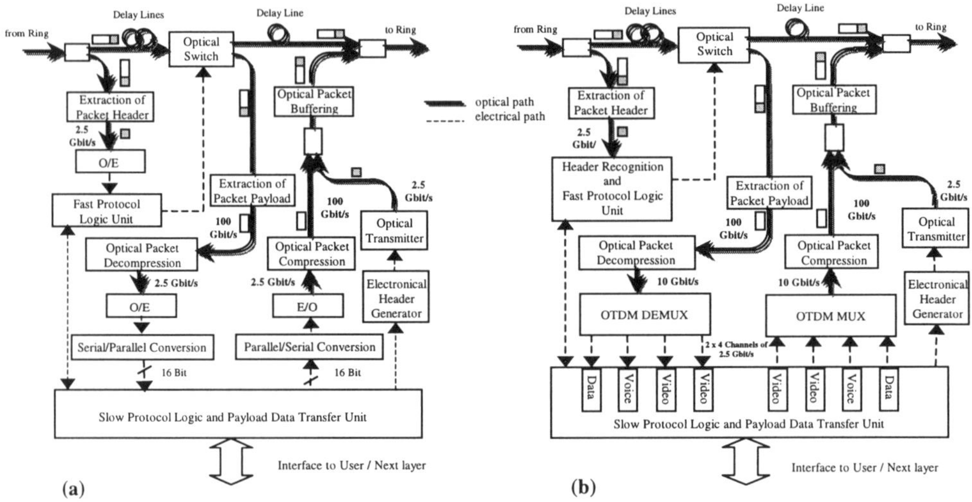

Figure 3: Two possible node architectures for a high-speed ring LAN with 100 Gbit/s medium access by use of the optical packet compression and decompression method

The functionality of the high-speed MAC node can be described as follows: the arriving slot is split into the *optical delay line path* and the *header processing path.* After the header has been recognized and extracted, the Fast Protocol Logic Unit processes it on the fly. The header processing, or its pipeline stage, must be fast enough so that it is able to finish header processing before the next header arrives. With a customary electronic technology it could be feasible to extract and process the main header information within the duration of 50 ns.

By the use of a $LiNbO_3$ or a Y-SOA switch (with a switching speed of the order 1 ns) we can design delay lines resulting in a header processing latency T_{hp} between 50 and 60 ns, corresponding to a packet length of 5kbit – 6kbit at 100 Gbit/s. Alternatively, by deploying a silica-based thermo-optic switch the delay lines must be larger than 1 ms.

If the destination address in the slot header matches the address of the node, the payload is extracted and converted down to 2.5 Gbit/s. If it is not the case, the packet simply propagates through the node. The Optical Packet Decompression Unit converts 100 Gbit/s (HDR-High Data Rate) packets to the lower bit rate 2.5 Gbit/s (LDR-Low Data Rate) for electronically saving and processing. Consequently, a new packet cannot be received until the previous packet has been completely converted. Due to this restriction, every 40th (100/2.5) slot of the slotted ring may be received. Note that the data rate conversion time is given by: $T_{cv} = l_s/\text{LDR} = l_s/2.5$.

Inserting a new packet into the ring-LAN can be realized by synchronically unloading the optical packet buffer if an empty slot arrives. The optical packet must be prepared (payload compressed and attached to its own header) and loaded into the optical packet buffer. Since optical buffers (optical storage loops) can store a full packet size for more than hundreds of microseconds, the insertion time T_{in} could be simply determined by the header processing time and the synchronization constraints, i.e., $T_{hp} - l_s \leq T_{in} \leq T_{hp} + l_s$.

Alternatively, in Fig. 3(b), the nodes' traffic is composed of four distinct service classes (channels), which have to be time-division multiplexed in a *bit-interleaved* manner (besides the packet interleaving on the ring). Due to the shorter data conversion time, $T_{cv} = l_s/\text{LDR} = l_s/10$ in this case, the time constraints for the corresponding access protocols can be relaxed. Consequently, each 10th slot is accessible for reception here.

4. Access Control Design

The medium access of the considered ring network is based on the Slotted Ring protocol employing *destination release* and *spatial slot reuse*. By allowing destination release and reuse of slots, the network throughput significantly exceeds the line rate of the medium. In contrast to alternative multiplexing methods such as WDM or bit-interleaved OTDM, the slotted OTDM scheme enables sharing a fast medium by access nodes capable of operating at maximally 100 Gbit/s rates.

However, due to the strict time constraints associated with the presented node architectures, the basic Slotted Ring scheme is not applicable any more. The limitations in terms of the packet reception given in section 3 lead to a static (fixed-assignment) Slotted Ring based access protocol, where each node is assigned a certain slot (or slots) for the reception of packets. In the case of node architecture (a), i.e. without OTDM facilities, a node can use every 40th slot for reception, whereas concerning the node configuration (b) deploying OTDM multiplexer/demultiplexer every 10th slot is available for reception.

Therefore, in our proposed access control scheme, the slots are successively numbered in the ring initialization phase and assigned to the individual nodes in an interleaved manner. That means, that for instance slots having the numbers 1, k + 1, 2k + 1, ... are uniquely dedicated to node 1, slots with the numbers 2, k + 2, 2k + 2, ... to node 2 and so forth, given that every kth slot can be emptied. As a result, we may speak of k virtual channels on the ring.

Consequently, a source node willing to send a packet to a specific destination node waits for the appropriate slot to arrive in order to ensure that the corresponding packet can safely be received by the destination node. In order to increase the efficiency of this protocol, i.e. to alleviate the head-of-line blocking problem in case of a single transmit buffer, k transmit buffers are deployed at each node equaling the number of virtual channels in the network. Thereby, the incoming packets are placed into those buffers according to their destination addresses, i.e. a packet destined to node d is put into buffer b, where

$$b = d \text{ modulo } k \qquad (b = 1,2,\ldots,k \; ; \; d = 1,2,\ldots,N)$$

Thus, when a backlogged node detects an empty slot with slot number s it then checks whether there is a packet awaiting transmission in the buffer corresponding to the current slot number, meaning in buffer $b = s$ modulo k. The considered buffer configuration is shown in Fig. 4.

In summary, this access protocol guarantees that the destination nodes meet their reception requirements, however, it suffers from the fact that some slots are wasted when there is nothing to send to a certain destination node. In order to make the medium access more flexible, a *dynamic* channel (slot) allocation may be envisaged where destination node occupancies could be recorded at each node via tables and accordingly different channels be used for reception.

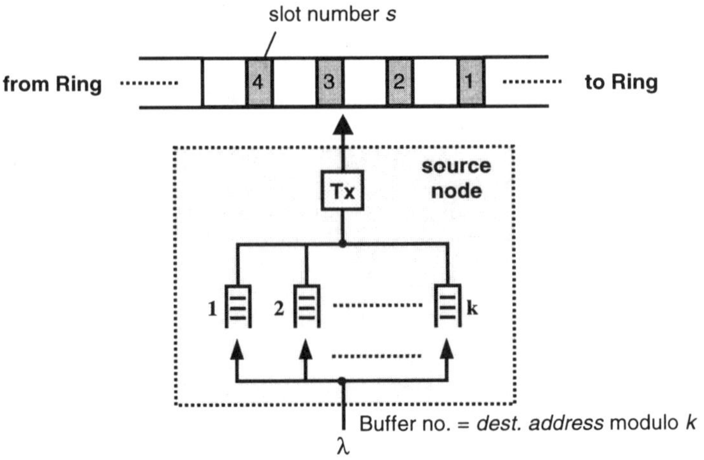

Figure 4: Buffer configuration for the proposed fixed-assignment access protocol

5. Performance Study

A performance study is carried out by means of discrete-event simulation in order to gain deeper insight into the behavior of the discussed protocol. The network throughput, the message transfer delay, and the packet transfer delay have been taken as performance measures. The *message transfer delay* is defined as the time period between the message generation and the moment when the packet has been completely received by the destination. The *packet transfer delay* is defined as the amount of time elapsed since a packet (acquired from a message) has reached the first place of the transmit queue until the packet has been completely received at the destination.

Concerning the traffic pattern we assumed a homogeneous load derived from a Poisson process with mean arrival rate λ. The message length is exponentially distributed and its mean value is here set to $l = 5$ slots. Furthermore, the nodes attached to the network are equidistantly spaced along the ring. In terms of the node parameters, we have adopted the values derived in section 3, i.e., the slot period T_s is set to 86 ns (corresponding to a payload length of 8.5 kbit and a header length of 64 bits for 100 Gbit/s), while assuming the header processing latency to be $T_{hp} = 55$ ns. Additionally, the number of network nodes was chosen to be $N = 20$ for our considerations.

In this performance study, particular attention is paid to the comparison of the 100 Gbit/s ring with a system operating at a medium rate of 2.5 Gbit/s in order to find out whether the implementation of the 100 Gbit/s access nodes with the processing delays as bottlenecks is justified regarding the additional costs in terms of the optical packet compression/decompression and OTDM multiplexer/demultiplexer units.

Fig. 5 illustrates the delay/throughput behavior for both considered delay types and a number of slots on the ring being $n = 150$ (corresponding to a ring length of $L \approx 2.6$ km). Four different scenarios are considered here, namely the proposed ring network with the processing restrictions corresponding to $k = 40$ and $k = 10$, respectively, compared to a conventional Slotted Ring scheme, where each empty slot can be used for reception, i.e. $k = 1$, operating at 100 Gbit/s (assuming that the attached nodes could achieve 100 Gbit/s without processing overhead) and 2.5 Gbit/s. Concerning the packet transfer delay, which is particularly decisive for real-time transmission it can be observed that the system with $k = 10$, i.e. the node configuration (b), behaves close to the "ideal" schemes with $k = 1$, whereas the node architecture (a) leads to significant degradations compared to the other cases. Apparently, the processing overhead associated with deploying solely optical compression at the nodes is too large to obtain an efficient access scheme.

In terms of the message transfer delays, where the queuing delay is included, it may be clearly seen that a 100 Gbit/s slotted ring shows a superior behavior compared to a 2.5 Gbit/s ring network only when node architecture (b), i.e. $k = 10$, is considered (partly deviations of about 200 slots !) even though the situation slightly changes for heavy loads. Therefore, it can be claimed that the implementation of an access node operating at 100 Gbit/s by means of optical packet compression and optical time division multiplexing is really justified compared to its 2.5 Gbit/s counterpart. A similar behavior can be observed in Fig. 6, when $n = 300$ ($L \approx$ 5 km) is considered while keeping the remaining parameters unchanged.

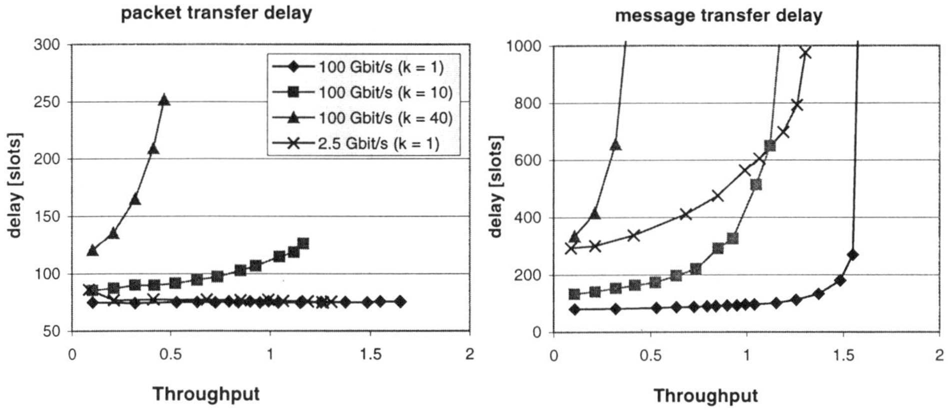

Figure 5: Delay-throughput behavior for $N = 20$ and $n = 150$

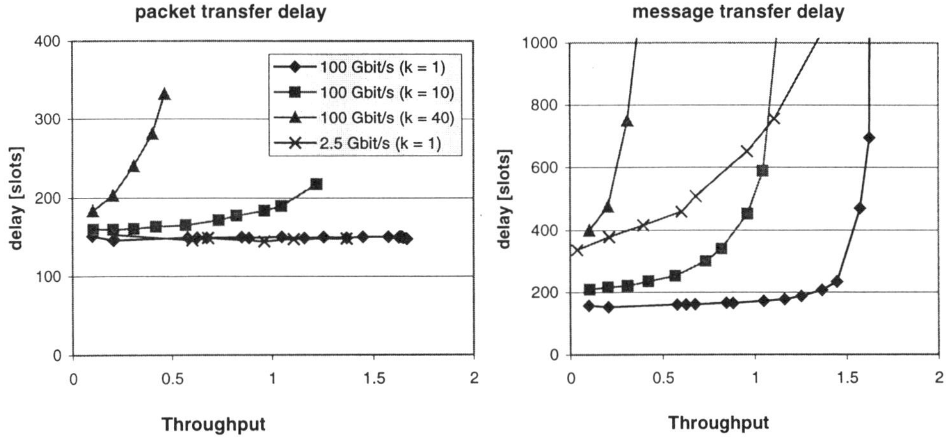

Figure 6: Delay-throughput behavior for $N = 20$ and $n = 300$

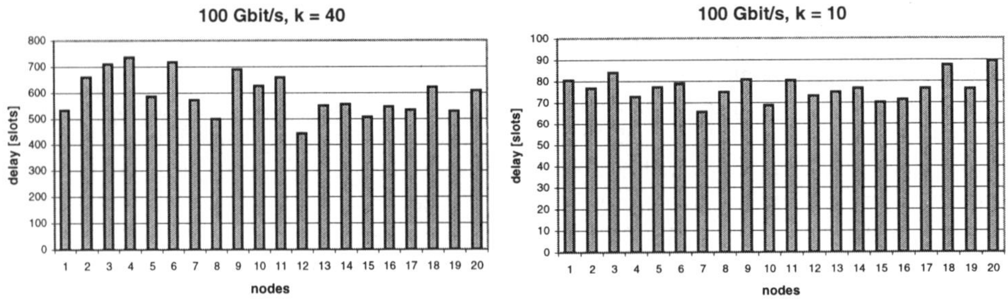

Figure 7: Access delays of the individual nodes for $N = 20$, $n = 150$, and $\lambda = 0.003$

Furthermore, the individual mean message access delays (excluding the propagation delay) are plotted against the nodes in Fig. 7, in which an almost fair access among the nodes (maximum difference $\approx$ 20 slots) under the above mentioned assumptions can be recognized for the $k = 10$ case, in contrast to $k = 40$.

6. Conclusion

In this paper, we proposed two different node architectures for a high-speed packet-switched ring-LAN employing optical packet payload compression and decompression. Taking the processing overheads of these architectures into account a fixed-assignment access protocol based on the Slotted Ring scheme with destination release has been designed using several virtual channels. The performance study clearly revealed that the implementation of such a 100 Gbit/s ring network with access nodes also operating at 100 Gbit/s speed at the expense of additional processing latency is justified compared to its 2.5 Gbit/s counterpart without this processing bottleneck.

Acknowledgement

This work is supported by the Austrian FWF (*Fonds zur Förderung der wissenschaftlichen Forschung*) under contract P13144-INF.

References

[1] M. Nakazawa, "Toward terabit/s single-channel transmission", *Conference Proceedings OFC/IOOC'99*, San Diego, Feb. 1999, pp. 132 / FI1-1 – 134/FI1-3.

[2] B.Y. Yu, P. Toliver, R. Runser, I. Glesk and P.R. Prucnal, "Towards terabit/s TDM Photonic Systems", *Conference Proceedings ECOC '97*, Edinburgh, Sept. 1997, **3**, pp.19 – 21.

[3] K.Bengi, G.Remšak, H.R.van As, "An optically transparent ultra high speed LAN-ring employing OTDM", *Working Conference on Optical Network Design and Modeling ONDM'98*, Rome, Feb.1998, pp. 35-41.

[4] K.Bengi, G.Remšak, "MAC protocol analysis for an optical packet switched ultra high-speed LAN-ring", *European Conference on Networks and Optical Communication NOC'98*, Manchester, June 1998, pp. 16-23.

[5] F.S. Choa, H. Chao, "All-Optical Packet Routing-Architecture and Implementation",*Photonic Network Communications* **1** (1999), pp. 303-311.

[6] Data Sheet, "CY7C43683 – 16K x 36 Unidirectional Synchronous FIFO w/ Bus Matching" *Cypress Semiconductor Corporation*, September 13, 1999.

[7] Data Sheet, "VSC7146 2.5 Gbit/s, 20 Bit Transceiver" *Vitesse Semiconductor Corporation*, June 13, 1999.

[8] H. Toda, F. Nakada, M. Suzuki, A. Hasegawa, "An Optical Packet Compressor Using a Fiber Loop for a feasible all optical TDM Network", *Conference Proceedings ECOC'99*, Nice, Sept. 1999, pp. I-256 – I-257.

[9] N. S. Patel, K. L. Hall, K. A. Rauschenbach, "Optical Rate Conversion for High-Speed TDM Networks", *IEEE Photonics Technology Letters* **9** (1997), pp.1277 – 1279.

[10] J.D. Moores et al., "80-Gbit/s 9-kbit optical pulse storage loop", *OFC'97 Technical Digest* (1997), pp. 88 – 89.

[11] C. Xie, P. Ye, "A novel scheme of optical packet-compression and decompression for all-optical packet switching networks", *Conference Proceedings ECOC'99*, Nice, Sept. 1999, pp. I-258 – I-259.

Broadband Access, WDM Metro and Network Management
D.W. Faulkner and A.L. Harmer (Eds.)
IOS Press, 2000 © AKM

A Simple Spread Time Optical Code Division Multiplexing System

Masanori HANAWA, Midori WATANABE, Mikio TAKAHARA
Faculty of Engineering, Yamanashi University,
4-3-11 Takeda, Kofu-shi, Yamanashi-ken 400-8511, Japan

Abstract. We propose a simple Spread Time Code Division Multiplexing (ST-OCDM) system which uses Chirped Fiber Bragg Gratings as encoder and decoder. The operation principle and performance of the simple ST-OCDM system are discussed based on computer simulation results. Our simulation results show that the proposed system is able to realize two channel CDM in spite of its simplicity. The proposed system may double the capacity of WDM systems at very low cost.

1 Introduction

Spread Time-Optical Code Division Multiplexing (ST-OCDM)[1] is one of the interesting optical multiplexing scheme. It realizes CDM using optical signature codes in frequency domain. Pulses modulated by data stream are encoded to represent an optical signature code in frequency domain. By this encoding process, pulses are spread in time and they may restore their original pulse shape only when they are decoded by a suitable decoder.

So far, the ST-OCDM system using the Arrayed Waveguide Gratings(AWG's) as the encoder and decoder have been reported[2][3]. This type of encoder/decoder consists of two AWG's and phase modulators. The number of phase modulators and branches of the AWG's depends on the code length used. In addition, as the frequency resolution of the AWG's is limited, the ultra wide band optical pulses (i.e. very short pulses like femto-seconds pulses) are needed to represent a long code. Thus the more the code length is, the more complex the system structure is.

On the other hand, Chirped Fiber Bragg Gratings(CFBG's) have been actively studied as dispersion compensation device[4]. A CFBG has a linear group delay characteristic in a limited bandwidth. This means that if a pulse has a bandwidth which matches with the one of the CFBG and it is launched into the CFBG through optical circulator, the output signal from the circulator is spread in time due to the group delay characteristic of the CFBG. When this spreading signal passes through a CFBG which has an inverse group delay characteristic for the first one, the original pulse shape may be restored.

In both operations described above, a pulse is once spread in time and restored again. We noticed a similarity between these operations. Thus, in this paper, we discuss applicability of CFBG's to encoder/decoder of the ST-OCDM system. A simple ST-OCDM system using CFBG's, that is able to multiplex two channels, is proposed. The system to be proposed is rather simple compared with the ST-OCDM system using AWG's.

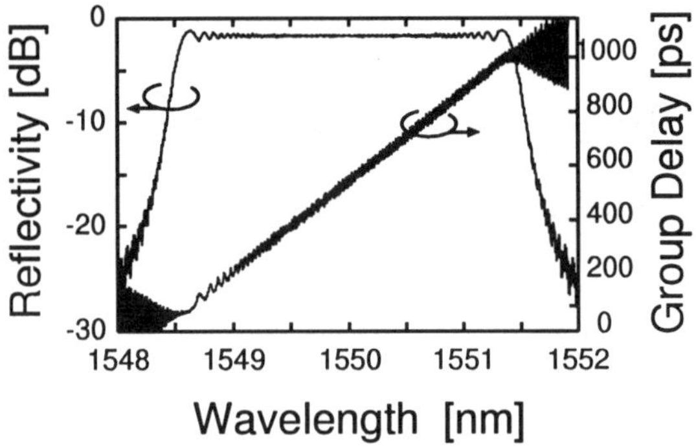

Fig. 1: Reflectivity and group delay of a CFBG.

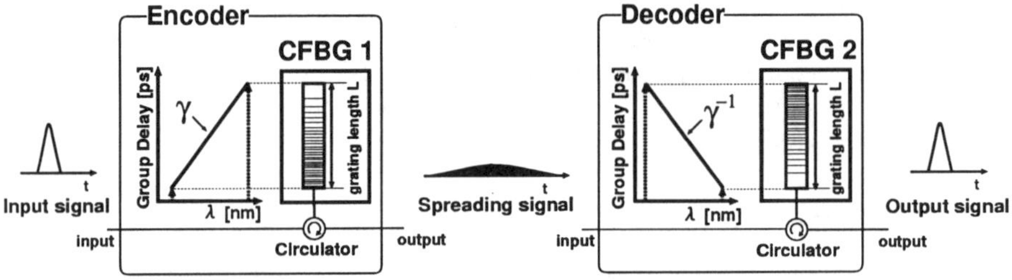

Fig. 2: Basic concept of the ST–OCDM system using CFBG's as encoder/decoder.

2 CFBG's as encoder/decoder of the ST-OCDM

A Fiber Bragg Gratings (FBG) is a fiber type device of which refractive index of the core changes periodically along the grating length[5]. This index change composes a diffraction grating in the core and a specific wavelength component is reflected in the FBG. A CFBG is a variant of the FBG. In the CFBG, the period of the index change varies along the grating length, so that the reflection wavelength also varies along the length. Thus the group delay characteristic of the CFBG linearly changes in its reflection bandwidth as shown in Fig.1.

Let's consider to use the CFBG as encoder/decoder of the ST-OCDM. Fig.2 demonstrates basic operation principle of the encoding/decoding process using CFBG's. When a pulse having some bandwidth matching to a reflection bandwidth of a CFBG are launched into the CFBG through an optical circulator, different wavelength components of the pulse receive different group delay by the CFBG. It makes the output signal from the circulator spread in time. If the spreading signal is launched into the second CFBG having an inverse group delay characteristic for the first one through another circulator, each wavelength components receive relatively inverse group delay. Here, the term 'relatively inverse' means that the absolute amount of delay is not canceled by passing these two CFBG's but only equalized. Namely, after passing through these two CFBG's, all the wavelength components receive equal amount of group delay. Thus the output signal from the second circulator takes the original pulse shape back.

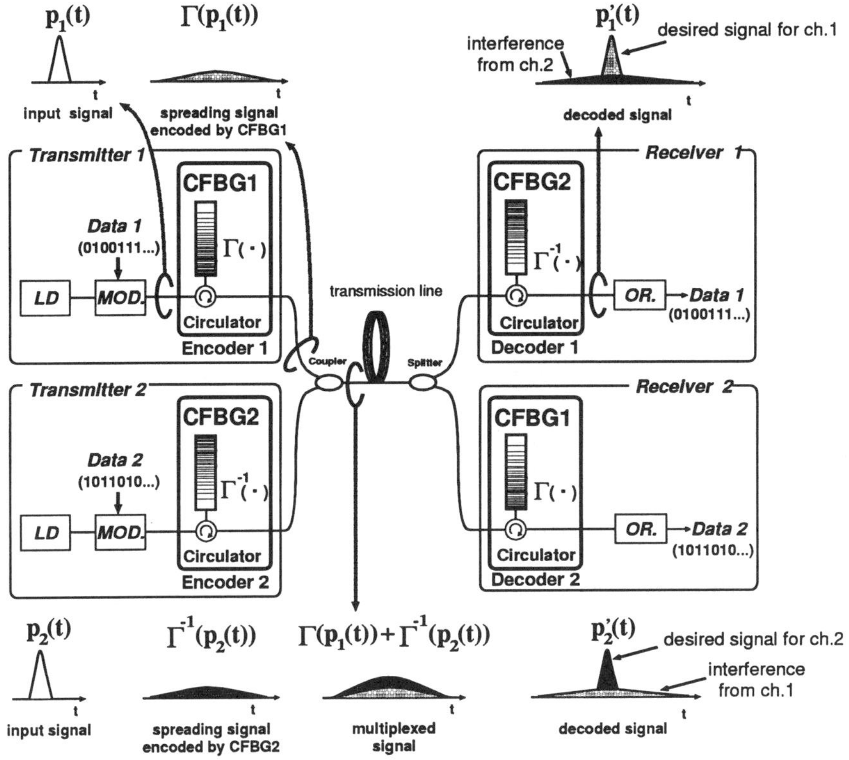

Fig. 3: Proposed ST–OCDM system using CFBG's as encoder and decoder. Tx: Transmitter, Rx: Receiver, LD: Laser Dioad, MOD: External Modulator, OR: Optical Receiver.

Two channel OCDM is easy to realize using this type of encoder/decoder. In the next section it will be demonstrated and a simple ST-OCDM system using CFBG's will be proposed.

3 A proposal of simple ST–OCDM systems

Here, we represent codes corresponding to the CFBG's shown in Fig.2 as γ (the first CFBG) and γ^{-1} (the second CFBG). In addition we show the coding process using the code γ and γ^{-1} as $\Gamma(\cdot)$ and $\Gamma^{-1}(\cdot)$, respectively. That is $\Gamma(\cdot)$ and $\Gamma^{-1}(\cdot)$ are corresponding to CFBG1 and CFBG2 in Fig.2, respectively. According to the discussion in the previous section, $\Gamma^{-1}(\Gamma(\cdot))$ and $\Gamma(\Gamma^{-1}(\cdot))$ should have no effect to their arguments. Using these notations, a simple ST-OCDM to be proposed is easily explained.

Consider a two channel OCDM system shown in Fig.3. As the encoder and decoder, the channel 1 uses $\Gamma(\cdot)$ and $\Gamma^{-1}(\cdot)$, while the channel 2 uses $\Gamma^{-1}(\cdot)$ and $\Gamma(\cdot)$, respectively.

A pulse $p_1(t)$ generated in the transmitter 1 is encoded as $\Gamma(p_1(t))$ by CFBG1 while a pulse $p_2(t)$ of channel 2 is encoded as $\Gamma^{-1}(p_2(t))$. They are coupled at a coupler and the multiplexed signal $\Gamma(p_1(t)) + \Gamma^{-1}(p_2(t))$ is transmitted to the receiver side. At the receiver side, the arriving signal is firstly split by a splitter. Thus the received signal at each receiver is

$$r(t) = \frac{1}{\sqrt{2}} \left\{ \Gamma(p_1(t)) + \Gamma^{-1}(p_2(t)) \right\}.$$

If the received signal $r(t)$ is decoded by a decoder $\Gamma^{-1}(\cdot)$ in the receiver 1, the decoder output $p_1'(t)$ is given as follows,

$$\begin{aligned} p_1'(t) &= \Gamma^{-1}(r(t)) \\ &= \Gamma^{-1}\left(\frac{1}{\sqrt{2}} \left\{ \Gamma(p_1(t)) + \Gamma^{-1}(p_2(t)) \right\} \right) \\ &= \frac{1}{\sqrt{2}} \Gamma^{-1}(\Gamma(p_1(t))) + \frac{1}{\sqrt{2}} \Gamma^{-1}\left(\Gamma^{-1}(p_2(t))\right) \\ &= \frac{1}{\sqrt{2}} p_1(t) + \frac{1}{\sqrt{2}} \Gamma^{-1}\left(\Gamma^{-1}(p_2(t))\right). \end{aligned} \tag{1}$$

Here, the first term of the last row is the desired signal from the transmitter 1 and the second one is the interference from the transmitter 2. Note that $\Gamma(\Gamma(\cdot))$ and $\Gamma^{-1}(\Gamma^{-1}(\cdot))$ mean passing through the same CFBG twice. Therefore, $\Gamma^{-1}(\Gamma^{-1}(p_2(t)))$ is spread wider than $\Gamma^{-1}(p_2(t))$. Spreading a pulse wider makes its amplitude be smaller. Thus, as interference signal from the transmitter 2 is lowered enough compared with the desired signal from the transmitter 1, the desired signal can be extracted by thresholding the decoder output. In the receiver 2, the principle is the same but vice versa. Therefore, we may double the capacity at the cost of some degradation of the extinction ratio between the desired signal and the interference. The extinction ratio is mainly decided by the CFBG's used as encoder/decoder. In the next section, selection of the CFBG, that satisfies the requirements of the extinction ratio, will be discussed.

4 Selection of CFBG's for encoder/decoder

As described above, the extinction ratio between the desired signal and the interference is the key issue for the proposed system and it mainly depends on the selection of the CFBG's used as encoder/decoder. Generally speaking, the wider the interference spreads in time, the better the extinction ratio is. Thus we briefly calculated the extinction ratio based on the theoretical equation representing the group delay of CFBG. It is

$$v_g = D_g \times \lambda_{chirp}, \tag{2}$$

where v_g is the maximum group velocity, D_g is grating dispersion and λ_{chrip} is the chirped bandwidth. The basic strategy for CFBG selection is as follows,

1. Assume a gaussian pulse.
2. Select a pulse width to be used.
3. Select a parameter set corresponding to a specific CFBG. The parameter set consists of a grating length, an reflection bandwidth, a coupling coefficient, a profile of refractive index change and so on.
4. Calculate a maximum group delay in the reflection bandwidth of the CFBG specified above by Eq.(1).
5. Calculate a peak amplitude of the interfering signal after passing through the CFBG twice. This is because the interfering signal passes through a same CFBG twice as described in the previous section.

Table 1: Parameters used for numerical calculation.

Pulse shape	Gaussian
Pulse FWHM	$\tau = 1 \sim 10$ps, 20ps, 30ps
Refractive index modulation	$\Delta n = 1.0 \times 10^{-4}$
Sections	N = 200

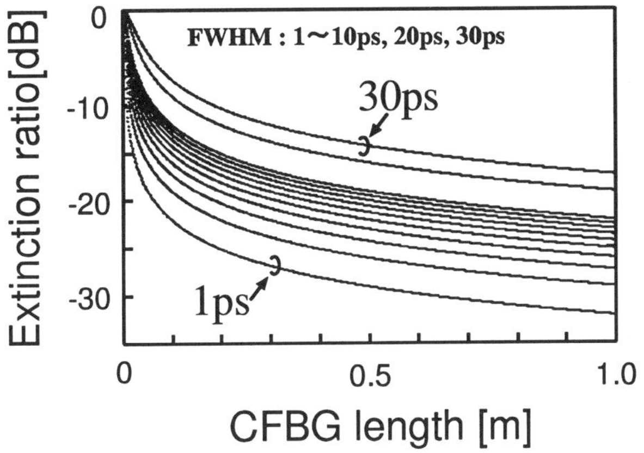

Fig. 4: The extinction ratio vs. CFBG length.

6. Calculate the extinction ratio from amplitudes of the original pulse and the interfering signal.
7. Repeat 3 to 6 until a parameter set which satisfies required extinction ratio are found.

Fig.4 shows the numerical results obtained by above method. Parameters used for numerical calculation are also shown in Table1. If the maximum permissible extinction ratio for the proposed system is defined, a parameter set (pulse width and CFBG length) enabling the requirement can be selected from Fig.4. For example, if the permissible extinction ratio is -25dB or less and the pulses having 1ps of the FWHM are used, we can determine the CFBG length as 0.2m from Fig.4.

5 Confirmation of the operation principle by computer simulation

Here, the operation principle of the proposed system is confirmed by computer simulations. According to the discussion described in the previous section, we selected 0.2m of the CFBG length for the pulse having 1ps of the FWHM. In this case, the reflection bandwidth of the CFBG is 8nm. Although this bandwidth may be difficult to realize in practical situation, implementation of such a CFBG is out of scope for this paper. Here we concentrate to possibility of the basic idea of the proposed system.

Fig.5 shows the simulated waveforms for the proposed ST-OCDM system. Parameters used to obtain the waveforms are shown in Table2. In this simulation, a back-to-back system was assumed.

Fig.5(a) shows a pulse sequence representing a data sequence '0100111' in the transmitter 1. This pulse sequence is passed through CFBG1 and spreads in time as shown

Table 2: Parameters used in computer simulations.

CFBG length	$L_g = 0.2$ m
Refractive index modulation	$\Delta n = 1.0 \times 10^{-4}$
Sections	N = 200
Pulse shape	Gaussian
Carrier wavelength	1550 nm
Pulse FWHM	1.0 ps
Launching peak power	1 dBm
Bit sequence	$2^3 - 1$ PRBS

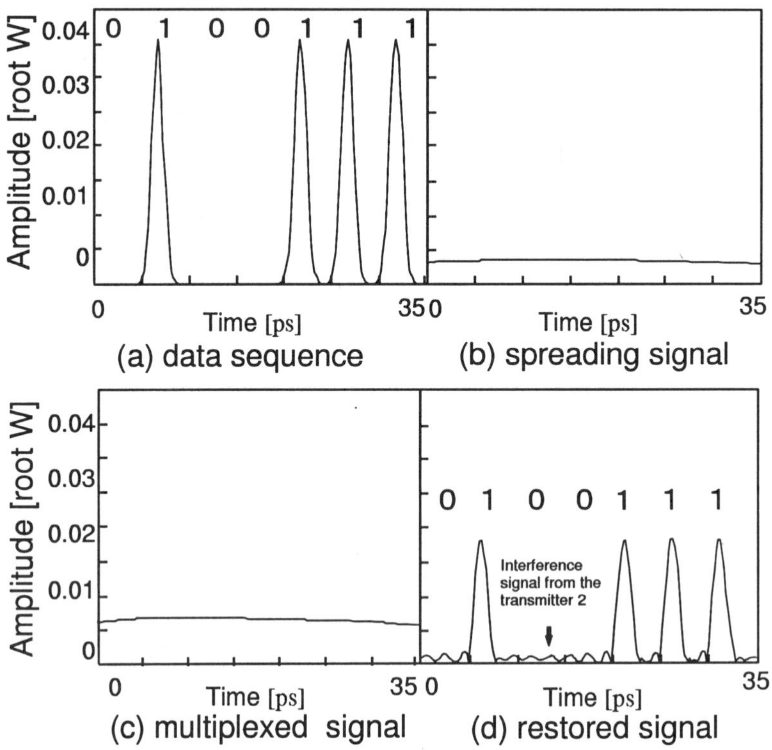

Fig. 5: Simulated waveforms. (a) Data sequence ' 0100111 'in the transmitter 1. (b) Spreading signal encoded by CFBG1. (c) Multiplexed signal. (d) Restored signal decoded by CFBG2.

in Fig.5(b). It is clearly seen that the spreading signal is spread enough compared with the original pulse sequence. Those spreading signals are multiplexed by a 3dB coupler and the multiplexed signal shown in Fig.5(c) is obtained. This multiplexed signal is then split by a 3dB splitter and the signal power are equally split to both receivers.

Fig.5(d) is the restored signal decoded by CFBG2 in the receiver 1. There are two points to notice. One of them is the small interfering signal lowered enough compared

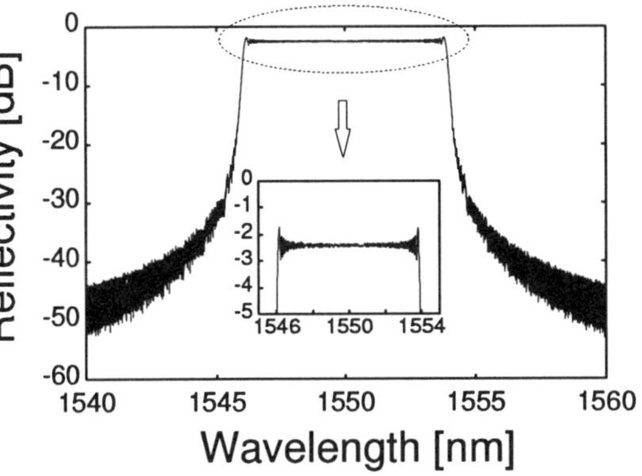

Fig. 6: Reflectivity of the CFBG used in the simulation. CFBG length: 0.2m. Reflection bandwidth: 8 nm. Refractive index modulation : 1.0×10^{-4}.

with the desired signal and the other is the smaller amplitude of the restored pulses compared with the original ones. The former means selection strategy of the CFBG is sufficient for the proposed system. The latter is caused by two factors, i.e. the power splitting at the 3dB splitter and the loss in the reflection bandwidth of the CFBG's used.

Fig.6 shows the reflection spectrum of the CFBG used in the simulation. The attenuation in the reflection bandwidth is only -2.5dB. This attenuation due to the CFBG, however, can be reduced by fine tuning of the CFBG design. Thus, the amplitude drop by the 3dB splitting is the only theoretical problem of the proposed system. Anyway, Fig.5(d) clearly shows that the proposed system would work as expected.

6 Conclusion

In this paper, a simple ST–OCDM which uses CFBG's as encoder and decoder has been proposed. From computer simulation results, it has been found that CFBG's can be used as encoder/decoder of the ST–OCDM system and two channel multiplexing is possible. More precise evaluations are further works.

References

[1] Jawad A. Salehi, Andrew M.Weiner, and Jonathan P. Heritage, "Coherent Ultrashort Light Pulse Code-Division Multiple Access Communication Systems," J.of Lightwave Technology, vol.8, No3, pp.478-491, March 1990.

[2] Koichi Takiguchi, Katsunari Okamoto, Yasuji Ohmori and Steven Mechels, "Preliminary Experiments on Coherent Optical Code Division Multiple Access Using Arrayed-Waveguide Grating Pairs," 1998 IEICE General Conference, C-3-315, March 1998.

[3] Masanori Hanawa, Takehiro Yanagi, and Mikio Takahara, "DWDM transmission by time spreading signals," Proc. of OECC Vol.1, pp.333-336, Oct 1999.

[4] K.O.Hill, F. Bilodeau, B. Malo, T. Kitagawa, S. Theriault, D.C.Johnson, and J. Albert, "Chirped in-fiber Bragg Gratings for compensation of optical-fiber dispersion," Opt.Lett, vol.19, pp.1314-1316, 1994.

[5] Raman Kashyap, Fiber Bragg Gratings, Academic Press, Inc., 1999.

Broadband Access, WDM Metro and Network Management
D.W. Faulkner and A.L. Harmer (Eds.)
IOS Press, 2000 © AKM

Code Acquisition of Optically Orthogonal Codes in a Fibre-based Optical CDMA Network Using Dual-Threshold Sequential Detection

Mazlyn M. MUSTAPHA and Richard F. ORMONDROYD
Communications and Wireless Networks Group, Cranfield University, Royal Military College of Science, Shrivenham, Swindon, SN6 8LA, United Kingdom
Tel: +44 1793 785201 Fax: +44 1793 785902
e-mail: mustapha@rmcs.cranfield.ac.uk; r.f.ormondroyd@rmcs.cranfield.ac.uk

Abstract. This paper compares the performance of a dual- and single-threshold sequential detection acquisition system in the presence of multi-access interference and receiver noise for a non-coherent fibre-based optical CDMA system using optically orthogonal codes. The synchronisation performance is analysed and derived using a Markov flow graph method. By varying the sequential detection threshold, the synchronisation performance can be optimised. The results show that a dual-threshold sequential detector outperforms the single-threshold sequential detector. Also, when both systems are designed for a specific number of users, it is found that the synchronisation performance deteriorates when the number of users varies from the design value.

1. Introduction

In any code-division multiple-access (CDMA) system, synchronisation is required to align the timing of the CDMA decoder with the wanted signal received from the network in order to recover the data. The existing synchronisation methods for wireless coherent CDMA cannot be used directly for the non-coherent fibre-based optical CDMA network considered in this paper because this optical CDMA system has a different signal structure to a coherent CDMA system. As a result, new synchronisation methods must be devised.

The majority of the work in the literature regarding optical CDMA system performance assumes synchronisation has been achieved earlier by some unknown method [1],[2],[3],[4]. There has been very little work on developing synchronisation techniques for optical CDMA systems and this aspect is addressed here. Code acquisition for an on-off keyed spread spectrum system with coherent pseudonoise codes over an optical channel was considered by Kiasaleh [5] in 1990. More recently, synchronisation for coherent optical CDMA systems was considered in [6] and [7]. A basic synchronisation scheme for non-coherent optical CDMA has been considered by Yang [8]. Following this, the authors have considered a more sophisticated synchroniser for a non-coherent optical CDMA network that uses a serial-search acquisition method [9] that provides performance improvements over [8]. This paper describes further enhancements to the performance of a synchroniser in [9] by using a synchronisation system based on a sequential detector [10],[11].

In this paper, the optical CDMA system model and both the single- and dual-threshold sequential detector models are presented and analysed. The synchroniser mean acquisition time is then derived by using the Markov flow-graph method [12]. The mean acquisition time depends on the sequential detection thresholds, the multi-access interference (MAI), receiver noise and coupler losses. Also, the effect of changing both the sequential detector thresholds and the number of active users on the synchroniser performance is investigated for a star network configuration. The paper also considers the performance degradation that results when the true MAI varies from the design MAI that the sequential detector has been optimised to accept. A comparison is then made between the single- and dual-threshold sequential detector performance.

2. System Description

The system considered here is a non-coherent, asynchronous, fibre-based optical CDMA star network using matched-filter optical tapped delay-line encoders and decoders [13], as shown in Figure 1 and Figure 2.

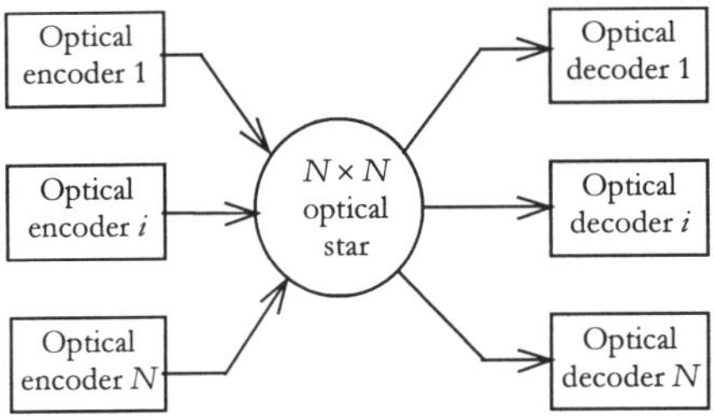

Figure 1: Star Topology Optical CDMA Network

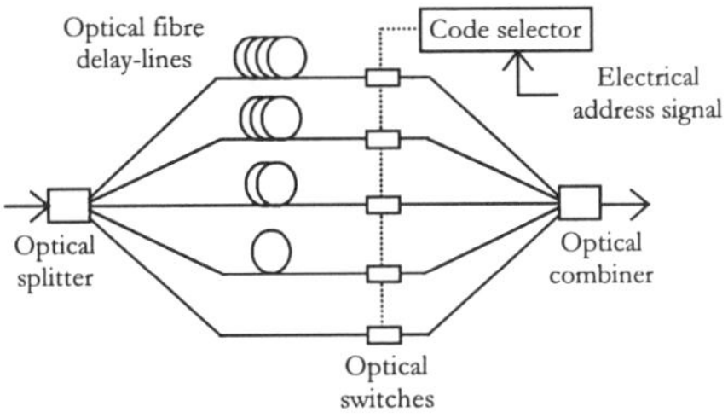

Figure 2: Optical Fibre Tapped Delay-Line Encoder

In this system, optical orthogonal codes (OOC) [14] are used for the spreading sequences. The OOCs are F chips long with K '1' chips and $(F - K)$ '0' chips. The uniqueness of each code is determined by the location of the K '1' chips. The code is effectively ANDed with the data bits so that each '1' data bit is represented by an OOC and each '0' data bit is represented by F '0' chips. It is assumed that there are N active users on the network that continuously transmit data. Each user is bit asynchronous and chip synchronous, with a relative random time delay of $\tau_d \in \{0, T_C, \ldots, (F-1)T_C\}$ with respect to each other, where T_C is the chip width.

At the receiver, the optical fibre delay-line matched filter converts the K wanted chips distributed over the F chip code to a single wanted output which has a duration of T_C located at some position, w, within the F possible output positions. This output is corrupted by MAI and noise. The other $(F - 1)$ matched filter output positions are MAI and noise. In the context of this paper, synchronisation is the process of locating output position w.

In a very simple synchronisation strategy, the receiver matched filter output is detected by an APD and threshold detected. The matched filter output position corresponding to the signal that exceeds the threshold is chosen as w. However, due to MAI and noise, it is possible that the wanted signal is missed or a large extraneous signal at a wrong output location is selected instead. To improve synchronisation reliability, a synchronisation preamble that is made up of several data bits is sent. This preamble must be known in advance by the receiver. Threshold detection of the APD output is performed as before. The synchroniser then locates output position w by finding the sequence of outputs at positions w, $(w + F)$, $(w + 2F)$, ... that matches the synchronisation preamble. This process can be described as a correlation between the matched filter outputs and the synchronisation preamble and the search strategy is essentially a serial-search [9]. The longer the preamble, the more accurate the detection of w will be. The disadvantage of the serial-search method is that if one or more of the preamble bits are decoded incorrectly, output position w could be missed.

When the noise and interference statistics are known, this *a priori* knowledge can be exploited to improve the synchronisation performance. Instead of making a hard decision on the state of the data bit using a simple threshold detector, the *likelihood* that the output of the matched filter corresponds to the preamble bit is computed by sequential detection [10], [11].

This process is performed for each chip position (0 to $F-1$) and for each bit of the synchronisation preamble. For each preamble bit, F likelihoods are obtained. Each of the F likelihoods is accumulated in an F-element array corresponding to its chip position, e.g. the likelihood for chip ℓ is accumulated in element ℓ. By accumulating the likelihoods for several synchronisation preamble bits, synchronisation is now indicated when one of the F accumulated likelihoods exceeds an acceptance threshold. Compared with serial-search, sequential detection does not require each preamble bit to be received correctly, and therefore this improves the synchronisation performance.

3. System Model

The non-coherent optical CDMA system used here is shown in Figure 1 and Figure 2 [15]. For every '1' data bit, the input optical power launched into each optical CDMA encoder is $P_{in} = \langle n_{in} \rangle \cdot hc/\lambda$, which corresponds to a mean number of photons $\langle n_{in} \rangle$, where h is Planck's constant, c is the velocity of light, and λ is the operating wavelength. Each of the N active users is assumed to transmit at this input power level. At the star coupler, the power splitting loss is N. Also, the optical splitters in the optical tapped delay-line encoder and decoder yields a power splitting loss of K^2 for an OOC of weight K. Furthermore, each coupler has an intrinsic loss of L_C. Hence, the total coupler loss between any source and any receiver is $L_C^5 NK^2$ [16], and this is suffered by each chip through the network when measured at the tapped delay-line decoder. At this point in the receiver, each '1' chip is represented by $\langle n_c \rangle = \langle n_{in} \rangle / \left(L_C^5 NK^2\right)$ photons.

Using the optical disc analogy [14], the optical interference level at the output of the tapped delay-line decoder when all the users' signals are received at the same power level is $l \cdot \langle n_c \rangle$ for $l \in \{0,1,\ldots,N-1\}$ where l is a discrete random variable due to overlapping chips from other users which cause MAI. Thus, the output of the tapped delay-line decoder becomes $\langle n_s \rangle = (l+K) \cdot \langle n_c \rangle$ for a data bit '1' and $\langle n_s \rangle = l \cdot \langle n_c \rangle$ for a data bit '0'. This value is then converted to an electrical output by an avalanche photodiode (APD).

In an ideal chip-synchronous optical CDMA network that uses OOCs, the probability of l interfering chips overlapping a wanted chip is defined as [15]:

$$P_I(l) = \binom{N-1}{l} \left(\frac{K^2}{2F}\right)^l \left(1 - \frac{K^2}{2F}\right)^{N-1-l} \tag{1}$$

In general, the photocurrent probability density function (PDF) at the output of the APD for a '1' or '0' data bit, assuming the APD photocurrent PDF is Gaussian [17], is then:

$$p(i) = \sum_{l=0}^{N-1} \left(\frac{P_I(l)}{\sqrt{2\pi\sigma^2}} \exp\left[-\frac{\left(i - I_{ph}\right)^2}{2\sigma^2} \right] \right) \tag{2}$$

where σ^2 is the photocurrent variance given by:

$$\sigma^2 = 2eB\left(I_{ph} + I_D\right) M^{2+\gamma} + \left(4kTBF_n / R_L\right) \tag{3}$$

Here, e is the electron charge, I_{ph} is the generated photocurrent, M is the APD multiplication factor and k is Boltzmann's constant. The other parameters are defined in Table 1. The generated photocurrent, $I_{ph} = \eta e \cdot \langle n_s \rangle$, is dependent upon the optical power incident upon to the APD and the APD quantum efficiency, η. All the optical losses are accounted for in I_{ph}, *via* the $\langle n_s \rangle$ term. From equation (2), the PDF conditional on a data bit '1' being expected in the preamble is $p(i_{\ell n}|1)$. Likewise, $p(i_{\ell n}|0)$ is the PDF conditional on a data bit '0' being expected in the preamble.

3.1 Sequential Detection

The APD-detected output of the optical matched filter is passed to the sequential detector synchroniser. At the sequential detector synchroniser, the likelihood that the received data for each chip position is a '1' or '0' is calculated with *a priori* knowledge of the statistics of the received signals when either a '1' or a '0' synchronisation preamble bit is expected by the synchroniser. For example, if the synchroniser expects the *n*th preamble bit to be a '1', the log-likelihood value is calculated from:

$$\Lambda_{\ell n}^{1} = \ln \frac{p\left(i_{\ell n}|1\right)}{p\left(i_{\ell n}|0\right)} \tag{4}$$

$\ell \in \{0,1,\ldots,F-1\}$ represents the ℓth chip location within the F chip code. However, if a preamble bit '0' is expected instead, the log-likelihood value becomes:

$$\Lambda_{\ell n}^{0} = \ln \frac{p\left(i_{\ell n}|0\right)}{p\left(i_{\ell n}|1\right)} = \ln p\left(i_{\ell n}|0\right) - \ln p\left(i_{\ell n}|1\right) = -\Lambda_{\ell n}^{1} \tag{5}$$

The log-likelihood values are accumulated individually for each chip location 0 to $(F-1)$ for each preamble bit, *n*. The accumulated value is then:

$$v_{\ell j} = \sum_{n=1}^{j} \Lambda_{\ell n} \tag{6}$$

where $0 < j < T_{tr}/\left(FT_C\right)$ and T_{tr} is the pre-set truncation time. Typical running accumulations of the log-likelihood, $v_{\ell j}$ are shown in Figure 3(a) and (b) for the single-threshold and dual-threshold sequential detectors, respectively as the number of preamble bits checked, j increases. Both figures show $v_{\ell j}$ for four different matched filter output locations (i.e. different values of ℓ). Using the dual-threshold sequential detector as an example, the sequential probability ratio test compares $v_{\ell j}$ to the acceptance and rejection thresholds, th_a and th_r respectively. This is illustrated in Figure 3(b) and it can be seen that after a number of preamble bits have been checked the accumulated log-likelihoods for two of the matched filter output positions fall below the rejection threshold. This means that it is extremely unlikely that those matched filter output positions have provided outputs that correspond to the training preamble received so far. These output positions are then eliminated from further consideration. Ultimately, the accumulated log-likelihood for one of the matched filter output positions exceeds the acceptance threshold. This indicates that it is highly likely that this output corresponds to the wanted synchronisation preamble received so far. This output is then selected as the chip position *w* of the code sequence, and the search is terminated. For the single threshold case, it is not possible to eliminate unlikely output positions early and these must still be tracked until the wanted output position is found.

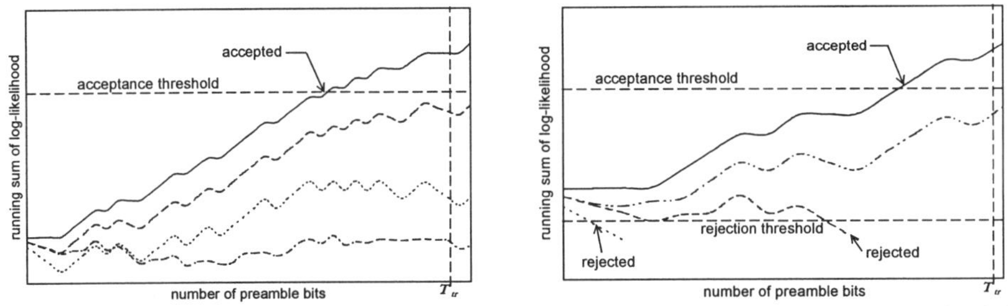

Figure 3: Running sum of log-likelihood function vs. dwell-time (number of preamble bits) for (a) single-threshold and (b) dual-threshold sequential detector

In both the single- and dual-threshold cases, only when $v_{\ell j} \geq th_a$ exists will the process be terminated. Otherwise, if $v_{\ell j}$ does not exceed th_a, the synchroniser continues to check the next preamble bit. This process is continued until either $v_{\ell j} \geq th_a$, or until the truncation time, T_{tr}, is reached.

Once $v_{\ell j}$ reaches the acceptance threshold, ℓ is selected as the wanted chip position w. The probability that the selected chip, ℓ, is the true wanted position, w, is P_D. However, the probability that ℓ is not w, i.e. a false alarm has occurred, is P_{fa}. Once a false alarm has taken place, a penalty time, T_{fa}, is incurred. Furthermore, it is possible that the acceptance threshold is not exceeded after the truncation time, T_{tr}, has elapsed. The probability of this event taking place is called P_{tr}, and $P_D + P_{fa} + P_{tr} = 1$. If synchronisation has not occurred by the truncation time, a decision must be made regarding the wanted chip location, w. This is resolved by selecting ℓ that yields the largest value of $v_{\ell j}$ as the wanted chip position w. Consequently, there are only two possible outcomes, either that the correct chip position was chosen, with probability P_{Dtr}, or an error (false alarm) was made, with probability $(1 - P_{Dtr})$.

3.2 Derivation of Mean Acquisition Time

The performance metric of the synchroniser is its mean acquisition time. This is the mean time taken for the synchroniser to select the true wanted chip position w when the signal is subject to random noise and interference. In a conventional CDMA sequential detection system, the acquisition time can be obtained by estimating the time needed to search the 'uncertainty' region, where the synchronisation signal is absent, plus the time required to search the cell where the synchronisation signal is present. This is possible because the search and dismissal processes are performed serially. In the optical CDMA sequential detection system considered here, the accumulated log-likelihoods for each matched filter output position are compared with the acceptance and rejection thresholds in parallel after the completion of each bit period. Therefore, a flow graph method is more suitable to derive the mean acquisition time than the conventional CDMA technique. In this paper, it is derived using the generating function flow graph of a discrete time-invariant Markov process [12].

In a process with several states, where transitions from one state to another is governed by probabilities and time delays, the mean time to terminate the process can be obtained by using a Markov flow graph. The flow graph is a topological representation of simultaneous equations that make up a system transfer function. After a process has been represented as a flow graph, the system transfer function can then be obtained using reduction techniques [12] to minimise the original flow graph to its minimal form - two nodes and one branch. The mean acquisition time can then be calculated from this transfer function.

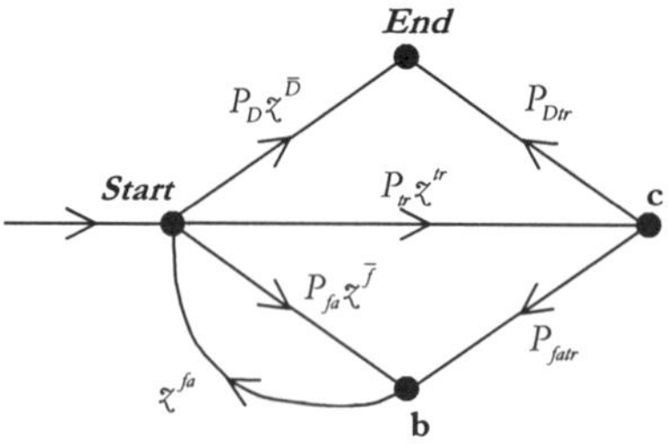

Figure 4: Generating Flow Graph for Mean Acquisition Time

We have derived the signal flow graph for the optical CDMA system sequential detector and this is shown in Figure 4. This signal flow graph applies to both the single-threshold and dual-threshold sequential detectors. The signal flow graph of Figure 4 can be reduced to give the sequential detector transfer function:

$$\Gamma(z) = \frac{P_D z^{\bar{D}} + P_{tr} P_{Dtr} z^{tr}}{1 - z^{fa}\left(P_{fa} z^{\bar{f}} + P_{tr} z^{tr} - P_{tr} P_{Dtr} z^{tr}\right)} \tag{7}$$

$\bar{D}$ is the mean number of chips to detect a true wanted chip position, $\bar{f}$ is the mean number of chips to acquire a false chip position, $fa = T_{fa}/T_C$, $tr = T_{tr}/T_C$, and z represents a unit delay. The partial derivative of the transfer function, $\Gamma(z)$, gives the mean acquisition time:

$$\bar{T}_{acq} = \frac{\partial}{\partial z}\Gamma(z)\Bigg|_{\substack{(T_C)\\ z=1}} = \frac{T_C \cdot \left(\bar{D}P_D + \bar{f}P_{fa}\right) + T_{tr} \cdot P_{tr} + T_{fa}}{P_D + P_{tr} P_{Dtr}} - T_{fa} \tag{8}$$

This requires that P_D, P_{fa}, P_{Dtr}, $\bar{D}$ and $\bar{f}$ are obtained for each system configuration. Note that these parameters are generally different for the single- and dual-threshold sequential detectors.

4. Discussion of Results

A simulation of the optical CDMA receiver and the single- and dual-threshold sequential detector synchroniser was implemented using a Monte-Carlo method with different values of acceptance and rejection threshold to obtain P_D, P_{fa}, P_{Dtr}, $\bar{D}$ and $\bar{f}$ for K=3, F=1000, and N=20. The Monte-Carlo simulations were repeated until any variations in the results had evened out. Then, these values were substituted into equation (8) to calculate the mean acquisition time. The APD multiplication factor, M, was obtained by finding the value of M that produced the optimum bit error probability for the optical CDMA network with N users. The values of the other parameters used in the simulation are shown in Table 1.

Parameter	Value	Parameter	Value
Optical input power, P_{in}	80 μW	Wavelength, λ	1.55 μm
Chip period, T_C	10 ns	Detector dark current, I_D	1 nA
Truncation time, T_{tr}	0.2 ms	APD material constant, γ	1
Penalty factor, T_{fa}	0.1 ms	Ambient temperature, T	300 K
Receiver bandwidth, B	100 MHz	Receiver noise figure, F_n	2.0
Detector quantum efficiency, η	0.8	APD bias resistor, R_L	50 Ω

Table 1: System parameters

Figure 5 shows the single-threshold sequential detector $\bar{T}_{acq}$ characteristics for different values of th_a for N=20 and N=50. It is seen from the graph that there exists a value of th_a that minimises $\bar{T}_{acq}$ for both cases of N. This is because when th_a is too small, a false w may be accepted for synchronisation, whereas if th_a is too large, the true w may be missed. For larger values of N, the optimum synchroniser performance deteriorates by a factor of approximately 2 and the optimum occurs at a higher threshold.

In Figure 6, the graph shows the effect of varying th_a with $\bar{T}_{acq}$ for a dual-threshold sequential detector, again for N=20 and N=50. In effect, each point on the graph represents a th_a-th_r pair. For each value of th_a, the value of th_r that optimises $\bar{T}_{acq}$ is obtained. It can be seen that when the number of users increases, the optimum performance deteriorates. However, at higher values of th_a this degradation is not very pronounced and not as significant as for the single-threshold case. Comparing Figure 5 with Figure 6, it can be seen that the performance of the dual-threshold sequential detector outperforms the performance of the single-threshold sequential detector by a factor of approximately 1.5 for N=20, and by a factor of approximately 2 for N=50.

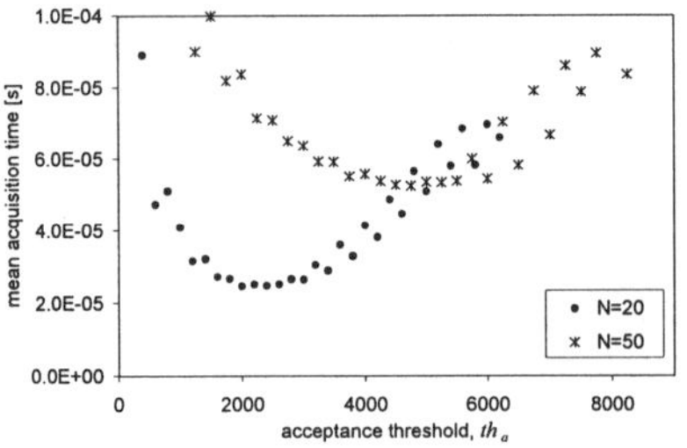

Figure 5: Effect of th_a on $\overline{T}_{acq}$ for a single-threshold sequential detector

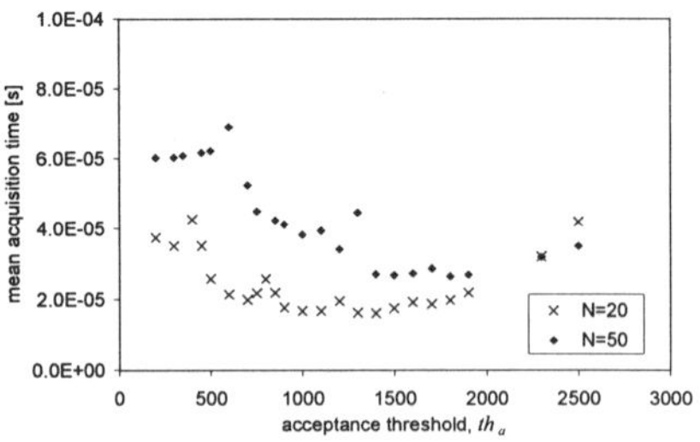

Figure 6: Effect of th_a on $\overline{T}_{acq}$ for a dual-threshold sequential detector

The sequential detector synchroniser is designed on the basis of the known statistics of a specific number of users. When the actual MAI differs from the number that the synchroniser is designed for, the actual likelihood function differs from the assumed likelihood function and the performance generally degrades. This is illustrated in Figure 7. Here, the single- and dual-threshold sequential detectors are both optimised for N=20. $\overline{T}_{acq}$ is then obtained for several different values of the actual number of users. The graph in Figure 7 shows that the optimum $\overline{T}_{acq}$ is only obtained when the actual number of users is the same as the design value, which is N=20, and if the actual number of users varies from this design value, $\overline{T}_{acq}$ deteriorates. This condition holds even when the actual number of users falls below the design value.

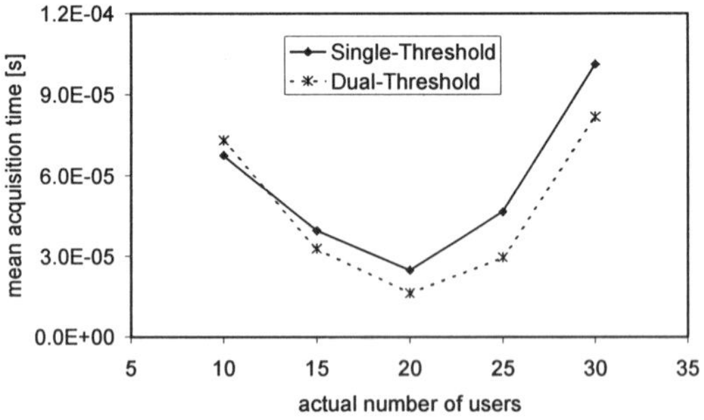

Figure 7: Effect of varying the true number of users for a dual-threshold sequential detector designed for 20 users

5. Conclusions

The synchronisation performance of a single-threshold and dual-threshold sequential detection synchronisation system for a non-coherent optical CDMA system with MAI has been analysed. The minimum mean acquisition time was obtained after optimising the acceptance threshold for the single-threshold case, and both the acceptance and rejection thresholds for the dual-threshold case. The results show that the dual-threshold sequential detector outperforms the single-threshold sequential detector, and the performance improvement increases as the number of active users increase. Also, it is demonstrated that a large number of simultaneous active users has an adverse affect on the optimum mean acquisition time. The optimum mean acquisition time, however, can only be obtained if the number of simultaneous users is known *a priori*, so that the likelihood functions can be computed correctly.

6. Acknowledgements

The authors are pleased to acknowledge the support of Tenaga Nasional Berhad, Malaysia, for their kind provision of a studentship for M. M. Mustapha.

7. References

[1] H. M. H. Shalaby, Chip-level receivers in optical overlapping PPM-CDMA, *IEEE International Symposium on Spread Spectrum Techniques & Applications* **3** (1998) 907-911.
[2] C. L. Lin and J. Wu, Synchronous fiber-optic CDMA system using adaptive optical hardlimiter, *Journal of Lightwave Technology* **16** (1998) 1393-1403.
[3] J. G. Zhang, A. B. Sharma and W. C. Kwong, New fiber-optic ATM switching networks using optical code-division multiple access for broadband communication applications, *IEEE Transactions on Consumer Electronics* **44** (1998) 952-962.
[4] T. Ohtsuki, Channel interference cancellation using electrooptic switch and optical hardlimiters for direct-detection optical CDMA systems, *Journal of Lightwave Technology* **16** (1998) 520-526.
[5] K. Kiasaleh, PN code acquisition for the spread-spectrum optical on-off-keying communication system, *IEEE Transactions on Communications*, **38** (1990) 1879-1885.
[6] W. Huang and K. Kitayama, Code acquisition in optical pulse CDMA utilizing coherent correlation demodulation, *IEEE Global Telecommunications Conference* **1** (1997) 265-269.
[7] W. Huang and I. Andonovic, Code tracking in optical pulse CDMA through coherent correlation demodulation, *IEEE International Conference on Communications* **3** (1998) 1264-1268.
[8] G. C. Yang, Performance Analysis for Synchronization and System on CDMA Optical-Fiber Networks, *IEICE Transactions on Communications*, **E77B** (1994) 1238-1248.
[9] M. M. Mustapha and R. F. Ormondroyd, Performance of a Serial-Search Synchronizer for Fiber-based Optical CDMA Systems in the Presence of Multi-access interference, *Proceedings of the SPIE* **3899** (1999) 297-306.
[10] A. Wald and J. Wolfowitz, Optimum Character of the Sequential Probability Ratio Test, *Ann. Math. Statist.* **19** (1948) 326.
[11] J. J. Bussgang and D. Middleton, Optimum Sequential Detection of Signals in Noise, *IEEE Transactions of Information Theory* **IT-1** (1955) 5-18.
[12] J.K. Holmes, Coherent Spread Spectrum Systems. ISBN: 0 89464 468 8. Robert E. Krieger Publishing Company, 1990.
[13] P. R. Prucnal, M. A. Santoro and T.R. Fan, Spread Spectrum Fiber-Optic Local Area Network using Optical Processing, *Journal of Lightwave Technology* **LT-4** (1986) 547-554.
[14] J. A. Salehi, Code division multiple-access techniques in optical fiber networks - Part I: Fundamental principles, *IEEE Transactions on Communications* **37** (1989) 824-833.
[15] J. A. Salehi and C. A. Brackett, Code division multiple-access techniques in optical fiber networks - Part II: Systems performance analysis, *IEEE Transactions on Communications* **37** (1989) 834-842.
[16] R. F. Ormondroyd and M. M. Mustapha, Optically Orthogonal CDMA System Performance with Optical Amplifier and Photodetector Noise, *IEEE Photonics Technology Letters* **11** (1999) 617-619.
[17] R. J. McIntyre, Multiplication noise in uniform avalanche diodes, *IEEE Transactions on Electronic Devices* **ED-13** (1966) 164-168.

Broadband Access, WDM Metro and Network Management
D.W. Faulkner and A.L. Harmer (Eds.)
IOS Press, 2000 © AKM

Next Generation Building Backbone Technologies

James Donovan
Lucent Technologies, 121 St. Stephens Green, Dublin, Ireland

Abstract. Network planning. It never has been an easy process, and in fact it can often lead to many a sleepless night for network managers. But it seems that as of late this task has become even more daunting. The prognostications of literally exploding demand for network bandwidth have stimulated the development of LAN technologies promising to deliver anywhere from 10 Mbps up to Gbps+ data rates. But is this fervor surrounding Gbps networking founded and justifiable? What are the needs of the End-user and what are the solutions? Are there issues surrounding the new technology solutions and how do new developments solve them?
In defining the requirements for the next generation building backbone, several key objectives must be accounted for to provide the best solution for the customer. This paper explores the options and will provide an up to date account of the standards activities in this arena.

Over the past decade, the LAN industry has developed numerous LAN protocols such as Token Ring, FDDI, ATM and Ethernet. All share the common characteristic of pushing data rates ever higher to serve the increasing demands of users. Of these, the Ethernet solution suite has become the most widely accepted application set. The most prevalent application of this suite today is 10-megabit Ethernet, but the 100-megabit upgrade, known as Fast Ethernet, is now displacing its lower speed predecessor for desktop applications. Because switches aggregate bandwidth from many users, the backbone network that interconnects these switches must support even higher rates. Thus, as 100- megabit speeds serve desktops, 1000-megabit speeds are needed in the backbone. Fortunately, the industry has responded to this need by developing the latest Ethernet solution, known as Gigabit Ethernet. But even gigabit speeds are not just contained within the backbone. Today' s "super users" are demanding gigabit speeds at the workstation. Past trends indicate that today' s super user speeds become tomorrow' s mainstream applications. It is in this light that IEEE has now undertaken the development of the next generation of the Ethernet suite, commonly referred to as Ten-Gigabit Ethernet.

Recently, the cabling industry standards bodies of TIA and ISO/IEC agreed to develop specifications for a multimode fibre that supports a 10 Gb/s over 300 metres objective for 850 nm systems. Also, in light of these cabling standards developments, the Higher Speed Study Group of IEEE 802.3 decided to develop a 300-metre next-generation multimode fibre solution for 10 Gigabit Ethernet. These developments signal an exciting improvement in applications support for multimode fibre, breaking through the limitations of current technology.

The IEEE 802.3 committee is beginning the development of the 10G Ethernet (10GbE) standard. The 802.3 Higher Speed Study Group (HSSG), which investigates the feasibility and industry interest in developing a 10GbE standard prior to the application for a standards project, has established objectives to provide solutions for five combinations of distance and fibre type, which include three singlemode and two multimode targets. The singlemode objectives are as follows:

1. at least 40 km on singlemode fibre
2. at least 10 km on singlemode fibre
3. at least 2 km on singlemode fibre

These different distance targets allow for optimisation of the cost and complexity of the optical components to best serve the needs of the customer. This optimisation allows the customer to select the lowest cost technology that fits their needs. For example, the lowest cost singlemode lasers will likely serve the 2 km target, which fits the distance requirements of most campus applications. The 10 km distance target serves those customers with multiple locations within a town, but will require more expensive lasers. And the 40 km target is aimed at metro-ring applications using sophisticated laser technology.

The two multimode targets are:

1. at least 100 m on installed-base multimode fibre
2. at least 300 m on multimode fibre

The 100 m distance of target 1 reflects the committee' s recognition that it will be difficult to support the installed base of multimode fibre to lengths sufficient for most building backbones. Solutions have been proposed that use parallel transmission schemes with multiple transmitters and receivers to extend the installed base fibre support beyond 100 m, but even these complex solutions will be hard pressed to achieve the 300 m needed to properly serve in-building distribution. The second target reflects the committee' s desire to enable extended distances on new multimode fibre. While it does not disallow the standardisation of a solution for the installed base, supporting the installed base to 300 m is not required to fulfil this objective. This gives the committee the directive to define a 300 m multimode solution that can take advantage of the emerging improvements in multimode fibre performance, such as offered by the LazrSPEED solution.

In defining the requirements for the next generation multimode fibre, several key objectives were established, designed to provide the best solution for the customer. These objectives are to:

1. Meet the needs of in-building network applications with a single universal medium,
2. Support lowest cost and complexity system solution for LAN applications to a minimum of 10 Gb/s,
3. Provide a smooth migration path for applications from 10 Mb/s to 10 Gb/s using achievable technology,
4. Remain friendly for cable manufacturer and installer.

Figure 1 - In-Building Fiber Backbone Distance Distribution
7/96 IEEE survey of 107 U.S. companies of diverse size, location, and industry

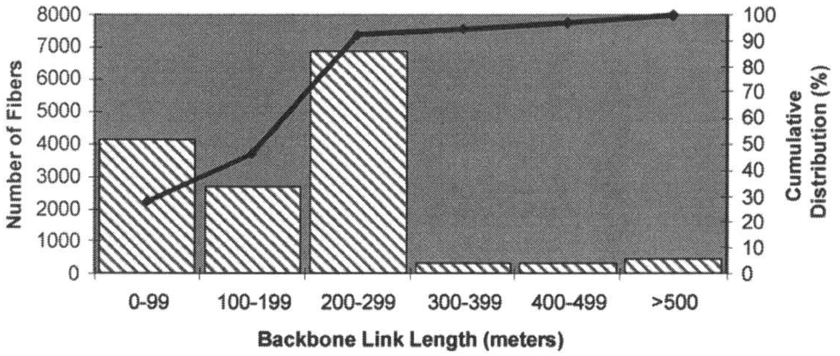

Several surveys have consistently shown that the vast majority of building backbones are less than 300 metres in length. Data from one such survey, gathered by the IEEE during the development of Gigabit Ethernet, is shown in Figure 1.

The new cable is designed specifically to support 300-metre link lengths for 10 Gb/s applications. Its industry standard 50 μm core size couples sufficient power from LED sources to support legacy applications like Ethernet, Token Ring, FDDI, Fast Ethernet and ATM for in-building network distances. The 50 μm core size is also directly compatible with laser-based applications like Gigabit Ethernet. Together, these capabilities satisfy the first objective of meeting the needs of in-building networks with a single universal medium.

Bell Laboratories examined the cost of various 10 Gb/s alternatives to determine the lowest cost solution. The study considered the use of coarse wavelength division multiplexing (CWDM) on the installed base of multimode fibre, upgrade paths using singlemode fibre and long wavelength (1300 nm) lasers, and upgrades to an advanced next generation multimode fibre using short wavelength (850 nm) vertical cavity surface emitting lasers (VCSELs). Figure 2 shows the relative costs associated with each of these alternatives for the upgrade of a building riser backbone from 1 Gb/s Ethernet (GbE) to 10 Gb/s. The costs are estimated. The exact costs will differ in each case depending on the specific configuration and costs actually charged to the customer at the time of purchase. In each of the cited cases it is evident that the cost of the electronics far exceeds the cost of the cabling.

Figure 2 - 1 GbE Riser + Upgrade Riser to 10 GbE
Relative Cost Projection for Typical Configuration

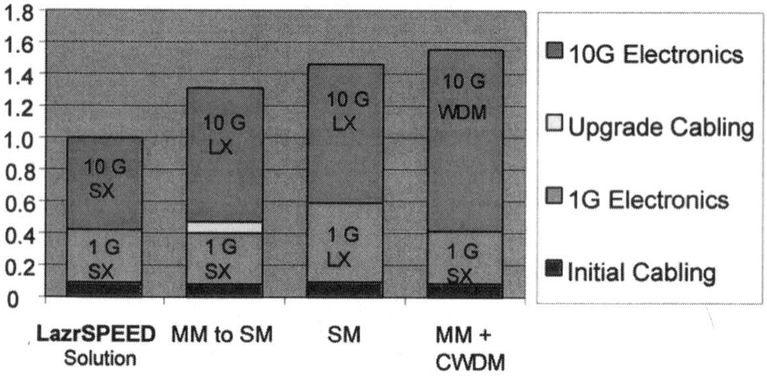

(Electronics include LAN switch plus labour for installation. Cabling includes cable, apparatus, and labour for installation)

The right-most bar, labeled "MM + WDM", shows the projected costs of using the installed base of multimode fibre. At 1 Gb/s, the electronics are cost-effective short wavelength (850 nm) Gigabit Ethernet devices, known as 1000BASE-SX, or SX for short. But, migration to 10 Gb/s requires use of CWDM electronics that have the highest projected cost due to the need for multiple lasers, combiners, splitters, filters and detectors.

The next bar, labeled SM, shows the projected costs of using singlemode fibre for both 1 Gb/s and 10 Gb/s applications. At 1 Gb/s, the electronics are long wavelength (1300 nm) Gigabit Ethernet devices, known as 1000BASE-LX (abbreviated LX in the figure). Initially at 1 Gb/s, LX devices are more expensive than SX. But, the upgrade to 10 Gb/s would allow the use of single-channel singlemode electronics (10G LX) that are simpler than CWDM electronics. So overall this solution is less expensive than the first.

The bar labeled "MM to SM" shows the cost of upgrading an existing multimode infrastructure to singlemode for 10 Gb/s. In this case, use of SX devices instead of LX at 1 Gb/s more than offsets the cost of the additional singlemode cable installation, making this upgrade path less costly than the previous two.

The left-most bar, labelled "LazrSPEED", shows the projected cost of using the new multimode fibre cabling system. Because this solution allows the use of single-channel short wavelength electronics for both 1 and 10 Gb/s, it has the lowest projected cost. Thus, the new solution satisfies the second objective by providing the lowest cost and lowest complexity solution for 10 Gb/s applications.

Work is underway to develop and perfect low-cost 10 Gb/s VCSEL and detector technology operating at 850 nm wavelength, within the laboratories of several companies. Bell Laboratories worked with these devices and leading-edge test and measurement equipment to develop a fibre cabling system that supports these technologies as they emerge into real-world applications. Because the new connectivity solution supports existing applications as well as future 10 Gb/s applications without changing any of the cabling or connecting hardware, it automatically provides a ready upgrade path simply by plugging in new electronics endpoints. The emerging 10 Gb/s VCSEL technology, combined with this total support flexibility, satisfies the third objective of providing a smooth migration path from 10 Mb/s to 10 Gb/s with achievable technology.

The last objective is essential to industry acceptance. The new fibre must remain easy to cable and install. The cable uses the same termination and test procedures already familiar to installers. But the fibre is vastly superior to conventional 50 μm fibre in one very important aspect – performance.

So what' s is the secret of the new solution capability?

Simply put, the cable is highly optimised for 10 Gb/s 850 nm VCSELs while retaining the performance required to fully support legacy applications. The critical performance parameter needed for 10 Gb/s transmission is laser bandwidth. Modelling and experiments prove that the laser bandwidth of the fibre is sufficient to provide robust link performance over 300 metres of cable.

This remarkable performance is attributed to careful attention to the light propagation properties of the fibre. In travelling from one end of a multimode fibre to the other, the light pulses can take many different paths, called modes. Like cars travelling on a multi-lane highway, each lane can travel at different speeds.

In order for the fibre to support very high speed digital transmission, the modes must all travel at nearly the same speed so that one pulse does not mix with the next. The allowable difference in the propagation time between modes over the entire length of the link must be much less than the duration of one bit. As the data rate increases, the pulses become shorter and the propagation time between modes must be nearly equal. This effect can be graphically depicted by plotting the difference in propagation time across all lanes of the multimode highway, as shown in figure 3.

Figure 3 - Received laser pulse Vs position in core

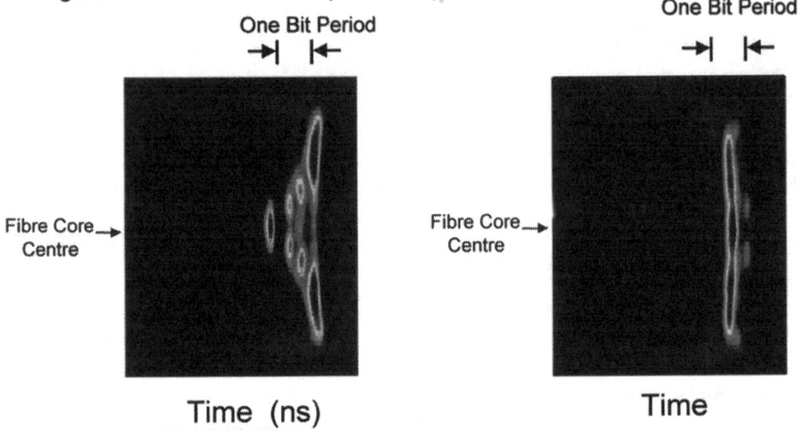

The fibre characteristics for conventional multimode shown on the left cause the light pulse to arrive at different times so that the signal energy is spread out over many bit periods causing interference and errors. The characteristics of the new fibre shown on the right keep the signal concentrated in a single wave that doesn't cause interference between bits allowing error-free transmission.

In summary, the key points of this new building backbone technology are:
1. Fast – Unprecedented 10 Gigabit/second performance using 850 nm electronics.
2. Flexible – 10 Megabits/second through 10 Gigabits/second, all on the same fibre!
3 Cost Effective – Allows network designers to use lower cost short wavelength electronics, providing lowest system cost for a 10 gigabit solution.

Broadband Access, WDM Metro and Network Management
D.W. Faulkner and A.L. Harmer (Eds.)
IOS Press, 2000 © AKM

All Optical Bit Skew Compensation Schemes for High-Speed, Long Distance Bit Parallel WDM Systems

M. E. V. Segatto[1], R. Kashyap, G. D. Maxwell, D. Moodie
Corning Researsh Centre, Martlesham Health, Ipswich IP5 3RE, United Kingdom
e-mail: segatto@ele.ufes.br
J. R. Taylor
Femtosecond Optics Group, Physics Department, Imperial College, Prince Consort Road, London SW7 2BZ, United Kingdom
V.A. Bhagavatula, G.E. Berkey, and A. Evans
Technology Division, Sullivan Park, Corning Inc., Corning, NY 14831, USA
email: bhagavatva@corning.com

ABSTRACT

Two different schemes for all optical bit skew compensation in bit parallel WDM systems are presented in this paper. In the first scheme, Dispersion Managed Fibers (DMF) are used as a replacement for the standard or DS fibers. In the second scheme, fiber gratings are used as compensator elements. An expression for the total bit skew as a function of the fiber dispersion and system bandwidth is derived and compared with experimental results. A 4 bit x 10 Gbit/s-per bit BP-WDM transmission experiment over 30 km DMF is used and an aggregate bite rate×distance (BL) product of 1.2 Tbit/s-km is obtained. The total bit skew of the system is reduced to a half of the bit period.

I - Introduction

High-performance computing systems such as high-speed mainframes, distributed systems, and supercomputers require expensive electronic circuits to perform the serial-to-parallel (S2P) and parallel-to-serial (P2S) conversions in serial communications. A different approach is to transmit all of the data in parallel form. Parallel transmission has several advantages over serial transmission since it requires no S2P and P2S conversions. The time needed to transmit one single digital word in series is ***n*** (number of bits in the word) times that required for parallel transmission, and consequently requires transmission rate that is ***n*** times larger. The drawbacks of present parallel systems are their limited short distance range and low bit rates but can be easily overcome with the help of Wavelength Division Multiplexing (WDM) technology.

Parallel optical communication may be achieved in two different ways as shown in Fig. 1a and 1b: (i) *parallel-per-fiber*, where each bit of the digital word is sent in a different fiber; and (ii) *parallel-per-wavelength*, where all bits are sent in the same fiber but using different wavelengths. Several research projects and consortia have been working with parallel-per-fiber systems [1-4] and some commercial products are already available. At high bit-rates, temporal alignment is difficult to maintain since each bit has a different optical path. The lack of alignment is the main restriction of this system and its application is limited to very short distances and low bit-rates [5].

In the parallel-per-wavelength system (also known as Bit-Parallel WDM system, or BP-WDM), each bit of the digital word is used to modulate one different optical wavelength and these wavelengths are then launched into the same fiber. Loeb and Stilwell [6] proposed the first BP-WDM system in 1988. In that work, they showed that it is possible to transmit up to 1 Gb/s per channel for a few kilometers using standard Single Mode (SM) fiber. Ten years later, Bergman *et*

[1] Federal University of Espírito Santo, Brazil. Also on leave of absence at Imperial College. This work was partially supported by CAPES/Brazil.

al [7] proposed a new system working at 1Gbit/s, 2 bits over 32-km of Dispersion Shifted (DS) fiber.

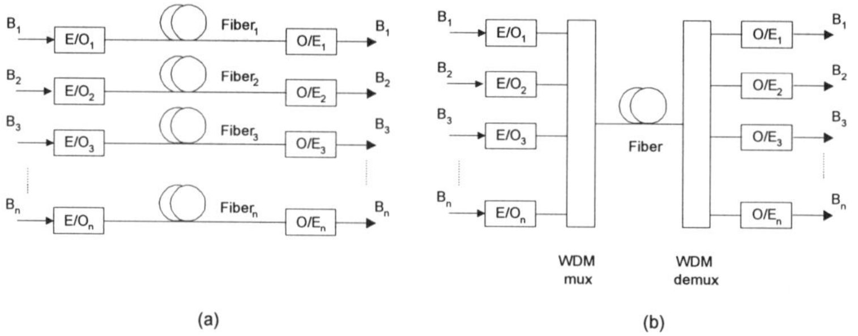

Fig. 1. Parallel optical systems. (a) Parallel-per-fiber, and (b) parallel-per-wavelength systems. E/O and O/E are respectively, the Electrical/Optical and Optical/Electrical converters.

In this paper, we study the advantages of using BP-WDM for long distance and high-speed systems as an alternative to optical serial and to optical parallel-per-fiber transmission. As each bit is represented by a different wavelength, each bit suffers a different group delay due to the dispersion characteristics of the fiber and bit skew is the main degradation parameter for this system. We propose two different schemes to optically compensate for bit skew in the system. In the first scheme, a Dispersion Managed Fiber (DMF) is used as the transmission media. The low residual dispersion and slope of this fiber greatly reduces the bit skew. In the second scheme, fiber gratings are used as delay and dispersion compensators.

The BP-WDM system is presented in Section II. An expression for the fiber bit skew is derived as a function of the fiber parameters and the spacing between wavelengths. Section III discuss all optical bit skew compensation schemes. In Section IV results are given for an experimental 4 bit, BP-WDM system at 10 GB/s. Conclusions are given in Section V.

II – The Bit-Parallel WDM System

The main difference between parallel and conventional WDM is that in parallel systems all channels are used to transmit information of the same binary word and the bits must arrive at the receiver end at the same time. The transmitter, fiber and receiver contribute to the bit skew. In the transmitter, bit skew is mainly due to the differences in the laser turn-on delay times. In the fiber, dispersion is the major contributor. The receiver contributes to bit skew due to the optical filters used in the demultiplex process. This contribution can be neglected compared to the fiber contribution for bit rates up to 10 Gbit/s [8-9].

The bit skew in the fiber can be written as a function of the dispersion as follow: The total dispersion in a single mode fiber, expressed as a function of the dispersion slope S_0 the zero-dispersion wavelength λ_0 and wavelength λ, is given by [10]

$$D = \frac{\lambda S_0}{4}\left[1 - \left(\frac{\lambda_0}{\lambda}\right)^4\right] \tag{1}$$

The difference in the group delay between the fastest and the slowest channel determine the maximum bit skew. For the sake of simplicity, we assume that all channels are on the same side of the zero-dispersion wavelength. Then, the delay between the first channel and the last channel gives the maximum bit skew. If we define a system with n channels (i.e. a digital word with n bits) equally spaced and total bandwidth $\Delta\lambda$, the wavelength of the first channel and of the last channel can be expressed as a function of the central wavelength λ_c as

$$\lambda_1 = \lambda_c - \Delta\lambda/2, \tag{2.a}$$

$$\lambda_n = \lambda_c + \Delta\lambda/2 . \tag{2.b}$$

The group delay τ_g in a fiber of length L is determined by integrating the total dispersion over the wavelength, and

$$\frac{\tau_g}{L} = \int D(\lambda')d\lambda'. \tag{3}$$

The maximum bit skew per unit of length is given by

$$\frac{\Delta\tau_g}{L} = \int_{\lambda_1}^{\lambda_n} D(\lambda')d\lambda'. \tag{4}$$

Substituting (1) in (4) and using (2.a-b), we have

$$\frac{\Delta\tau_g}{L} = \Delta\lambda \frac{\lambda_c S_0}{4}\left[1 - \frac{\lambda_0^4}{\left(\lambda_c^2 + \Delta\lambda^2/4\right)^2}\right]. \tag{5}$$

Assuming that $\lambda_c >> \Delta\lambda$, then

$$\frac{\Delta\tau_g}{L} \cong \Delta\lambda \frac{\lambda_c S_0}{4}\left[1 - \left(\frac{\lambda_0}{\lambda_c}\right)^4\right] = \Delta\lambda D(\lambda_c). \tag{6}$$

Eq. (6) shows that the maximum bit skew is a function of the total bandwidth and of the dispersion of the fiber at the central wavelength and it is still valid even if different expressions are used for the dispersion in (1) [11]. The only assumptions made here are that the central wavelength must be much greater than the total bandwidth and that all wavelengths are in the same side of the dispersion curve.

III Bit Skew Compensation

A. THE DISPERSION MANAGED FIBER SCHEME

Recently, a new kind of fiber has been reported [12]. Alternating sections of positive and negative dispersion and slope are pulled in a single continuous fiber without splices, forming a Dispersion Managed Fiber (DMF). Fig. 2 shows the fiber, the first section is a single mode fiber with positive dispersion, positive slope and length L_1 and the second section has negative dispersion and slope and length L_2. The dispersion, slope and length of each section are chosen in such a way that the total dispersion and the slope are virtually zero and a flat dispersion fiber is obtained. As the fiber is formed by sections of single mode fibers with high positive and negative dispersion, nonlinear effects such as Four-Wave-Mixing and Cross-Phase-Modulation are reduced [12].

Fig. 2 The Dispersion managed fiber. Type 1 is positive dispersion and slope and Type 2 is negative dispersion and slope.

For the DMF, the total bit skew can be computed by adding the contribution of each section of the fiber

$$\Delta\tau_g\Big|_{TOTAL} = L_1\Delta\tau_{g1} + L_2\Delta\tau_{g2} \tag{7}$$

where L_i, $\Delta\tau_{gi}$ (i = 1,2) are, respectively the length and the maximum bit skew in the sections 1 and 2 of the DMF.

It is easy to prove that if [11] $L_1S_{01} = -L_2S_{02}$ and $\lambda_{01} = \lambda_{02}$, where S_{0i}, and λ_{0i} (i = 1,2) are respectively, the dispersion slope and zero-dispersion wavelength of each section of the fiber, the bit skew is nulled in the DMF.

B. THE FIBER GRATING SCHEME

Fiber gratings have received considerable attention lately [13] because of their low-insertion loss, passive operation and versatility. Applications range from filters, pulse compression to dispersion compensation [14].

Fiber gratings also provide an alternative for Optical Bit Skew Compensation (OBSC) in BP-WDM systems. Fig 3a shows a proposed OBSC system using fiber gratings and a circulator. Light coming from the fiber enters into the circulator via port 1 and it is reflected by the grating in port 2. Two different kinds of gratings can be used:

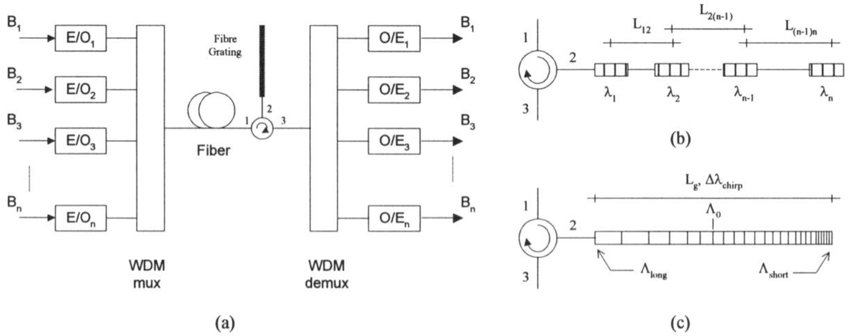

Fig. 3. (a) The fiber grating scheme used to compensate the bit skew. (b) One high-reflectivity grating per wavelength and, (c) one chirped grating.

B.1 One High-Reflectivity Grating per Wavelength

The simplest method to compensate the bit skew is to use a number of high reflectivity gratings equal to the number of wavelengths transmitted. Each grating reflects one wavelength (Fig. 3b). The gratings work as selective mirrors and the bit skew in the gratings are defined by the separation among the gratings.

B.2 One Chirped Grating

A single chirped grating could be used to compensate the bit skew (Fig. 3c). The reflection from a chirped dispersion compensating grating is a function of wavelength, and therefore, light entering into a positively chirped grating suffers a delay τ on the reflection that is approximately

$$\tau(\lambda) \approx \frac{(\lambda_0 - \lambda)}{\Delta\lambda_{chirp}} \frac{2L_g}{v_g}, \qquad \text{for } 2n_{eff}\Lambda_{short} < \lambda < 2n_{eff}\Lambda_{long}, \tag{8}$$

where λ_0 is the Bragg wavelength at the center of the chirped bandwidth of the grating, L_g the grating length, $\Delta\lambda_{chirp} = 2n_{eff}(\Lambda_{long} - \Lambda_{short})$, v_g is the average group velocity of the light in the fiber, and n_{eff} is the effective refractive index. One great advantage of the use of chirped grating is that we can compensate for bit skew and dispersion at the same time.

IV. Experimental Demonstration of a 4 bit, BP-WDM system at 10 GB/s

In this section, a 4 bit x 10 Gbit/s BP-WDM transmission experiment over 30 km of DMF is presented. Fig. 4 shows the experimental setup. The transmitter had 4 DFB lasers diodes (LD) with wavelengths 1557.03 nm (bit 1), 1558.74 nm (bit 2), 1560.87 nm (bit 3), and 1561.77 nm (bit 4). For simplicity, the outputs were multiplexed (using a 4x1 coupler) and simultaneously modulated by a $LiNbO_3$ Mach-Zehnder modulator with 10 Gbit/s $2^{15} - 1$ PRBS, although in a field transmission system each bit would be modulated separately and then multiplexed. In order to simulate a truly independent pseudo-random bit stream for each of the wavelengths, a simple scheme using a 4.8 km long SM fiber was implemented. The high dispersion of the SM fiber creates a different delay for each wavelength and at the end of the fiber each bit stream is separated by at least one bit period.

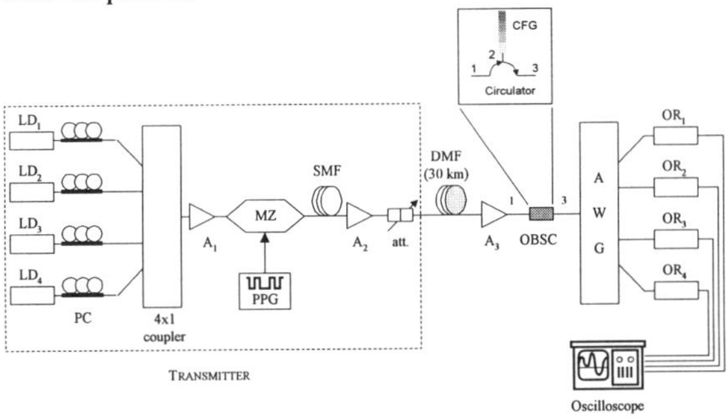

Fig. 4. The experimental setup. LD: DFB laser diode, PC: polarization controller, A: EDFA optical amplifier, MZ: $LiNbO_3$ Mach-Zehnder modulator, PPG: pulse pattern generator, SMF: single-mode fiber, att.: variable optical attenuator, DMF: dispersion managed fiber, OBSC: optical bit skew compensator, CFG: chirped fiber grating, AWG: arrayed-waveguide-grating demultiplexer, OR: optical receiver.

After the SM fiber, the 4 optical signals were transmitted over the DMF with optical power of 0 dBm per wavelength. The dispersion parameters of the DMF (L_1, S_{01}, L_2, and S_{02}) were not optimized. The residual dispersion slope was 0.03 ps/nm^2·km, the zero-dispersion wavelength was 1535.8 nm and the loss was 0.4 dB/km. An arrayed-waveguide grating (AWG) [15, 16] (fabricated at Corning Research Centre/UK using the flame-hydrolysis process), with inter-channel cross-talk of 25 dB and insertion loss of 10 dB was used as a demultiplexer. Optical amplifiers were used to compensate for the losses in the coupler, fibers and AWG. An optical variable attenuator also was used after the SM fiber to minimize the nonlinear effects by limiting the input optical power in the DMF. In the future, the 4x1 coupler will be substituted by an AWG with the same characteristics of that used as demultiplexer. The inset in Fig. 4 shows an optical bit skew compensator (OBSC) using a three-port circulator and a fiber grating.

Fig. 5 shows a picture of the four transmitted and received pulses. A long sequence of zeros followed by a single "one" was used to modulate the optical signal. Fig. 5a shows the four bits after the SM fiber. The minimum and maximum delay generated by the fiber were, respectively, 110 and 500 ps. After the DMF (Fig. 5b), the maximum delay has increased to 600 ps including a maximum bit skew of 100 ps.

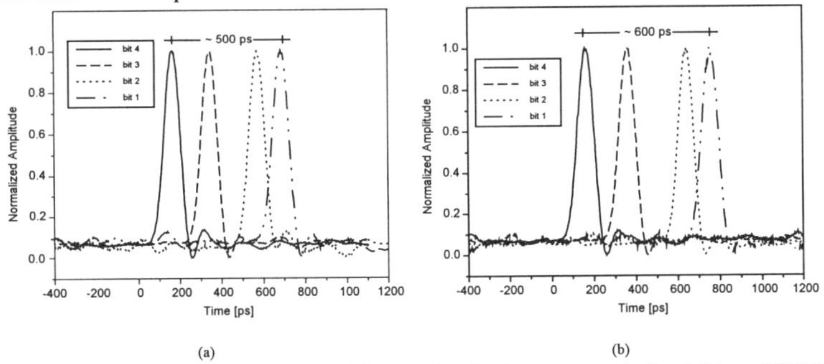

Fig. 5. The four transmitted and received pulses: (a) transmitted pulses measured after 4.8 km of SM fiber and (b) received pulses after 30 km of DMF.

The same experimental setup was used to measure the bit skew as a function of the total bandwidth ($\Delta\lambda$) for the DMF and for a DS fiber (S_0 = 0.08 ps/nm^2·km and λ_0 = 1537 nm). In this experiment, one fixed wavelength laser and one variable laser are used. As the wavelength changes, $\Delta\lambda$ and λ_c also change. The results for both fibers, shown in Fig. 6, show a good agreement with those obtained theoretically with Eq. 6 (continuous lines in Fig. 6). A 100 ps maximum bit skew represents a system bit skew of ±50 ps, i.e. a half of the bit period. Those values can be easily compensated by the receiver circuits or optically.

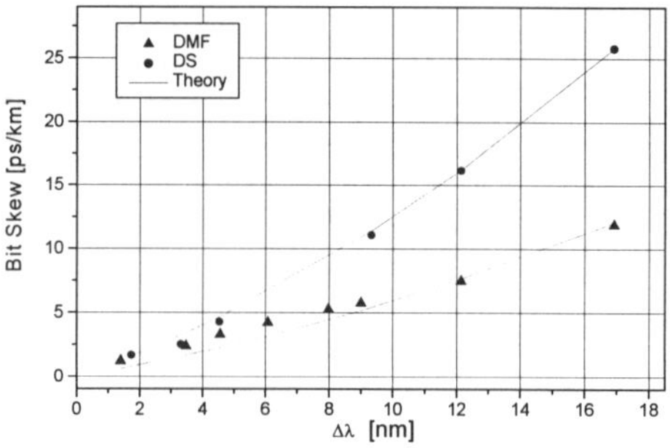

Fig. 6. Measured maximum bit skew as a function of the total bandwidth ($\Delta\lambda$) for the DMF and a DS fiber with S_0 = 0.08 ps/nm^2·km and λ_0 = 1537 nm. The straight lines were obtained by Eq. (6).

Fig. 7 shows demultiplexed 10 Gb/s waveforms for back-to-back and after DMF and the first and the last bits. The eye closure after the fiber is due mainly to noise in the optical amplifiers used in the system. The bit error rates are shown in Fig. 8 for Back-to-Back (BB) and afterthe fiber (AF) measurements. The power penalty is 0.8 dB and 2.9 dB at 10^{-9} BER for the best and the worst case, respectively. The power penalty is mainly due to crosstalk in the AWG.

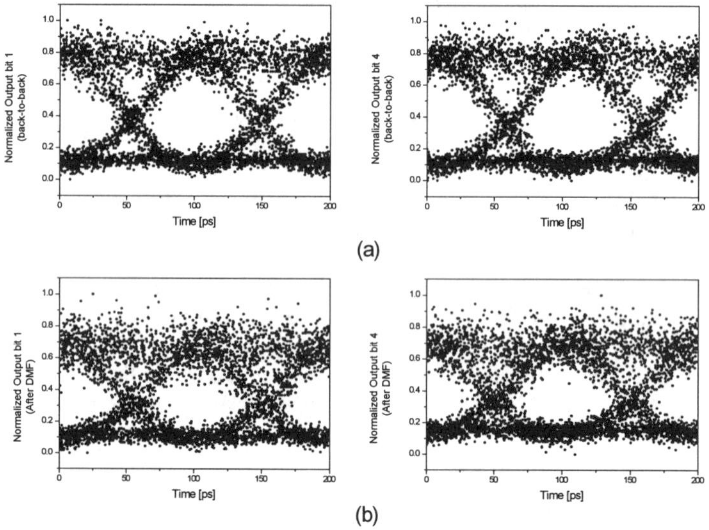

Fig. 7. Eye diagrams for the first and last bits: (a) back-to-back and (b) after DMF measurements.

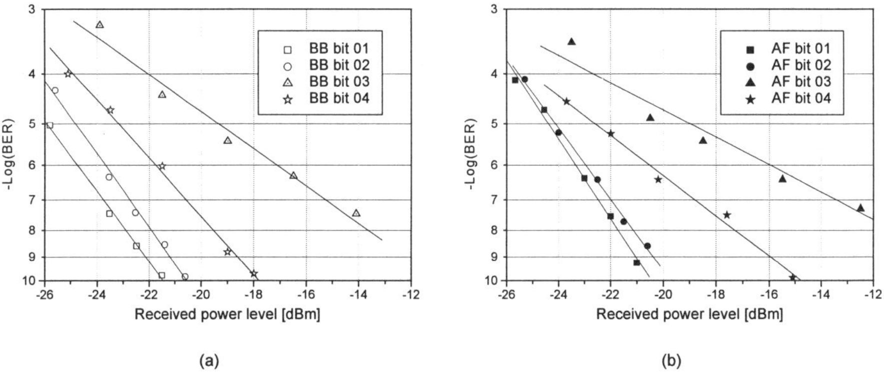

Fig. 8. Bit error curves for (a) Back-to-Back (BB), and (b) After fiber (AF) measurements.

V. Conclusions

In this paper, we have proposed two different schemes for all optical bit skew compensation in bit parallel WDM systems. In the first scheme, dispersion managed fibers are used as a replacement to standard or DS fibers. The characteristics of these fibers have been briefly presented, and expressions for the minimum bit skew have been given. The advantage of the DMF over DS fibers for this kind of application also has been shown. In the second scheme, fiber gratings are used as compensator elements.

A 4 bit x 10 Gbit/s-per bit BP-WDM transmission experiment over 30 km DMF has been used, and an aggregate BL product of 1.2 Tbit/s-km was obtained. The system bit skew was ±50 ps, which represents a half of the bit period. This bit skew can be easily compensated optically or electronically. An OBSC using fiber gratings was presented and is being currently implemented. A higher BL product can be achieved if a denser WDM system is employed.

We believe that systems using BP-WDM will be useful for computer interconnects in high-speed parallel systems and ring networks.

References

[1] N. Fujimoto, K. Wakao, M. Yano, "Intelligent parallel optical link - Parallel optical transmission system and LSIs," *FUJITSU Sci. Tech. J.*, 30, 2, pp. 214-23, December 1994.

[2] K. H. Hahn, "POLO-parallel optical links for gigabyte data communications," *Proceedings. 45th IEEE Electronic Components and Technology Conference*, pp. 368-75, May 1995.

[3] J. D.Crow, et. al., "The Jitney parallel optical interconnect," *Proceedings 46th IEEE Electronic Components and Technology Conference*, pp. 292-300, May 1996.

[4] Y. M. Wong, et. al., "OptoElectronic Technology Consortium (OETC) parallel optical data link: components, system applications, and simulation tools," *Proceedings 46th IEEE Electronic Components and Technology Conference*, pp. 269-78, May 1996.

[5] N. Fujimoto, A. Ishizuka, H. Rokugawa, and K. Mori, "Skew-Free Parallel Optical Transmission Systems," *IEEE J. Lightwave Technol.*, pp.1822-1831, vol. 16, October 1998.

[6] M. L. Loeb, and G. R. Stilwell, Jr., "High-speed data transmission on an optical fiber using a byte-wide WDM system," *IEEE J. Lightwave Technol.*, vol. 6, pp.1306-1311, August 1988.

[7] L. Bergman, J. Morookian, and C. Yeh, "An all-optical long-distance multi-Gbytes/s bit-parallel WDM single-mode fiber link," *IEEE J. Lightwave Technol.*, vol. 16, pp.1577-1582, September 1998.

[8] M. E. Vieira Segatto, et. Al, "High-speed transmission and dispersion characteristics of an AWG," *in Proc. WFOCP2000*, Pavia-Italy, June 2000.
[9] G. Lenz, B. J. Eggleton, C. R. Giles, C. K. Madsen, and R. E. Slusher, "Dispersive properties of optical filters for WDM systems," *IEEE Journal of Quantum Electronics*, vol. 34, no. 8, pp. 1390-1402, August 1998.
[10] J. M. Senior, *Optical Fiber Communications: Principles and Practice*, London: Prentice Hall, 2nd ed., p. 129, eq. 3.55, 1992.
[11] M. E. Vieira Segatto, J. R. Taylor, and R. Kashyap, "Use of Dispersion Managed Fibers in Parallel WDM Systems," *in Proc. IMOC99, Ref. I312*, Rio de Janeiro-Brazil, August 1999.
[12] V. A. Bhagavatula, G. Berkey, D. Chowdhury, A. Evans, V. Srikant, and M. Cain., "Novel fibers for dispersion managed high-bite-rate-systems," *in Proc. OECC98, Ref. 15C1-2*, pp. 362-363, July 1998.
[13] R. Kashyap, "Fibre Bragg Gratings, " Academic Press, 1999.
[14] J. A. R Williams, K. Sugden, L. Zhang, I. Bennion, and N. J. Doran, "In-fiber grating systems for pulse compression and complete dispersion compensation," *IEE Colloquium on 'Optical Fibre Grating and Their Applications* (Digest No.1995/017), 30, pp. 9/1-6, January 1995.
[15] H. Takahashi, K. Oda, K. Toba and Y. Inoue, "Transmission characteristics of arrayed waveguide N *x* N wavelength multiplexer," *IEEE Journal of Lightwave Tech.*, vol. 13, no. 3, pp. 447-455, March 1995.
[16] C. Dragone, A. Edwards, and R. C. Kistler, "Integrated optics N x N multiplexer on silicon," *IEEE Photonics Tech. Letters*, vol. 3, pp. 896-899, October 1991.

Broadband Access, WDM Metro and Network Management
D.W. Faulkner and A.L. Harmer (Eds.)
IOS Press, 2000 © AKM

Fibre Channel 100-TP-EL-S: Gbit/s transmission via Cat.7 cable

Prof. Dr. Albrecht M. Oehler
Fachhochschule Reutlingen, University of Applied Sciences,
School of Electronic Engineering, Alteburgstr.150, 72762 Reutlingen, Germany

1. Fibre Channel New Work Item

The international subcommittee ISO/IEC JTC 1 SC25, which is responsible for the "Interconnection of Information Technology Equipment", just started a new project on the physical fibre channel interface 100-TP-EL-S:

This stands for

100 MByte/s, i.e. 1,0625 Gbit/s with 8B/10B endconding,
Twisted **P**air transmission medium,
Electrical transmission and
Short length, i.e. 100 m in this case.

The scope of the New Work Item is as follows:

Specify 100 MB/s (1,0625 Gbit/s) physical layers that will operate over 100 meters of Category 7/class F balanced cabling, as defined in ISO/IEC SC 25/WG 3 N 552.

Type 1:
The physical interface will tansmit 50 MB/s (503,125 Mbit/s) using NRZI (Non Retrun to Zero Inverted on Ones) Modulation on each of two pairs and receive 50 MB/s (503,125 Mbit/s) on each of the other two pairs of a 4-pair cable: This simplifies the endsystem as it eliminates the need of multi level encoding and the need to support duplex transmission on each pair.

Type 2:
The physical interface will transmit 100 Mbit/s (1,0625 Gbit/s) using MLT-3 encoding (Multi Level Transmit) on one pair and receive 100 Mbit/s (1,0625 Gbit/s) on the other pair, which eliminates the need to support duplex transmission on each pair. This interface enables the fibre channel to use one pair of balanced cable instead of one fibre.

Both types support full and half-duplex operation and provide a Bit Error Rate (BER) performance of better than 10^{-10}.

There was a great international support of this new work item. 15 countries voted in favour of it, only one country against it. Germany proposed the project editor.

2. Balanced class F link

The international standard for information technology customer premisis cabling ISO/IEC 11801 is under revision. The newest working draft reflecting the second edition is ISO/IEC JTC 1/SC 25/WG 3 N 568 from 6th October, 1999.

New cabling classes were added, i.e. class E with frequencies up to 250 MHz and class F with frequencies up to 600 MHz. In the new model the horizontal link contains a maximum of four connections and its length is maximum 100 m.

Furthermore new electrical characteristics will be included. Thus beside the return loss an “Attenuation to Return Loss Ratio” was discussed, similar to the Attenuation to Crosstalk (ACR) ratio of the actual version.

The cat.7 telecommunication outlet (TO) will be either a backward compatible EN 60603-7 or the fallback choice of the working group 3, which is a totally new connecting hardware presumable called IEC 61076-3-104.

Fig. 1 shows the horizontal link using only one connection at each end, so called interconnect - TO environment.

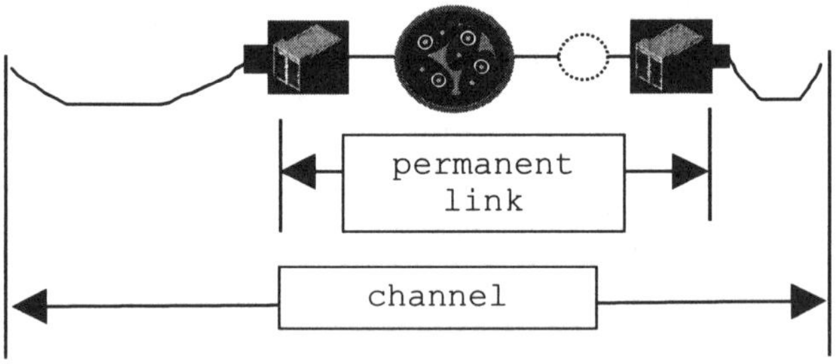

Fig.1 Horizontal link class F using interconnect and TO with fall back postion connecting hardware

3. Interface Type 1

This physical interface will tansmit using two pairs in parallel. The feasibility of this interface is proofen as given in [1]. The output voltage is 2 Volt peak to peak (see fig.2 of [1]). The input voltage after equalization is 0,12 V (see fig.3 of [1]).
Figures 2 and 3 show the eye pattern of a 600 Mbit/s signal instead of the 503,125 Mbit/s data stream. This is to prove the feasability and to show the safety margin of this interface.
The resulting BER of $0{,}5\cdot10^{-11}$ is significant below the requested BER.

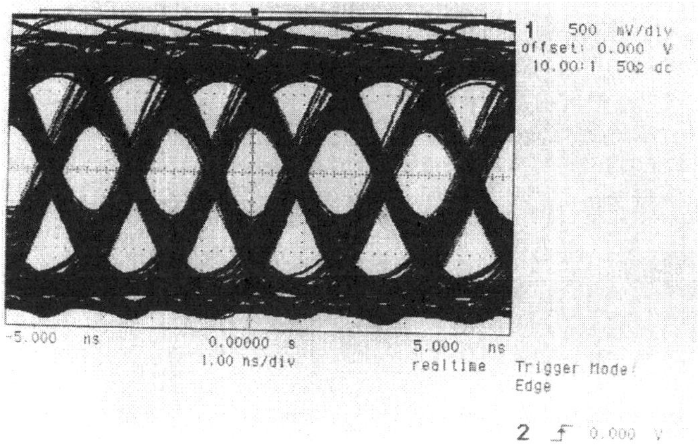

Fig.2 Output eye diagram at the near end

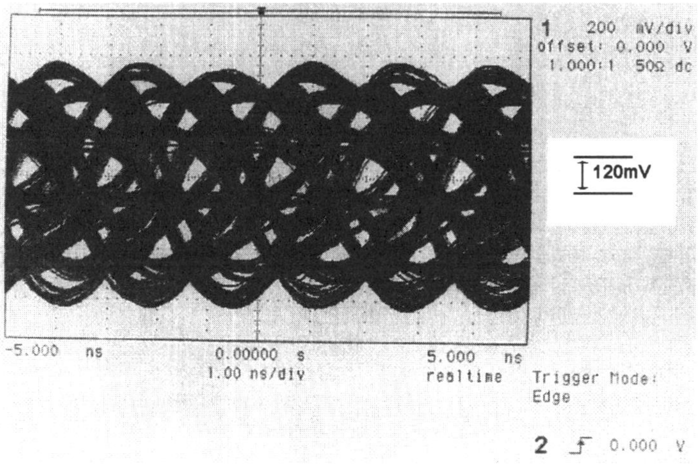

Fig. 3 Input eye diagram after the equalizer amplifier

4. Interface Type 2

The advantage of this interface type is the interchangeability of optical lines with copper lines, and vice versa. If only one pair of balanced class F cabling is needed for the transmission of 100 MB/s (1,0625 Gbit/s) in one direction, this corresponds to one fibre only per direction. Thus only the physical medium dependent (PMD) layer has to be changed, which allows easy adaption to the desired transmission medium.

Fig.4 shows the measurement setup for testing of this high speed interface. To realize the high data rate, a fibre channel digital multiplexer was used. This allows the BER measurement at 100 Mbit/s only, whereas the data rate on the balanced cabling can be up to twentyfive times higher, i.e. up to 2,5 Gbit/s.

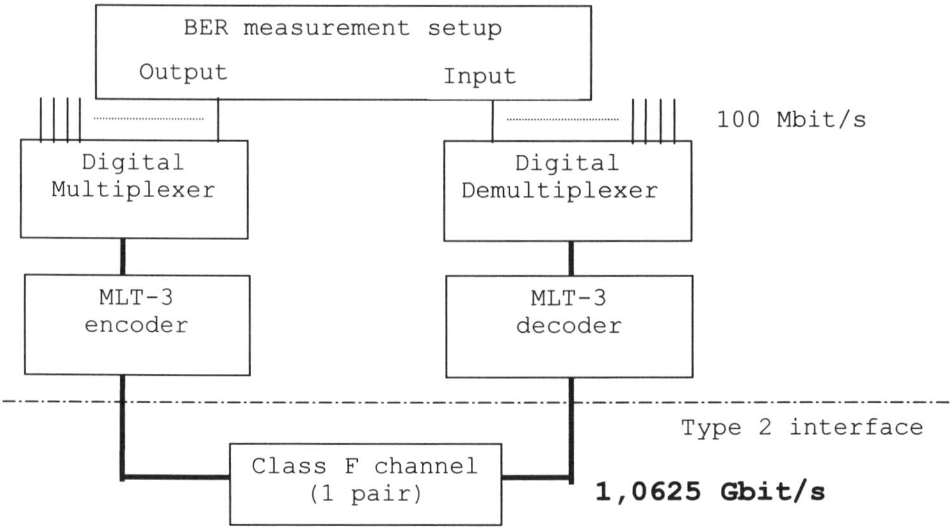

Fig. 4 Schematic of measurement setup for 1,0625 Gbit/s transmission

5. Conclusions

We realized 100MB/s (1,0625 Gbit/s) transmission via 100 Ohm class F cabling without additional digital signal processing. This approach for the 100-TP-EL-S fibre channel physical interface allows the interchangeability of optical fibre medium versus balanced cabling and vice versa. Only the PMD has to be adopted to the chosen transmission medium. This will increase the acceptance of fibre channel interfaces within customer premises.

6. Acknowledgements

The author thanks the consortium of cable manufacturers, co-ordinated by the ZVEI (Zentraler Verband der Elektrotechnik und Elektronikindustrie e.V.), which sponsored this research work.

References

[1] A. Oehler: 1,5 Gbit/s transmission in the horizontal area via two pairs of Class F/Cat.7 cabling; NOC'99 conference, 1999, Delft.

Broadband Access, WDM Metro and Network Management
D.W. Faulkner and A.L. Harmer (Eds.)
IOS Press, 2000 © AKM

Management of Multi-Client Optical Transport Networks

Results from EURESCOM Project P918

Georg Lehr

T-Nova Deutsche Telekom Innovationsgesellschaft mbH, Technologiezentrum
Am Kavalleriesand 3, 64307 Darmstadt, Germany
Tel: +496151-83-7495, Fax: +496151-83-4221, E-mail: Georg.Lehr@.telekom.de

Abstract. Owing to their property of transparency, optical transport networks are considered to be ideal candidates to leverage integration of the transport of different clients over a common platform. In order to be able to reasonably operate such OTN-based Multi-Client Networks it is necessary to develop a management concept that enables the integration of the client and the OTN server layer also from a management point of view. This contribution discusses the impact of client independence on the management concept. In this context, it is described how the existing functional architecture can be enhanced to include digital layer functionality. Furthermore, aspects of the interworking between OTN server layer and client layers are addressed, focussing on IP as a client.

1. Introduction

At this point in time, telecom network operators deploy a huge amount of WDM systems in order to satisfy the increasing need for bandwidth they have to cope with. This development has started with the deployment of WDM point-to-point systems to interconnect SDH nodes. As a next step, IP Routers are being interconnected with the same technology. From a network operator's perspective, using the same technology to transport different clients offers the opportunity to benefit from synergies with respect of network operation and management.

The prerequisite for this is that a common transport platform is available that is able to transport different clients and nevertheless can be operated and managed in a uniform manner. It is well known that optical transport networks have the potential to serve as a basis for such multi-client networks. Owing to the property of transparency, the transport of signals is independent to a large extent from the detailed characteristics of the signal. This is why they can support a lot of different data formats and, in a first approximation, the OTN can be considered client independent. Therefore, the management of the OTN and the client domain should be independent, too.

Imposed by the physics of transmission, however, there are some limitations to the property of transparency described above. These limitations necessitate to employ 3R regeneration under certain circumstances. In this context, the scheme of a digital wrapper or digital container is discussed in order to maintain client independence as far as possible. An appropriate inclusion of this new functionality into the management architecture is an important issue.

Furthermore, in order to optimise the multi-client network with respect of operation and maintenance, a certain level of interworking is required. For instance, in the layer stack for the transport of an IP client, similar or related functionality such as routing, switching, and crossconnection may be present. These closely related pieces of functionality have to be coordinated in an appropriate way to maximise synergies. For this aim, approaches such as Multiprotocol Lambda Switching [1] are being discussed for the transport of IP over OTNs.

This contribution describes aspects of the management of multi-client networks that have been studied by the EURESCOM P918 project. The focus of the contribution is on management aspects of transparency and layer interworking. Moreover, issues of the management of connectionless clients such as IP are addressed.

The paper starts with a description on how optical transport networks are modelled from a management point of view. In a second part, the issue of transparency is discussed and some of its impact on management is pointed out.
Finally, client interworking issues are addressed taking connectionless clients as an example and some preliminary considerations on the solutions to the problems discussed and to the modelling of connectionless clients are given.

2. Modelling Optical Transport Networks

The management concept in general comprises two different elements. As a first element, software systems are necessary to enable operational staff to configure the network and to retrieve information about its status. In the TMN architecture [2], agent software residing on the network element presents to a manager software via a management interface the information on the network elements needed for this purpose. In addition to this, a mechanism of signalling between network elements is implemented which carries information about fault situations encountered in the network. This OA&M (Operation, Administration, and Maintenance) signalling constitutes a second element of OTN Management. Both elements, the management interface between manager and agent and the OA&M signalling has to be specified exactly to enable interoperation of the systems procured from different vendors. Otherwise, the network operator looses the view on the network as a whole ("the end-to-end view") as it is very difficult then to retrieve information from different parts of the network.
For this specification work, the network elements need to be modelled in a way that is independent of the implementation chosen by a specific vendor. Furthermore, a scheme is necessary that allows to systematically arrange the vast bunch of information needed to control the network. ITU-T provides a methodology [3] that supports this goal. For this purpose, a network is divided into layers which describe the basic pieces of functionality implemented in the network. Figure 1 shows how this approach is applied to OTN's [4]. According to the functions implemented in the network elements (amplification, multiplexing, and establishment of end-to-end relationships, see upper part of the Figure), three layers are used to model the network. These layers are called Optical Transmission Section (OTS), Optical Multiplex Section (OMS), and Optical Channel (OCh) layer, respectively. The characteristic signal of the OCh layer, for instance, is a specific single channel signal, whereas the characteristic signal of the OMS layer is a full WDM signal. Within the layers, "adaptation process" and "termination processes", graphically represented by trapezoid and triangles, are defined to model the functionality and its supervision in more detail.

This functional architecture now allows to allocate the management information systematically to managed objects. In the TMN architecture [2], managed objects represent the "containers" for the management information, and the ensemble of all managed objects defines the management interface. In Figure 1, it is shown how the termination point managed objects TTP (trail termination point) and CTP (connection termination point) [5] of the different layers are mapped to the adaptation and termination functions. For instance, the adaptation function between OCh and OMS layer (OMSA) models the multiplexing and demultiplexing of single channel signals. Accordingly, information needed to control this function such as the channel number, or alarms indicating the loss of a single channel signal or misconnections are allocated to the CTP of the upper layer, in this case the sink instance of ochCTP.

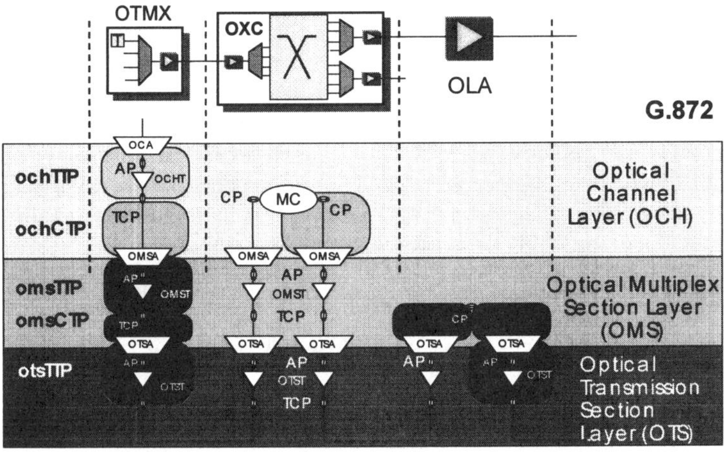

Figure 1: Functional Architecture and Information Model of Optical Transport Networks acc. to [2]

The product, that the network operator provides to his customers, is the end-to-end communication through the network. For this reason, the management of this resource is the major issue rather than the control of single network elements. In terms of modelling, the transparent end-to-end relationship in the OTN is modelled by an ochTrail managed object. As shown in Figure 2, this entity connects two ochTTPs. Following pointer and containement relationships, the network management system can access all information that is relevant for a specific ochTrail, for instance its connectivity, the status of all connections supporting the trail, the status of the termination points, current problems, detected OA&M signals and alarms, physical parameters, and the status of equipment that support the trail.

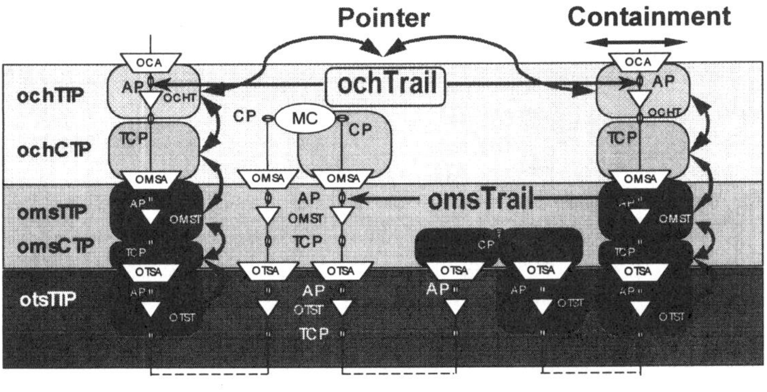

Figure 2: Network level view on functional architecture and information model

Considering this it can be said that the trail (in this case the optical channel trail managed object) presents whole management information "in a nutshell".

3. The Impact of Transparency on the Management of Optical Networks

Due to the analogue nature of transmission no full signal regeneration takes place between two nodes and signal quality is affected by analogue effects such as noise accumulation, fibre non-linearity, dispersion, and crosstalk. When a certain length called "transparency length" or maximum number of crossconnects is exceeded, the optical signal quality as well as the resulting electrical signal-to-noise ratio falls below the values required by the service. Consequently, the signal can be transmitted transparently only within a limited region of the network called "transparent domain". As soon as it is necessary to leave the transparent domain, a full 3R regeneration has to be applied (see Figure 3). 3R regeneration, how it is used today, is a function that is specific for client signal. This means, that transparent transmission, which forms the basis for the client independence, is no longer possible. So, the problem is to find a scheme for 3R regeneration which allows to maintain as much transparency with respect of the transmission format as possible.

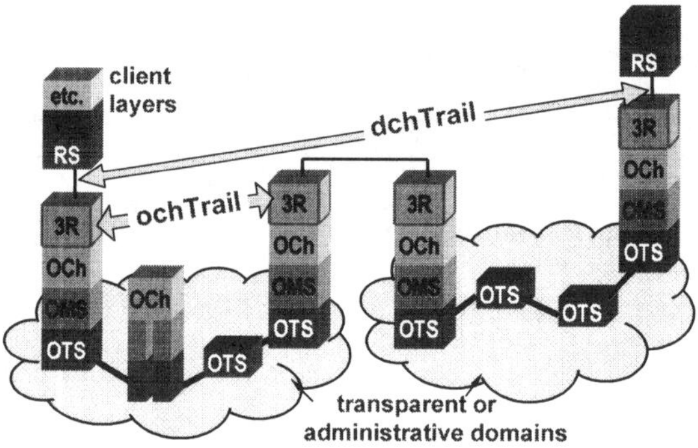

Figure 3: Optical Transport Networks with digital regeneration

A solution to this problem is to map the client signal to a digital frame that allows to perform digital regeneration. In principle, any digital frame can be used for that, but it is natural to use a common transport format such as SDH frames. Another solution, which is discussed intensively in standardisation is called "Digital wrapper" scheme [6]. For this approach (see Figure 4), a frame structure used for undersea cable transmission [7] is used. For this structure, mappings of some client signals into a payload envelope are already defined. In addition, forward error correction (FEC) can be performed to correct transmission errors in the signal. Furthermore, overhead to transport information about the client signal and information necessary to maintain the end-to-end view of this new type of characteristic information can be provided.

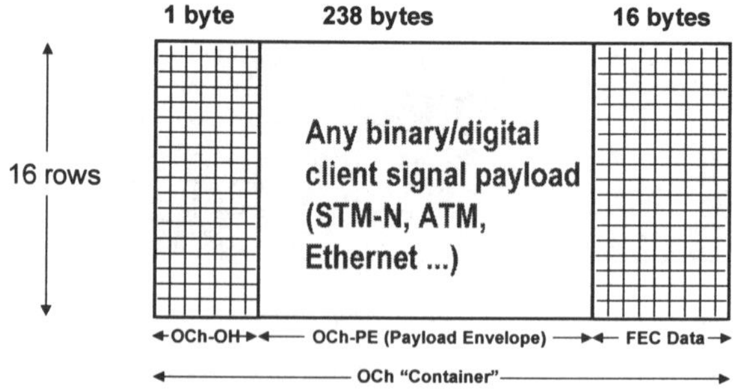

Figure 4: Digital Wrapper Approach

This shows that by introduction of the digital wrapper scheme we get a whole bunch of new functionality that needs to be described in the functional architecture and the OA&M mechanisms. As this is functionality that is basically different from the functions described in Section 2, it is proposed to include in the functional architecture a further layer or sublayer that describes this new functionality. Figure 5 shows how this impacts the different options to transport client signals over OTN and the resulting layering. The additional layer or sublayer is called "Digital Channel layer" and describes a digital signal in a frame with a certain bit rate and bit error rate. The optical channel layer then represents a single channel signal with its physical properties such as signal power level, signal frequency, OSNR rather than a logical signal as was the case in the G.872 approach.
It should be mentioned, however, that the access of the client via a digital interface should be only an option that does not preclude a fully transparent transport of the client signal within a transparency domain.

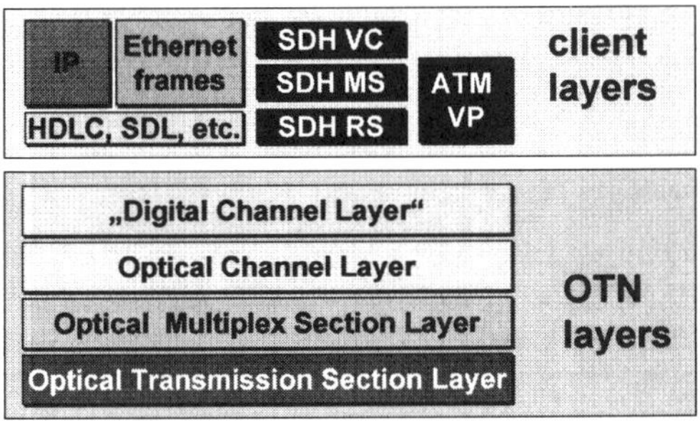

Figure 5: OTN with digital channel layer or digital channel sublayer

Figure 6 shows in some more detail how these functions can be modelled in terms of the functional architecture. On the left hand side the client signal enters the optical network for instance via a tributary port of an OADM or OXC, then passes through an OXC which performs 3R regeneration and goes back to the Client layer on the right hand side.

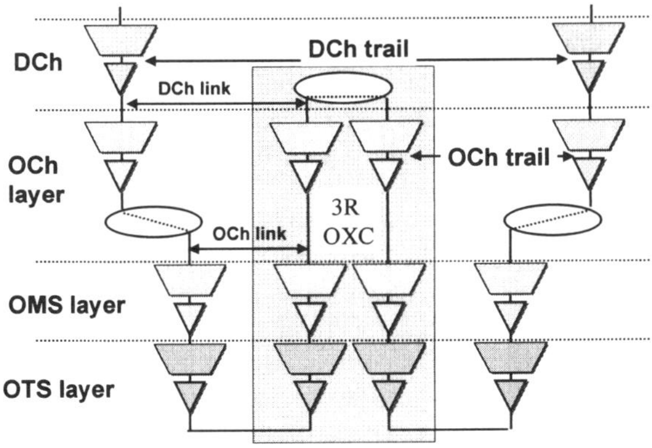

Figure 6: Modelling of the digital channel functionality

The functionality of the digital layer is allocated to the adaptation and termination functions in the following way: The mapping of the client signal into the frame is modelled by the DCh/Client adaptation (depicted as the uppermost trapezoids). The DCh termination (triangle below) is responsible for creating the end-to-end overhead (CRC, BDI, FDI, BQI,TTI, and unequipped signal). The OCh/DCh adaptation, finally, performs FEC coding/decoding (if used), frame alignment, line coding, and timing; besides FEC these are exactly the functions necessary for 3R regeneration.

Due to the new functionality two different modelling entity entities are needed, one for the transparent portion of the trail and one for the end-to-end relationship. As can be seen both from Figure 3 and Figure 6, the transparent section of the trail end-to-end trail is described by the ochTrail, whereas the dchTrail represents the end-to-end view through the whole OTN, including network elements that perform 3R regeneration.

From the perspective of the management interface, introducing a new layer in the functional architecture results in the specification of new managed objects in the management information model. Figure 7 gives a strongly simplified overview on the management information model, that has been specified in GDMO [8] within the P918 project. The Figure displays the containment relationship between the managed objects of the information model. Compared to existing information models [9]-[11], two managed objects have been added. These are the dchCTP and the dchTTP which contain management information needed to control and supervise the digital layer functionality as described above.

It is important to note that the clientCTP can be contained in the dchTTP as well as in the ochTTP. This accounts for the fact that model must describe both the digital access of the client and the fully transparent access.

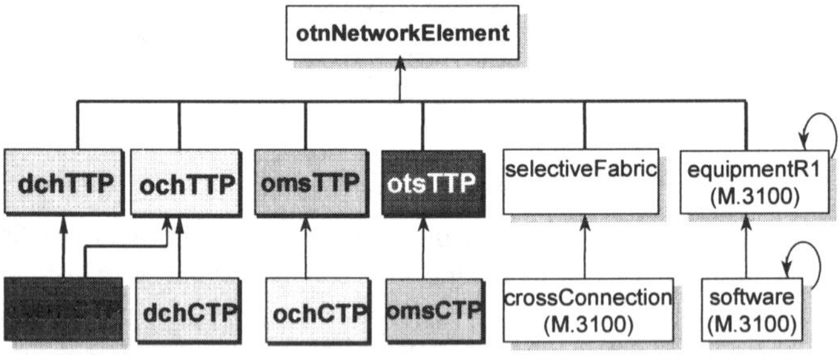

Figure 7: Containment diagram of P918 information model (simplified)

Up to now, the discussion was strongly focussed on modelling for the management interface. The remaining part of this Section is devoted to the description of the OA&M mechanisms. As already mentioned, these are mechanisms for exchanging signals between network elements to inform each other about fault conditions encountered. These OA&M signals are derived from processes which supervise the functions performed in the network elements. Figure 8 gives an overview over OA&M processes and signals, where the three lower layers correspond to the specification given in G.872 and the DCh layer has been included to account for supervision of functionality of the digital layer.

OA&M signalling takes place between network elements which perform the same functionality. This is why we find again here the layers shown before. The most important process is LOC (loss of continuity). This process which is present in each layer detects the interruption of the layer's characteristic signal. As far as the DCh layer is concerned, the functions of framing and FEC can fail causing a LOC. Associated to the LOC process are FDI (forward defect indication) signals to inform downstream network elements and BDI (backward defect indication) signals for the upstream direction.

Both FDI and RDI can be used for fault localisation; FDI, in addition, suppresses consequent alarms. G.872 proposes the MPP (measured performance parameter) process for the OCH layer and the OTS layer; for the OMS layer it is for further study. As we have an access to the digital signal on the DCh layer, we propose to introduce an MPP process for this layer. A degradation of the signal can be notified to the upstream network element by means of the BQI (backward quality identification) signal.

Layer	Process	OAM Signal
DCh	LOC MPP TTI	FDI, BDI BQI
OCH	LOC MPP TTI	FDI, BDI BQI
OMS	LOC (MPP)	FDI BDI (BQI)
OTS	LOC TTI MPP	FDI, BDI (BQI)

Figure 8: OA&M signals for an OTN including digital channel layer functionality

A TTI (trail trace identification) process is used to identify misconnections in the network. No OA&M signal is associated to this process, as misconnections are only notified directly to the management system. The TTI process is required in every layer that has flexibility. In addition it must be present in the uppermost layer of the network in order to enable end-to-end supervision. This is why we propose this for the DCH layer.

4. Connectionless Clients and Issues of Layer Interworking

Looking at the complicated stack of transport protocols depicted in Figure 5 it is evident that a lot of similar or related functionality is described in different layers. For instance, IP provides routing, ATM has switching functionality, SDH and the OCh layer, finally, both have crossconnection capability. Similar considerations apply for other areas of network functionality such as resilience. In order to optimise the transport of a particular client over OTN, it is necessary to decide which layer has to perform which functionality. Speaking in terms of the functional architecture, the aim is to allocate the different pieces of functionality to the elements of an overall functional architecture, including the client layers, and to specify potential schemes for the interworking of the layers.

As far as the classical transport network architectures are concerned, a lot of work has been done which can be used as a basis to describe the overall functional architecture. Functional architectures have been specified for SDH [12], ATM [13], and, as mentioned earlier, for OTNs [4]. For connectionless clients such as IP, however, no such architecture exists. The problem for defining a functional architecture according to [3] for connectionless clients has been addressed already some time ago by Sexton and Reid [14]. For explanation, Figure 9 tries to apply G.805 methodology to an IP network layer. The figure displays two IP access routers that transmit packets via intermediate routers, depicted by subnetworks (ellipses).

The basic problem now is that different datagrams could take different routes through the network. Due to this fact, it is not possible to correlate between the proper or failed delivery of a datagram and the status of any of the links between the routers. The existence of such a correlation, however, is a very fundamental assumption for transport networks which forms the basis for the whole OA&M philosophy. So, the basic difference between IP and classical transport network architectures is that best effort schemes are used instead of explicit configuration, reservation, and connection specific supervision of network resources by a management system. In terms of the functional architecture the problem is characterised by the fact that no such entity like a "connection" or trail exists in connectionless networks. In order to get an architectural analogy to trail and connection, the schemes of a "connectionless trail" and of a "flow" have been proposed (see Figure 9 and [14]). In this context, the flow is defined as "a set of "connections" all of which are routed towards the same transit group or access group".

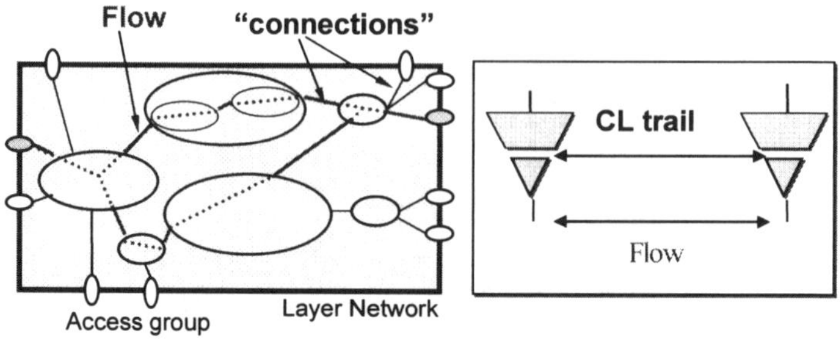

Figure 9: Modelling connectionless clients using G.805 methodology

The connectionless trail, however, still is basically different from a trail according to G.805 as the classical supervision, configuration, and reservation schemes do not apply.
Recently, however, an approach called Multiprotocol Lambda Switching [1] has been proposed which allows clients to dynamically request transport capacity from the OTN server layer. The basic idea is that, on request of the IP layer, a label switched path is established between label switch routers on the optical channel level. This example of a strong interworking between server and client layer is promising in particular for the transport of IP over OTN server layers as IP traffic can be very bursty. From the management point of view, the approach is very attractive as it brings, via the "label switched path" back something that is similar to a G.805 trail. This may, regarding management, ease the integration of IP over optical networks.

Conclusions

A common digital framing is needed to realize a „semi-transparent", open interface between OTN & client layers. It has been shown how the candidates for that, namely SDH frames or Digital Wrapper can be included in the management concept. Furthermore, aspects of the interworking between OTN server layer and client layers have been addressed, focussing on IP as a client.

References

[1] Daniel O. Awduche et al.: Multi-Protocol Lambda Switching: Combining MPLS Traffic Engineering Control With OXC's; IETF draft document: draft-awduche-mpls-te-optical-01.txt (November 1999).
[2] ITU-T Recommendation M.3010 (1992), "Principles for a telecommunications management network"
[3] ITU-T Recommendation G.805 (11/1995) "Generic functional architecture of transport networks"
[4] ITU-T Recommendation G.872 (2/99) „Architecture of optical transport networks"
[5] ITU-T Recommendation M.3100 (1995), Generic network information model
[6] WD16-3 (Lucent), ITU-T Q.16/15 Meeting, Feb. 99
[7] ITU-T Recommendation G.975 (11/96)"Forward error correction for submarine systems"
[8] ITU-T Recommendation X.722 (1992) " Guidelines for the Definition of Managed Objects"
[9] MOON, ACTS Project AC231, "Final MOON Information Model", Deliverable D1202, December 1998
[10] ITU-T Recommendation G.875, draft, July 1999
[11] PELICAN, ACTS Project 355, WP3, Working Document WD33, April 1999
[12] ITU-T Recommendation G.803 (6/97)"Architecture of transport networks based on the SDH"
[13] ITU-T Recommendation I.327 (3/93) I.327 - B-ISDN functional architecture
[14] M. Sexton, A. Reid: "Broadband Networking: ATM, SDH, and SONET", Artech House, Boston (1997).

Broadband Access, WDM Metro and Network Management
D.W. Faulkner and A.L. Harmer (Eds.)
IOS Press, 2000 © AKM

CAN YOU AFFORD ACCESS NETWORK MONITORING?

a customer and business process perspective

Andy Chattell
BT Adastral Park, Martlesham Heath, Ipswich, IP5 3RE, UK

Abstract

A question that is often considered by telecoms operators is "Will I get any benefit from proactive monitoring of my network?" For main core transmission networks, this is simple to answer – yes. The concentration of traffic in individual bearers means that the cost of monitoring a bearer is relatively small, whereas the impact on the company of hundreds or thousands of customers losing service is large. The cost benefit equation becomes indisputable.

Monitoring of the access network is a completely different problem. It has repeatedly proven to be extremely difficult to produce an accurate cost benefit analysis of monitoring systems for the access network. This has resulted in the use of such systems being constrained to where an overriding policy decision has been made, and not on a rigorous business case analysis.

This paper proposes a structured approach to determining the requirements and justification for proactive monitoring of access bearer networks. The approach can be broken into two components:

- Customer perspective
- Business processes perspective.

From a customer perspective, the major difficulty in quantifying the cost benefits of access bearer monitoring is that the "wrong" thing is being monitored. The monitoring systems are targeted at the bearer, whereas the impact on the customer manifests itself in their perception of the service that they are receiving. This paper considers the relationship between these two issues.

From a business process perspective, there are a number of different process areas that could potentially benefit from proactive bearer monitoring. This paper discusses these areas and proposes where and how benefits may be achieved.

It is not possible to give a single quantified answer to the question of the benefit of access bearer monitoring, as each operator is in a different business environment and has different operational processes. However, the approach discussed in this paper will allow an operator to quantify these benefits for their own individual circumstances.

Acknowledgements & EURESCOM Disclaimer

This document is based on results achieved in a EURESCOM Project. It is not a document approved by EURESCOM, and may not reflect the technical position of all the EURESCOM Shareholders. The contents and the specifications given in this document may be subject to further changes without prior notification. Neither the Project participants nor EURESCOM warrant that the information contained in the report is capable of use, or that use of the information is free from risk, and accept neither liability for loss or damage suffered by any person using this information nor for any damage which may be caused by the modification of a specification. This document contains material which is the copyright of some EURESCOM Project Participants and may not be reproduced or copied without permission. The commercial use of any information contained in this document may require a license from the proprietor of that information.

The author gratefully acknowledges the support of EURESCOM for carrying out this work. The author wishes to express a special thank to all P917 participants, since this paper reflects the results of their work, and in particular acknowledgement is given to the following people for their contribution and input to the work:

Paul Sadd	British Telecommunications plc
Angelantonio Gnazzo	Telecom Italia/CSELT
Andrew Kelleher	Eircom
Panayiotis Verrios	Hellenic Telecommunications Organization S.A (OTE)
Mike Balm	KPN Research
Marco Smit	KPN Research

1 Introduction

It has repeatedly proven to be extremely difficult to produce an accurate cost benefit analysis of bearer monitoring systems for the access network. This has resulted in the use of such systems being constrained to where an overriding policy decision has been made, and not on a rigorous business case analysis.

In order to calculate such a business case, we need to look at this problem from two perspectives, the customer and the business process.

2 Customer oriented benefit calculation

This section of the paper proposes a structured approach to determining the requirements and justification for proactive monitoring of access bearer networks, by taking a customer oriented view. It is presenting a way of calculating the benefits based on avoiding any impact on customers of network bearer failures.

2.1 Network/Service Hierarchy

The major difficulty in quantifying the cost benefits of access bearer monitoring is that the "wrong" thing is being monitored. The monitoring systems are targeted at the copper or fibre bearer, whereas the impact on the customer manifests itself in their perception of the service that they are receiving. It is therefore essential to consider the relationship between the customer and the bearer. This is illustrated in figure 1.

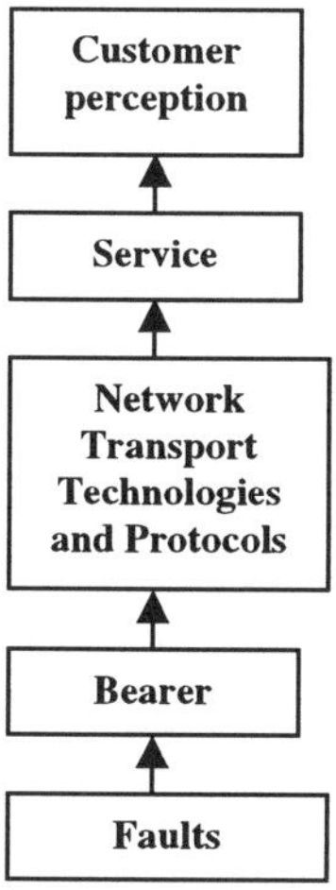

Figure 1. Fault impact hierarchy

2.2 The customer impact equation

Referring to the hierarchy illustrated in figure 1, the calculation of the impact on the customer can be represented as follows.

The set of possible bearer faults is taken to be **F**.

The transmission layer systems and protocols act as a filter (**T**) on these fault conditions. This filter may comprise more than one discrete step (T_1, T_2,...). For example, using IP over ATM, the IP layer will have error correction/handling features, in addition to those contained within ATM.

The service being carried over the network will determine how the network failure manifests itself to the customer. This is represented by a conversion function S. For example, a short network error could be translated into a "freeze frame" on a video service or a short delay for an internet service.

Finally, the way the customer perceives (P) the impact on their service will determine the overall significance. This will depend heavily on customer expectation. For example, if a customer is expecting a perfect picture on their video service, then they may not be happy with even light bursts of "snow", whereas if they simply expect better than what they had before, they may well ignore the problem.

The overall impact **I** on the customer of the potential bearer faults **F** can be represented as a simple application of a number of filter functions:

I = P.S.T(F)

2.3 Approach

The approach is broken into a number of stages, which are illustrated in figure 2.

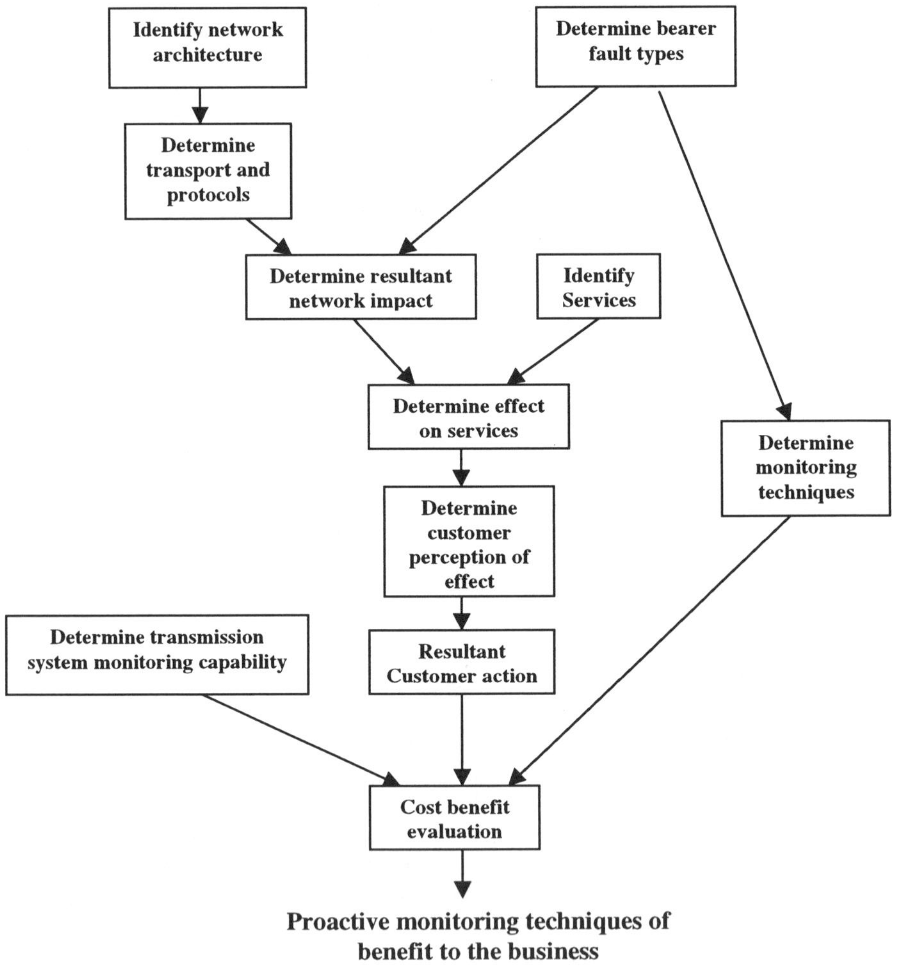

Figure 2. Analysis steps for calculating business benefit.

Identify network architecture. The first stage is to determine the network architecture, identifying the bearers and types of bearers. For example, copper pairs from exchange to customer premise, direct fibre, PON, etc.

Determine transport and protocols. Each of the transport layers running over the bearer(s) must be identified, e.g. IP over ATM over ADSL. The associated error handling functionality of each layer must also be identified and understood.

Determine bearer fault types. The different types and nature of all faults that affect the bearer must be identified. This will include the lifecycle of the degradation. For example, cable dig-up will result in immediate and total failure of the bearer, whereas a joint that is letting in water may not have any effect for several months and then lead to intermittent failure. The frequency of the types of failure must also be identified.

Determine monitoring techniques. For each of the bearer fault types, a technique needs to be identified that would enable the failure to be detected, preferably in advance of it becoming serious. If a failure mode cannot be detected then it may not be considered worthwhile accurately calculating the benefit of detecting the problem.

Determine transmission system monitoring capability. Digital transmission systems usually have some monitoring capability built in as it can be provided fairly cheaply and enhances the look of a manufacturers product specification. This capability needs to be understood as it may reduce or replace the requirement for bearer level monitoring.

Determine resultant network impact. Given the bearer faults and the error handling properties of the network transport and protocols, the resultant impact on the data transmission can be calculated. For example, an intermittent bearer fault could cause a bit error rate to exceed the manageable level in the higher layer protocol and lead to short interruptions in the data stream.

Identify services. The services being carried over the network need to be identified and understood. Only one service can be evaluated at a time as part of this approach. If multiple simultaneous services must be considered, then the benefits of each service must be combined in proportion to the percentage usage of the transmission path.

Determine effect on services. Given an understanding of the service and the nature of the impact on the network transport (caused by a given fault type), the resultant effect on the service can be deduced. For example, for an internet related service and short interruptions in data transport, the effect on the service would be to add a corresponding delay, e.g. a WWW page takes slightly longer to download. However, the same network problem for a real time video service could cause the video picture to freeze for a short time.

Determine customer perception of effect. We can calculate the quantitative effect of a bearer fault on a service, but this on its own is not enough. We must consider the customer perception of the impact on their service. This will be affected by the type of customer and their expectation of the service.

Resultant customer action. Given the customer's perception of a problem with their service, the likely action that they will take must be considered. If this action is to ignore (or not even notice) the problem, then there is little benefit in proactively monitoring for it. If, however, the customer response is to cancel their service, it is a completely different situation.

Cost benefit evaluation. A cost must be attributed to each course of action that a customer will take to a given failure condition. Given the frequencies and costs, it is possible to calculate the value to the telecomms operator of proactively monitoring for each type of bearer failure. This figure must then be adjusted by the cost of providing the monitoring capability.

Consideration must also be given to the capabilities of network monitoring at other levels. It may be that the transmission system itself can provide the capability for monitoring specific failure types. In which case, the benefit of additional bearer monitoring equipment is reduced. However, this is not the case where the transmission or service provider is a different company from the bearer provider, for example in the case of Local Loop Unbundling. In this case, there is no reduction in the benefit.

3 Business Process Benefit Calculation

The overall benefit accrued through improvements to business processes is simply given by the sum of individual benefits. This section identifies which processes will benefit and the nature of that benefit. The financial value of each benefit area is highly dependent on the detailed implementation of the process concerned. It is therefore down to the individual telecoms operator to calculate the value attributed to each component.

3.1 Provision

The provision process can benefit from bearer monitoring in two areas; link availability and quality.

For the scheduling of a provision job, the presence of a good bearer link should be known in advance of dispatching an engineer. If a link is either not present or is faulty, additional activity will be required as part of the provision process. Effort will be wasted if an engineer is dispatched to carry out an installation, which is not possible due to bearer link availability or faults.

The quality of the link becomes a more significant factor in the provision process as a link is being pushed to carry higher and higher bandwidths, for example ADSL, particularly where the quality can vary between bearers. By knowing the quality of a link, service requests can be accurately matched to network capability to minimise the number of provision failures.

3.2 Repair

3.2.1 Proactive identification of future fault conditions

A bearer monitoring system enables proactive maintenance, i.e. the detection of degradation and repair to the bearer before a service affecting fault occurs, or even before transmission is interrupted.

There are a variety of failure causes, each of which has a different possibility of being detected in advance of a fault causing a transmission loss.

It has been estimated [1] that a minimum of 17% of faults can be detected proactively, i.e. before they become service affecting. There are two factors that contribute to savings in this situation. One is the *cost saving* that occurs due to more effective scheduling of repair activities. The other is the *revenue recovered.* Reimbursement policies do not go into effect and the network is in use for a longer period of time. Repair scheduling improvements can be made on an individual link basis, but also monitoring data can be correlated across entire bearers to identify when degradation is such that a portion of the bearer must be replaced in advance of a major fault.

In addition to using a monitoring system in a proactive manner, data gathered by the system can be analysed statistically to reveal possible links to external factors which may result in a fault. This information can be used to plan around possible causes of failure and provide input to equipment design to avoid certain problems altogether.

In order to calculate the benefit of a proactive maintenance approach, the portion of outages repaired during regular working hours and the portion repaired during other hours must be evaluated. In addition, the fact that stand-by personnel are needed to support emergency situations in cases where the same work could have been routinely scheduled needs to be taken in account.

3.2.2 Repairing faults before they become bigger problems

For some types of bearer failure, e.g. water ingress into a cable, the longer they are left, the greater the damage that is caused. By repairing a bearer cable immediately, the typical activity is likely to be a re-make of a leaking joint and repair of the individual bearer link itself. If the failure is left until it becomes so severe that it starts to affect service, many bearer links could be affected, necessitating the replacement of a whole section of the bearer cable.

3.3 Unbundled metallic loop and QoS guarantees

In the situation that a telco operates services over the metallic access network that it owns, the process benefits of monitoring the metallic access network fall into the normal areas (Provision, maintenance, repair) as described elsewhere in this paper. However, for the case where the metallic access network is owned by one telco, but the transmission and services are owned and provided by a different company (i.e. unbundled loop), there are two specific process areas that must be looked at in addition. These are:

- Verification of contractual QoS guarantees
- Spectrum policing

It is likely that specific QoS parameters will be included in the contract between the metallic access network owning telco and the transmission/service providing company. Against these parameters will be compensation amounts defined for when the network performance is not up to the contractual limits.

Bearer monitoring would allow the telco to identify where specific circuits are degrading and have or are likely to drop below the acceptable performance limits set. Action can then be taken to rectify the problem before contractual penalties come into play.

It would be straightforward to calculate the value of this monitoring capability, given the appropriate network reliability information.

Cost to telco of not monitoring = Frequency of failure below QoS limits x Number of "unbundled" pairs x contractual penalty amount

This calculation would, however, by highly dependent upon the precise details of the contractual agreement. For example, the contract may allow for a certain amount of "down time" for a given circuit which means that a telco may be able to repair faults within the given time and avoid penalties. In this case, the frequency of failure below QoS limits would be small.

An additional area that must be considered is the case of intermittent or variable QoS. The quality of a line may be varying resulting in failures being reported, but could test OK. This would lead to long confused disagreements between telco and service provider, which could be very expensive. It is, however, very difficult to quantify this and hence the benefit of bearer monitoring which would remove any such disagreement.

The other aspect of the "unbundled loop" scenario is that of spectrum policing. As a different company is providing the transmission systems, a telco has no direct control over what their network is being used to carry. It is possible for an "illegal" transmission system to be used on an "unbundled pair" which would interfere with the digital transmission on other pairs in the cable that are controlled by the telco or even a third company. This type of situation could be extremely difficult and costly to correct. Interference could be at any time and would not necessarily be continuous. Bearer monitoring could be used to identify this type of interference quickly and effectively, and may even identify the type of transmission system that is causing the interference.

In order to quantify the benefits, the frequency of these types of problems occurring and the effectiveness of any manual identification techniques would need to be taken into account.

4 Conclusions

It has been shown that analysing the potential benefits of employing bearer monitoring for broadband access networks is a highly complex procedure which will require each operator to apply the techniques described within this paper on a per service, and per network topology, basis.

It has also certainly been shown that there are clear technical benefits to the use of bearer monitoring, but the difficulty is in determining whether the technical benefits can be justified on an overall cost benefit basis. It is only through applying the approach described in this paper that an accurate answer can be determined for a specific case.

This paper has mostly concentrated on the value of bearer monitoring to a single operator who is responsible for the network bearer and transmission system. It has touched on the issue of where the transmission system owner is a different company from the bearer owner, e.g. for the case of Local Loop Unbundling. In this case, it must be highlighted that the financial decision on whether bearer monitoring is implemented may be over-ruled by specific regulatory law which compels an operator to guarantee a QoS at all times.

References

[1] Calculations based on several information sources. Contact Mr Panayiotis Verrios - OTE Hellenic Telecommunications Organization S.A., 99 Kifisias Ave, GR-151 24 Maroussi – Athens, GREECE. e-mail: verrios@ote.gr

Quality of Service Techniques in Optical Packet Switches*

Carla Raffaelli
DEIS - University of Bologna, Viale Risorgimento,2, 40136 Bologna, Italy
e-mail: craffaelli@deis.unibo.it

Abstract. The paper proposes different techniques that can be adopted for service differentiation in an optical packet network used as an IP backbone. The main aim is to show how optical technology can be exploited to enhance performance and flexibility and to support different levels of quality of service within network nodes. Label switching is assumed as the basic switch functionality. Simulation results are given to evaluate the effect of two-level service differentiation techniques and to discuss the opportunity of their introduction in multi-service networks.

1. Introduction

The evolving telecommunication scenario is characterised by a significant growth of bandwidth-intensive services such as the World Wide Web, video on demand, distributed computing etc., and a wide varieties of networking solutions in the local and in the wide area environments, dominated by the IP protocol. The increase in performance that applications will need put severe demand on the networking solution. To meet this demand, optical networks are being deployed and are emerging as the next generation of high-capacity, multi-protocol networks. When speaking of optical networks, the quality of service is typically put in relation with the very high bandwidth offered by optical links that, at a first glance, can appear much greater than application needs. This can be true, at this moment, for link transmission capacity while the bottleneck typically move to switches and routers, even if implemented with optical technology and especially for services with QoS (Quality of Service) requirements. The IETF (Internet Engineering Task Force) has proposed two different approaches to improve the best effort services in IP networks, namely the Integrated Services (IS) and the Differentiated Services (DS) approaches. The IS provides QoS to the users according to specific application flow requirements. Each network node maintains information regarding application flow states through exchange of signalling information and for this reason has scalability limitations. On the other hand the DS model provides QoS to aggregated application flows by means of setting service bits at the network edges and then applying drop priority or delay priority algorithms to the aggregated flows in internal network nodes. The large aggregated bandwidth and the high number of application flows that is expected within an optical backbone call for a scalable scheme and for this reason the differentiated approach has been chosen.

Starting from previous experiences on optical networking [1], the diff-serv concept [2] is here applied to an optical Internet backbone [3]. In particular queuing techniques in the optical nodes, taking advantage of the intrinsic characteristics of photonic switch

* This work has been partially funded by the Italian Ministry for University and Scientific and Technological Research (MURST), in the framework of the National project "Techniques for quality of service guarantee in multi-service telecommunication networks".

implementation, are proposed. The networking scenario consists of an optical backbone, based on fixed length optical packet transfer mode, organised as a cloud of core routers and which is accessed by means of edge routers. This architecture matches with the IETF DS architecture in the sense that core and edge routers are responsible, as far as service differentiation is concerned, of the splitting of functions according to the general diff-serv architecture. Edge routers are responsible of traffic classification and shaping and core routers of packet forwarding, according to the Type of Service encoded in the IP header.

A broadcast-and-select switching scheme is considered for evaluation, where queuing is implemented by means of a set of fibre delay lines that realises delays multiple of the basic packet duration. Two-level differentiation is assumed where the first has drop preference indication with respect to the second. Threshold dropping techniques suitable for optical implementation based on enhancements of the conventional RED algorithm are proposed and evaluated by simulation in terms of packet dropping and latency time. The first is based on marking low priority packets and discarding them only in the presence of high priority packets in the node. This technique is expected to improve low priority packet dropping probability with respect to RED while maintaining the same performance for high priority traffic. Another technique has been considered for expedited forwarding (EF) applications, assuring very low latency and high time transparency. It also takes advantage of having a packet available in a broadcast and select FDL buffer with all possible delays. A precedence technique can be implemented with EF packets skipping AF/BE packets during congestion, thus relaxing the first in first out (FIFO) discipline. This happens only for packets belonging to different classes, while the FIFO discipline inside the same service class is maintained. EF traffic can take benefit of the modified FIFO discipline to obtain the required low latency in the queues.

In conclusion quality of service techniques are shown to be necessary also in the optical networking environment and some features peculiar to optical switch implementation are successfully exploited and evaluated in a service differentiation perspective.

The paper is organised as follows. In section 2 the employment of the OTP-N, defined in the KEOPS project, as an internet backbone is presented. In section 3 queuing inside optical nodes is described with reference to a switch based an the broadcast-and-select principle. In section 4 different quality of service management policies are introduced and their performance are analysed in section 5 for random and bursty traffic. Finally in section 6 conclusions are drawn.

2. The Optical Transparent Packet Network as a TCP/IP Backbone

The reference scenario is that defined within the ACTS Project KEOPS and named Optical Transparent Packet Network (OTP-N) [4]. The basic concept relies on the definition of fixed-length packet format and suitable network architecture that enable high speed transfer of information through optical links without any concern about original information format and speed. That is why the network is said to be transparent to services and bit rates. The transparency property is very important both for performance and integration reasons. The network is connection-oriented and the packet forwarding, on the basis of the virtual path identifier, read from the packet header, will be achieved by means of optical packet switches [4], [5]. To achieve full transparency the length of the packet is defined to be fixed-length in time and not in bits, that allows for the following achievements:

- It simplifies the switch design, especially in terms of delay lines dimensioning;
- It allows for scalability of the link speed without re-design optical switches;

- It reduces the number of accesses to the switch routing table since one OTP may contain several packets from the electrical domain (ATM cells for example).

Although the initial idea for OTP-N implementation was mainly for interconnection of electrically switched ATM networks, the rapidly evolving growth of IP networks required to devote research efforts to the application of OTP-N as a backbone for this kind of environment. The main manufacturers of IP routers and IP network providers are recently being considering advanced solutions to cope with the huge bandwidth need of aggregate IP traffic and the enhanced functions necessary to achieve the integration of data communication and telecommunication such as quality of service support. Optics is considered as a fundamental player in this environment and Gbit routers interconnection are considered by means of IP over SONET or IP over WDM. Both these solutions are circuit oriented so that although offering high bit rate transfer with commercially available technologies, they lack in scalability and in flexibility as circuit switching does. In order to enhance these network properties a solution based on optical packet switching can be adopted. Several possible application schemes have been considering for IP over OTP as described in [3]. In the framework of the development of standards for high speed IP networks, such as MPLS [5], it seems interesting to consider the OTP-N solution as a forwarding support for label switching technologies [5]. In fact label switching promises to enhance the speed of information transfer using fast hardware for forwarding purposes. Moreover better scalability is obtained running standard IP routing protocols on the top of this hardware thus reducing the number of adjacency that are bound to arise in large IP networks. Finally label switching overcome the control problems that are typical of IP over ATM networks relying on standard IP control functions. Quality of services may be also supported by means of label and interface management [5]. On behalf of these considerations the solution IP over OTP-N can be considered as an effective solution both for scalability and flexibility high speed implementations. In this framework the nodes of OTP-N play the role of label switch/routers with routing procedures based on available IP routing protocols and Gbit hardware for high speed transfer. In order to accomplish this task encapsulation of IP packets inside optical packets is necessary at the edges of the optical cloud together with label support in optical packet header [3].

3. Queuing in Optical Packet Switches

The operations of optical packet switches are considered with the assumption to work in a label switching environment.

Optical packet switch architectures have been developed based on wavelength routing or broadcast and select principles [6]. The broadcast and select switches, that are mainly considered in this paper, offer high degree of flexibility for a wider range of functions with respect to the wavelength routing solutions, among which the support of multi-point communications, although physical constraints arise in relation to power splitting [8].

Queuing in a broadcast and select switch is assumed to be implemented by means of a set of fibre delay lines that realise all possible delays which are multiple of the basic packet duration up to a maximum multiplying factor [7]. Packets arriving at the switching node are encoded with a wavelength related to the input link and split onto all delay lines. Packet header is converted to electronics to perform input to output logic association. Physical selection of optical information is performed for each output link by means of filters and optical gates thus achieving packet forwarding. It is worthwhile to underline that all optical information transfer is obtained with electronic control [6]. Figure 1 illustrates a block diagram of a broadcast-and-select switch.

An implementation of such a switching concept is described in [6], [7]. Basically it relies :

- on the use of wavelength encoding and selection to achieve packet routing,
- on the exploitation of optical Fibre Delay Lines (FDL) accessed by fast optical gates to perform packet buffering and time switching.

An NxN matrix, in relation to the sketch of figure 1, is composed of the following three main blocks:

- the packet encoder, which consists of a set of N all-optical wavelength converters (AOWC), one per input and each being assigned a fixed wavelength ;
- the packet buffer which includes K FDL and a space switch built up with semiconductor optical amplifier gates (SOA);
- the packet selector, based on a set of N wavelength-selectors (WS).

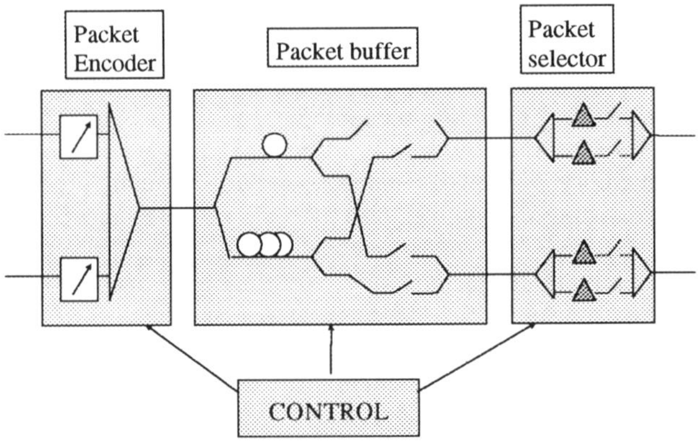

Figure 1- Scheme of a Broadcast-and-select optical switch

The switching operation is performed as follows. Each incoming packet is assigned to one wavelength by means of a wavelength converter identifying its input port, and then fed into the packet buffer. All packets experience all possible delays available within the buffer. The role of the set of optical gates associated to each output of the switch is to select the packet in the correct packet time corresponding to the appropriate delay, as determined from actual traffic conditions in the switch. All signals, at all available wavelengths are gated simultaneously. Finally, the wavelength selector only discriminates one outgoing packet at a time, on the basis of its encoding wavelength, i.e. its input address.

One of the main advantages offered by this architecture is that multicasting is provided easily, as the same wavelength-encoded packet is available at any time and at any output port. This should be of primary importance to meet the requirements of future applications and services. Moreover the remarkable flexibility of this kind of switch allows the implementation of different queuing disciplines for example in relation to performance and quality of service requirements.

In addition to the optical buffer which can be realised with optical fibres of calibrated lengths, three key photonic devices are required for the implementation of this architecture, namely the AOWC, the SOA and the WC. The AOWC can be based on a SOA operating under cross-saturation regime. More advanced devices relying on cross-phase modulation integrated in SOA-based interferometric structures will ultimately provide some signal regeneration (in particular, extinction ratio enhancement). Both the buffer and the SOAG operates in a multi-wavelength regime. To avoid inter-channel cross-modulation which might occur in conventional SOA, clamped-gain SOA are required, which maintain the gain at a constant value. The WS is realised by combining one optical de-multiplexer to separate the wavelengths, a set of SOA to achieve the packet selection and one multiplexer to recombine the selected packets towards one common output port and to remove the amplified spontaneous emission. This device takes advantage from the fast switching capability of the SOA.

4. Service differentiation techniques in optical nodes

With reference to the queue implementation described in the previous paragraph different strategies for service differentiation inside the switch can be realised. In this paper two-level differentiation is addressed with the aim to improve performance of the best effort Internet service.

Different techniques are proposed, taking into account the basic functions of optical queues. In particular being packets available at any time with a given wavelength, packet precedence may be implemented for the highest priority by simply taking the packet by a lower-delay delay line. Moreover being packets physically discarded at the head of the queues when they are out of timing for transmission, they are present inside the queues also in case of congestion and can be discarded only when really necessary. This allows to improve the classical RED (Random Early Discarding) technique as will be explained in the following [9].

Two-level differentiation is assumed that can be related to two different situations: *in-profile* traffic belonging to two different service levels, or, in alternative, to the same service level with both out-profile and in-profile traffic, where the first has drop preference indication with respect to the second.

The service differentiation that has been introduced is mainly evaluated in terms of packet dropping probability. The choice of this evaluation parameter has been done according to the peculiar nature of the OTP network that is characterised by packets with quite short duration (some microseconds) and nodes with small buffers (up to few tens of packets), that typically implies limited delays. On the other hand packet loss may be too high for some application, especially under heavy traffic load conditions. Also, delay and bandwidth usage are considered.

The first technique that is considered as a reference for comparison is dropping associated with RED functionality (figure 2). When the buffer occupancy reaches the threshold T, the discard probability for a best effort packet begins to be more than zero and its value is increased linearly up to 1, when the buffer occupancy reaches the value T + W. When the buffer occupancy exceeds this value only the high priority quality packets are buffered and all the best effort packets are discarded.

If this technique is applied some best effort packets may be discarded even if no high priority traffic needs transfer. Being in any case these packets present in the optical node for a time equal to the maximum delay introduced by FDL, an improved technique can be implemented that only marks these packets, and only discards them if any high priority packet arrives at the node during their life in the node. This technique is expected to improve best effort packet dropping performance while maintaining almost the same performance of the previous technique for high priority traffic, at least for percentage of the high priority traffic over the total not too high.

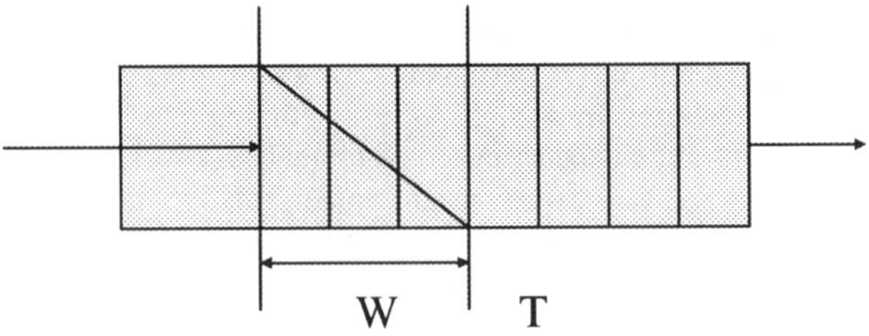

Figure 2 - Output buffer with drop priority threshold and RED functionality

A third technique is also considered for expedited forwarding applications, assuring low latency and high time transparency. It is based on the assumption of having a packet available in a broadcast and select delay line set with all possible delays so that a precedence technique can be implemented with high priority packets skipping best effort packets during congestion, thus relaxing the first in first out (FIFO) discipline but only for packets belonging to different classes, while maintaining the FIFO discipline inside the same service class.

5. Results

The evaluations were performed by simulation in terms of dropping probability for multiplexed bursty traffic and for random traffic as a particular case, assuming for this latter a two-state Markov chain representation and the value of burstiness, defined as the ratio between the average burst length and the average burst length of a Bernoulli traffic with the same load, equal to 2 and 1. The reference switch has 8 inputs and 8 outputs and performs output queuing with buffer length equal to 30. The main operating parameters that have been varied were the traffic load and the percentage of best effort and high priority traffic.

In figure 3 the effect of service differentiation operated by the dropping technique is analysed as a function of load and for different percentages of high priority traffic, and its effectiveness its effectiveness, resulting in a sensible difference between cell dropping for the two classes, is demonstrated. The high values of packet loss for best effort traffic is mainly due to the region of load considered and to the limited buffer size. In any case it is advisable to operate at lower traffic loads and in any case some techniques for buffer enhancement can also be applied to face temporary overload conditions [10].

In figure 4 the effect of the second technique based on packet marking is outlined for best effort packets. As expected the availability inside the switch of the best effort packets that would be lost in the discarding technique leads to a better exploitation of memory with consequent improvement of the related dropping performance. The packet loss probability for the high priority class remains unchanged unless high percentage of high priority traffic is present as shown in figure 5.

Figure 6 shows the results for the modified FIFO discipline, that combines delay transparency with low packet dropping. In fact dropping performance for assured forwarding traffic remains practically unchanged.

The effect of the presence of virtual connections is also evidenced. By assuming an output buffer with a single threshold, r virtual connections transmit data at a normalised rate λ , that is the fraction of link bandwidth required by a single virtual circuit, $r\lambda$ is defined as the *overbooking ratio* and for an overloaded link $r\lambda > 1$. Results are given here in the case of unbalanced load for 8 virtual connections with 50% of high priority traffic on each connection to show the effectiveness of the proposed techniques (Figure 7). It can be seen that the modified FIFO techniques offers the lower average delays while marking is less interesting for delay sensitive traffic. In figure 8 the usage of the output bandwidth is plotted as a function of the aggregate load showing that the mechanism guarantees the high priority traffic when overbooking occurs.

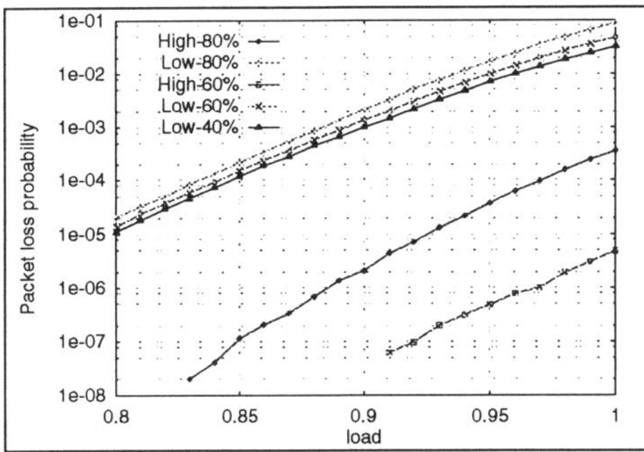

Figure 3 - Packet Loss probability as a function of load for the RED algorithm with threshold T=20 and W=5, random traffic.

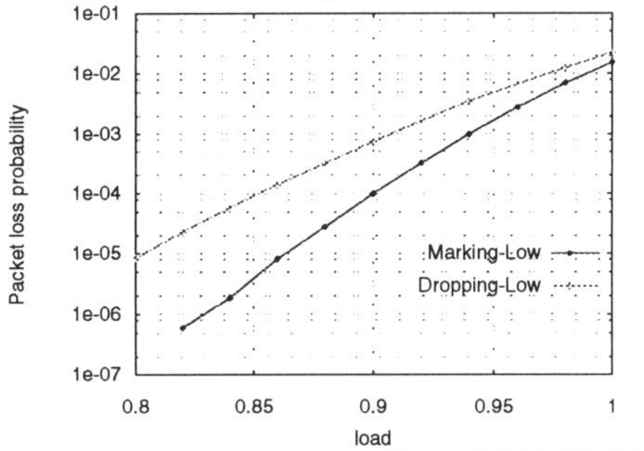

Figure 4 - Packet Loss probability as a function of load using the marking and the dropping techniques with 10% of high priority traffic, T=20 and W=5, random traffic.

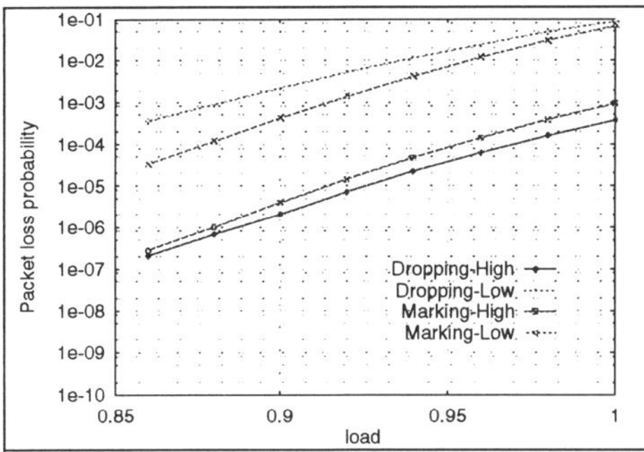

Figure 5 - Packet loss probability for high and low priority traffic as a function of load using the marking and the dropping techniques with 80% of high priority traffic, T=20 and W=5, burstiness 2.

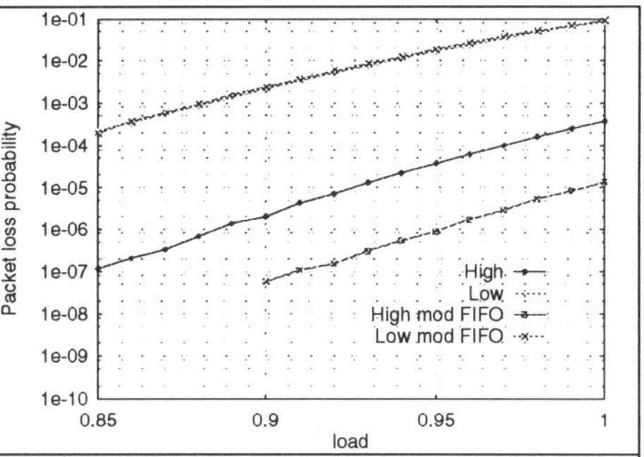

Figure 6 - Fixed length packets: packet loss probability as a function of the traffic load for high and low priority traffic. The discarding and the modified FIFO techniques in the presence of 80 % of high priority traffic, T=20 and W=5, burstiness 2

6. Conclusions

The aim of this paper was to show some achievements in service differentiation related to optical switches based on packet transfer mode. Different techniques that are feasible for broadcast and select switches based on fibre delay line queuing were presented and then evaluated by simulation. Packet dropping probability has been chosen as the main performance figure, being the queuing delay limited by queue size that is small for implementation reasons. The result of major interest is the possibility to improve the service differentiation in terms of packet dropping probability by exploiting packet availability inside switch queues and thus by marking them instead of discarding or by relaxing the FIFO discipline inside switch queues. This work has been developed within a particular network architecture based on packet transfer mode, but in any case suggest that optics offers the availability of new design degrees that can be suitably exploited for example in the field of service differentiation at the transport level.

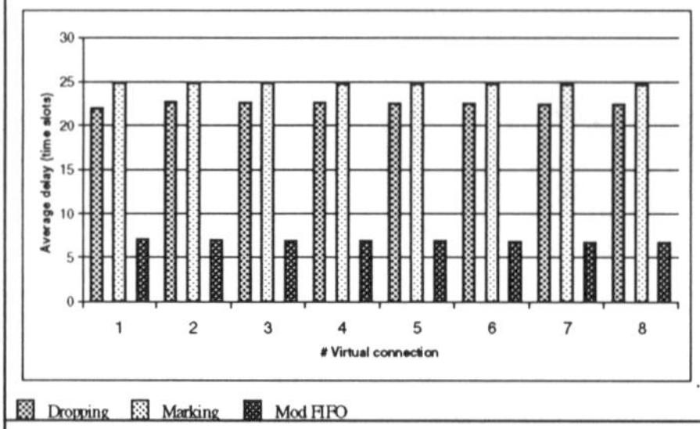

Figure 7 – Average delay in connection oriented transfer mode during overload (overbooking=1.3) for high priority traffic for the different techniques.

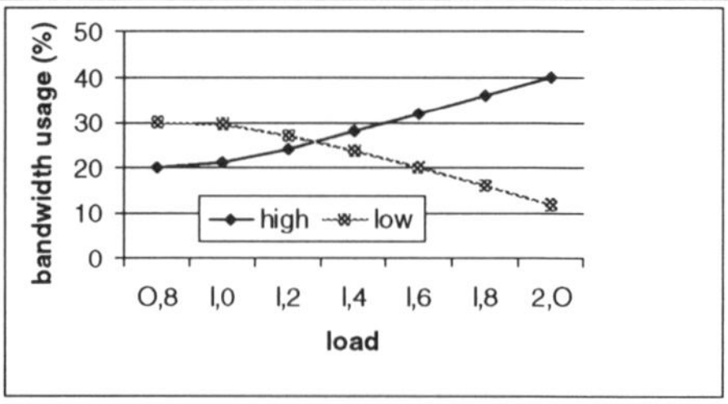

Figure 8 – Bandwidth usage on the output link as a function of load in overbooking condition for the modified FIFO discipline with a connection that requires the 50% of the offered load of which 40% high priority traffic and the remaining 60% low priority.

References

[1] M. Renaud, F. Masetti, C. Guillemot, B. Bostica, "Network and System Concepts for Transparent Optical Packet Switching", *IEEE Communication Magazine*, Vol. 35, No. 4, April 1997.

[2] S. Blake, D. Blake, M. Carlson, E. Davies, Z. Wang, W. Weiss, "An Architecture for Differentiated Services", RFC 2475, ftp://ftp.isi.edu/in-notes/rfc2475.txt, December 1998

[3] B. Bostica, F. Callegati, M. Casoni, C. Raffaelli, "Packet Optical Networks for High Speed TCP-IP Backbones", accepted for publication on IEEE Communication Magazine, 1998.

[4] P. Gambini et al, "Transparent Optical Switching: Network Architecture and Demonstrators in the KEOPS project", Journal on Selected Areas in Communications, invited paper, September 1998.

[5] B. Davie et al., "Switching in IP networks", Morgan Kaufmann Publishers, 1998

[6] F. Callegati, M. Casoni, G. Corazza, C. Raffaelli, D. Chiaroni, F. Masetti, M. Sotom, "Architecture and Performance of a Broadcast and Select Photonic Switch", Optical Fiber Technology, invited paper, Academic Press, 1998.

[7] F. Masetti et al., "Fiber Delay Lines Optical Buffer for ATM Photonic Switching Application'', Proc. of Infocom'93, pp. 935-942, San Francisco, CA, USA, March-April 1993.

[8] D. Chiaroni et al., " A Novel Photonic Architecture for High Capacity ATM Switching Applications ", Proc. of Photonics in Switching'95, Salt Lake City, UT, USA, March 1995.

[9] S. Floyd, V. Jacobson, "Random Early Detection gateways fo Congestion Avoidance", IEEE/ACM Transactions on Networking, vol. 1, no. 4, pp. 397-413, August 1993.

[10] M. Casoni, C. Raffaelli, "Photonic Switches with Enhanced Output Queuing", ICT 98, Chalkidiki, Greece, June 1998.

Broadband Access, WDM Metro and Network Management
D.W. Faulkner and A.L. Harmer (Eds.)

Comprehensive Fibre Management for the Evolving Broadband Network

David Tanis, Lucent Technologies, Dublin Ireland

Abstract

Recent advances in optical technologies have led to exponential growth in the amount of information that can be carried on a single fibre.

At the same time, very high fibre count cables are being deployed around the globe. As a result, an optical fibre cable installed today could carry Petabits (10^{15} or 1,000 Terabits) of traffic in the near future.

These two events have put a tremendous burden on telecom operators to protect their customers against the likelihood of a fibre cut. While protection and restoration mechanisms have been developed at the Optical, TDM, ATM and IP layers, these techniques do not explicitly address the physical fibre layer that actually incurs the fault.

This paper will explore recent trends in optical technology and will describe an automated fibre management system that provides a comprehensive set of configuration and fault location management capabilities, which will allow service providers to meet the challenges created by these trends.

Trends in the Evolving Optical Network

The past few years have seen unprecedented growth in the deployment of optical fibres coincident with tremendous advances in the optical technologies that utilise these fibres. And while this progress in achieving high capacity networks has helped facilitate such things as high-speed Internet backbones, it has also put a strain on the ability of service providers to manage the optical fibre cable plant. The following are a few examples:

+ The shear quantity of optical fibre installed around the world has grown dramatically over the past three years and this rate of growth is expected to continue beyond 2000.

+ New optical fibre packaging technologies such as AccuRibbon™ cabling have enabled the practical deployment of very high fibre count cables. While fibre cables deployed 3-5 years ago may have contained 24 to 48 fibres, cables being deployed today can consist of up to 864 fibres.

+ Continuous advances in Dense Wave Division Multiplexing systems have revolutionised the deployment of longhaul networks. For example, repeater spacings have been extended from 40-50km range with 1310 nm technology to several hundred kilometers, with optical amplifiers providing in excess of 120 km between spans. As shown below in Figure 1, this order of magnitude increase in the distance between electrical regeneration points can mean that the distance to isolate faults between end terminals can go up by a similar factor.

Figure 1: DWDM Technology Overview

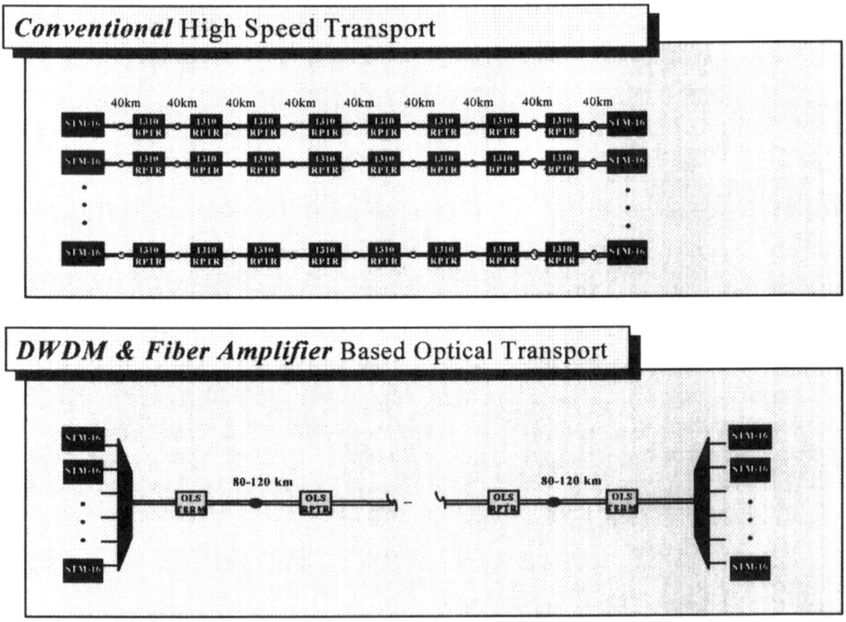

+ Similarly, the capacity per fibre that can be transported on DWDM systems is doubling on average every 9-12 months. While commercially available DWDM systems have the capacity for up to 400 Gb/s systems, it is anticipated that systems capable of supporting several Terabits per second on a single fibre will be available in the near future. New amplifier technologies, such as the enhanced Raman amplifier, along new fibre designs such as zero water peak fibre (AllWave™), serve to increase the amount of spectrum used to carry information over a single fibre.

DWDM technology has also begun permeating into the metropolitan network as well, and many believe that networks will evolve to an all-optical format where wavelengths, rather than circuits, will be the common denomination for carrying traffic.

In spite of the significant advances in optical technology, there has not been a corresponding improvement in technology to prevent fibre cuts. One large carrier reports an average of one fibre cut per day[1]. As the troubleshooting and fault location process is typically a manual process today, average fault isolation and repair times can range between 4 and 8 hours[2].

These trends all underscore the need for an automated fibre management system that is capable of continuously monitoring these very high capacity optical fibre cables.

Lucent has developed automated fibre test and surveillance system to specifically address these challenges. Known as the *Smart* LGX®, this system is a comprehensive fibre management system which integrates fibre configuration and fault management capabilities in conjunction with traditional fibre distribution frame technology.

This automated fibre management system is a frame-based product which incorporates test access and monitoring components, optical test equipment and fibre network database software. Through the combination of embedded surveillance sensors, Optical Time Domain Reflectometer (OTDR) and landmark database records, the *Smart* LGX can automatically detect AND locate a fibre fault within ***metres*** of the actual event. This feature enables service providers to rapidly dispatch their craft to repair the fibre fault, rather than undergo the time consuming process of trying to manually locate the fibre fault.

This feature becomes particularly beneficial in the event of an intermittent failure. While conventional SDH and DWDM terminals can provide alarm information on intermittent faults in the forms of Loss of Signal or excessive BER, they cannot provide fault isolation beyond identification of the fibre span experiencing the event.

A typical product configuration is shown on the following page:

Figure 2: SmartLGX Configuration

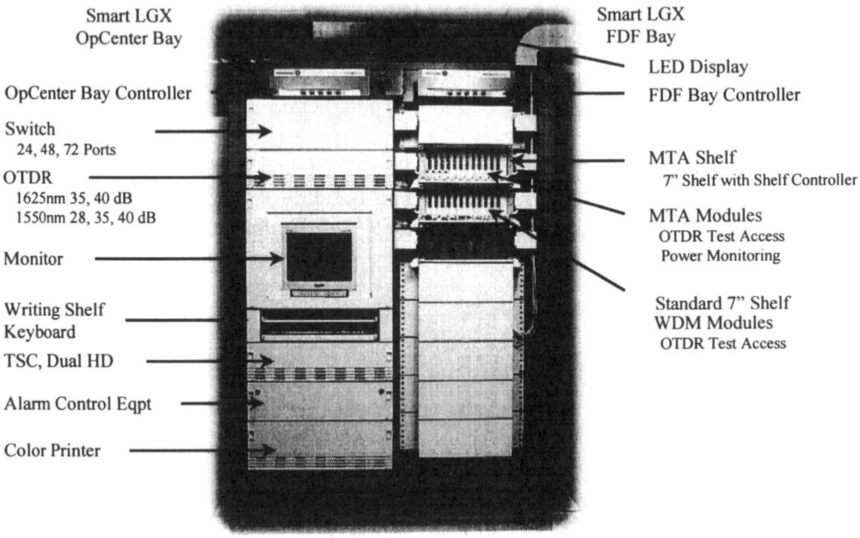

The major functions of the *Smart* LGX can be categorised into testing/fault management and configuration management functions. The following is an overview of the testing capabilities offered by the *Smart* LGX:

(A) Test/Fault Management Capabilities

- ***Test Access to both Dark and Active Fibres*** - This automated fibre management system can perform testing on in-service and dark fibres. When testing active fibres, the system employs an out-of-band test wavelength different from the operating wavelengths so as not to disturb the customer traffic. As current DWDM technology uses the 1550 nm band for transmission, a 1625 nm test wavelength can be used for in-service fibre testing. Wavelength Division Multiplexers (WDM) and filter components are used for combining and separating the different wavelengths. These components and are integrated into standard LGX (Lightguide Cross Connect) shelves. Similarly, dark fibre testing is available as a means of verifying the health of a fibre cable.

- ***Continuous Performance Monitoring*** – The system offers continuous monitoring with the Monitoring and Test Access (MTA) module. This module provides the highest level of protection and effectiveness in proactive maintenance, thereby reducing the occurrence and duration of service interruptions. The advantage of continuous power monitoring surveillance is that potential problems can be discovered prior to transmission alarms and prior to customer service interruptions for minimum down times. When power monitoring problems are detected, an OTDR test is automatically launched to locate the problem. Continuous power monitoring, in conjunction with test access, provides the most rapid fault location and detection possible.

- ***OTDR Batch Testing*** - The system provides a "batch" test mode that regularly performs OTDR tests on in-service or dark fibres. Batch testing will detect a problem in a fibre path utilizing user defined thresholds at user defined test intervals.

The following diagram outlines the capability of the automated fibre management system for monitoring active fibres through the use of the WDM module:

Figure 3: In-Service Fibre Testing with the *Smart* LGX

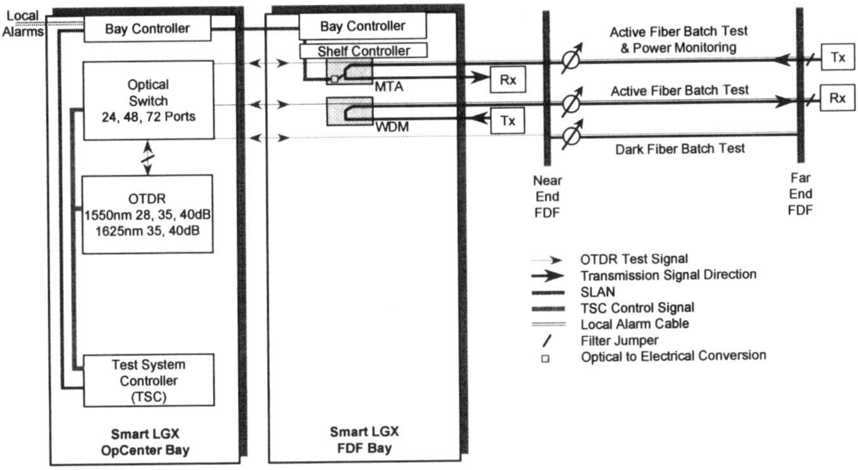

(B) Configuration Management

The advances in optical networking technologies have also had a profound impact on the importance of fibre configuration management. An example of some recent trends are listed below:

+ The line rates on long haul communications equipment continue to increase to the extent that much of the traditional twisted pair and coaxial interconnect cabling is being replaced with fibre jumpers. Many types of networking equipment are evolving to support fibre interfaces on the low-speed and high-speed inputs.

+ Similarly, the point of interconnect for customer traffic is migrating from the copper based Digital Cross Connect shelf (DSX) to the fibre based Lightguide Cross Connect Shelf (LGX). As such, fibre management within the central office is essential.

+ New 'applications-specific' singlemode fibres are emerging in the service provider network. In some cases, composite cables consisting of different fibre types are being deployed. Proper fibre identification must be made to ensure performance of the interconnected optical network equipment.

+ With the accelerated deployment of fibre cable around the globe, dark fibre is no longer the scarce commodity that it once was. As a result, service providers are required to effectively manage multiple dark fibre leases with multiple customers.

The *Smart* LGX addresses these configuration management issues through the use of the Fibre Administration System, or FAS. The FAS is a software database tool which provides comprehensive inventorying capabilities for both OSP and intra-office fibre facilities. The system utilizes a concept called "LightPaths" which allows the user to trace the optical signal through an optical network from the point of origin to the point of final destination.

The testing functions of the *Smart* LGX are integrated with the configuration management features of the FAS. By integrating configuration management with testing capability, the *Smart* LGX system also provides jumper selection and routing information required to cross-connect any two FDF terminations. Routing information is provided in graphical form, as well as specific written instructions for routing the jumper. The system incorporates extensive query and reporting capability for circuits, optical transmission equipment, outside plant cables/fibres, and frame terminations.

The FAS system also provides complete support for creating test circuits associated with fibre test and surveillance. The FAS offers seamless integration with the *Smart* LGX so the FAS user has a series of *Smart* LGX functions available directly from the FAS screens. Alarm messaging is facilitated with TL-1 over TCP/IP.

The following is a sample screen displayed by the FAS which would indicate a fibre event. In this example, the precise time, location, severity and associated frame and fibre information are displayed.

Figure 4: Sample Screen Depicting Alarm List

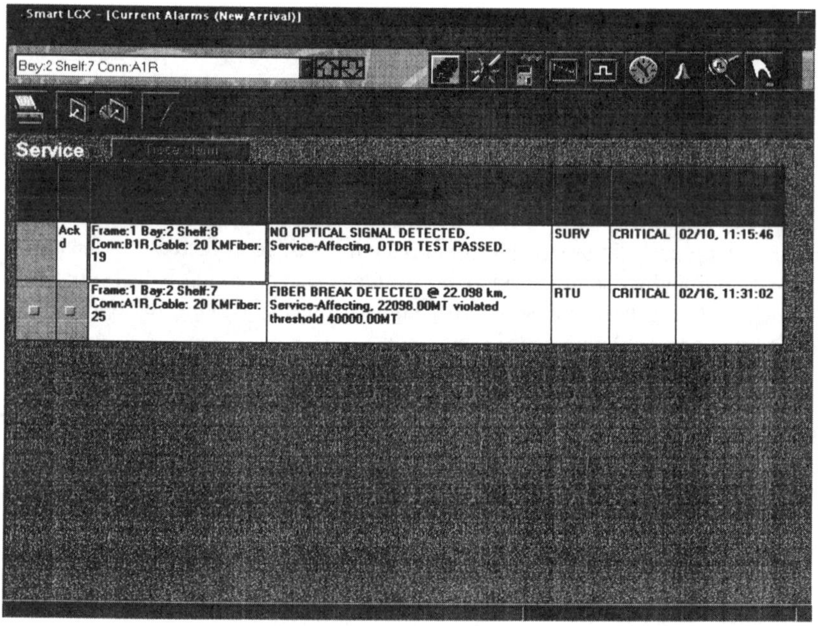

Summary

Ongoing advances in fibre deployment and optical technology have enabled tremendous growth in the deployment of high capacity networks. This unprecedented growth has created a challenge for service providers to effectively manage their investment in fibre plant. By utilising advanced, automated fibre management tools such as the SmartLGX, service providers can meet this challenge by delivering highly reliable, differentiated service to their end users.

---

[1] *'Optical Performance testing for the SmartLGX'* R.J. Pimpinella, F.S. Leone, and M.R. Jennings, LCET'97

[2] Ibid.

Broadband Access, WDM Metro and Network Management
D.W. Faulkner and A.L. Harmer (Eds.)
IOS Press, 2000 © AKM

Web-based Operation Support System for the Maintenance of Access Networks in Korea Telecom

Hyunmin Lim
Korea Telecom Telecommunications Network Laboratory
463-1 Junmin-dong, Yusung-gu, Taejeon, Korea
TEL : +82-42-870-8660, FAX : +82-42-870-8649, E-MAIL : hmlim@kt.co.kr

1. Introduction

Access networks are undergoing dramatic changes all over the world in the viewpoint of introduced new technologies and increasing new services. As a network & a service provider, KT(Korea Telecom) is actively taking part in this trend to offer high-quality services to customers and to scope with today's competitive business environment.

Among the diverse transport technologies KT has focused on the construction of optical fiber access networks in order to provide advanced services such as video or high bit-rate data and to enhance the conventional copper access networks. And KT has also been deploying DSLAM in order to build high-speed copper access networks.

In 1991, KT had established 3-phase evolution plan : FTTO(Fiber To The Office)-FTTC(Fiber To The Curb)-FTTH(Fiber To The Home) to complete optical fiber access networks by 2015. As the first step of this plan, KT has developed and deployed FLC-B(Fiber Loop Carrier-Building) system which can be used for business applications (FTTO) in 1996. Recently, KT is ready to deploy FLC-C(curb) system which was developed and has all been field-tested in 1999 for the purpose of application to the densely populated areas(i.e. apartment complexes) and small business buildings. And KT has also been using strategy to deploy DSLAM in copper access network since 1999.

In this circumstance of constructing new infrastructures in access network, it is indispensable for KT to prepare for business processes and operation support systems in order to satisfy customer's expectations of service order and maintenance. Without automatic business processes based on operation support systems, it will be evident for telecom operators to cost so much in maintaining resources and serving customers.

So KT has developed ANSWERS(Access Network Support & Warranty for End Resources) system which provides integrated maintenance services for the existing copper access network and broadband access networks such as optical access network and high-speed copper access network.

The first part of this paper describes the overview of ANSWERS system and presents the role, position & function of ANSWERS system. The second part of this paper describes ANSWERS system features in the viewpoint of architecture, environment and server configuration. And the last third part of this paper gives a short description of maintenance scenarios relating to ANSWERS system.

2. What is ANSWERS System?

2.1 Background

KT developed various OSS for the conventional copper access network and has been operating them. For example, TSIS(Telephone Subscriber Information System) deals with customer's request for PSTN service and manages customer's information such as customer name, address, directory number(i.e. telephone number) and so on. TIMS(Telecom service Integration Management System) activates Local Exchange's subscriber ports and manages wiring related to MDF through the work order. The above OSS has taken the role of service provisioning.

After service provisioning or activation, KT has to perform maintenance activities. SLMOS(Subscriber Line Maintenance & Operation System) and LCR(Line Conditioning Report) system has been charged with that mission, that is, receive customer's complaints about telephone services and carry out testing the subscriber line and responds to them. Until now, SLMOS or LCR has contributed to improving the quality of service for subscriber lines in Korea.

But as mentioned before, the above OSS focuses on the conventional copper access network. So it is difficult for existing OSS to accommodate new access networks such as an optical fiber access network and an advanced copper access network which consists of intelligent network elements, facilities and standardized interfaces to the LE.

Therefore KT had decided to develop new OSS for the maintenance of access networks in 1998. At that time KT conceived a separate system which is called ANOSS(Access Network Operation Support System) different from the existing OSS. But after a workshop on ANOSS in September in 1998, our Business Unit set forth it's views on adopting existing operation functions and processes into a new OSS. That was because the existing OSS needed to improve in the viewpoint of UI(User Interface), DB(Data Base), processing performance & flexibility. And also we couldn't afford additional human resources and organizations.

So KT had decided to develop ANSWERS system which provides integrated maintenance services for access networks to cover functions of the existing OSS and the newly required OSS early in 1999. And now KT has completed ANSWERS version 1.0 and has been field-testing during the first half year of 2000.

2.2 Role, Position & Function

When customers in trouble contact our maintenance center by telephone or internet (in a near future), our operators can response their trouble reports and performance complaints by using PC connected to ANSWERS Web Server. In addition to that, ANSWERS can provide proactive maintenance services, that is, before customers report their complaints, status reports & planned maintenance scheduling can be notified to customers by using the function of service affected customer identification, performance/fault trend check, performance/fault analysis and QOS assurance.

Figure 1 shows end-to-end maintenance diagram over the entire network. As a matter of fact, if we resolve customer's complaints fully, we have to handle overall management information relating to the specific customer service. But most problems tend to occur in AN, so ANSWERS system focuses on AN in the viewpoint of NM(Network Management), SM(Service Management) & CC(Customer Care). If ANSWERS system needs to cooperate with other OSS relating to CN, ANSWERS system will co-work to resolve the problem.

Figure 2 shows ANSWERS system in the present operational environment. We are in the development phase of ANSWERS Version 2.0(targeted for ADSL maintenance) and ANSWERS Version 3.0(targeted for accommodating xDSL tester). ANSWERS system makes use of the management information based on EMS of FLC system and DSLAM system. And ANSWERS system also accommodated existing copper line test equipment of SLMOS and LCR system.

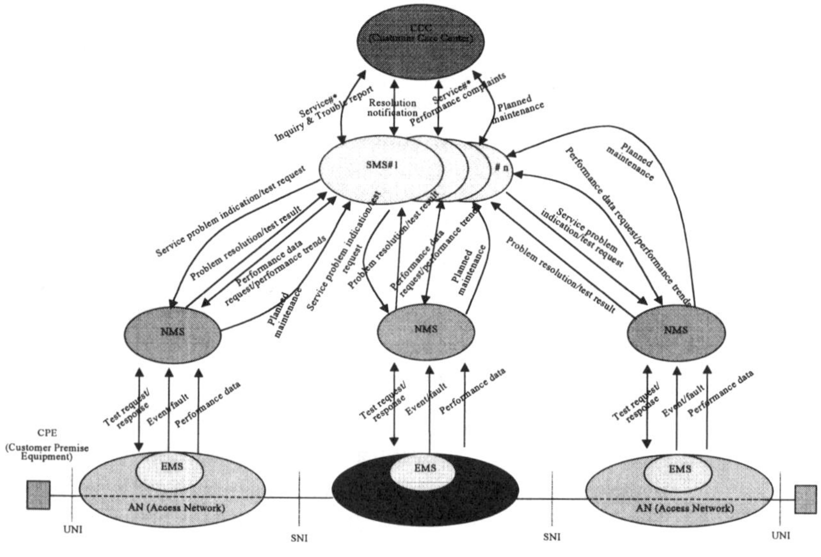

Figure 1. End-to-End Maintenance Framework

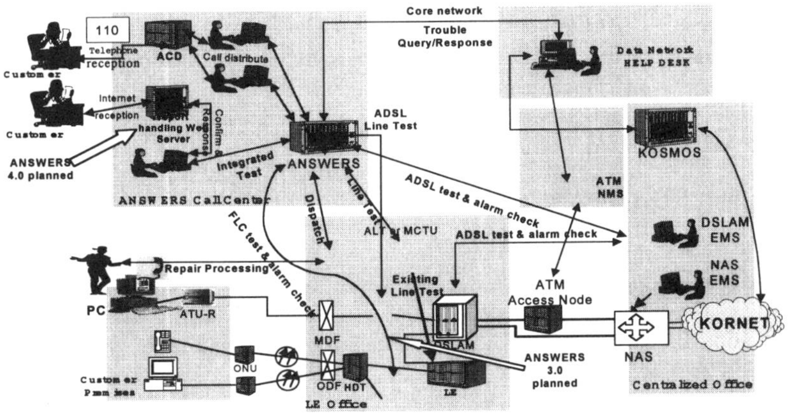

Figure 2. ANSWERS system maintenance flows in the operational environment

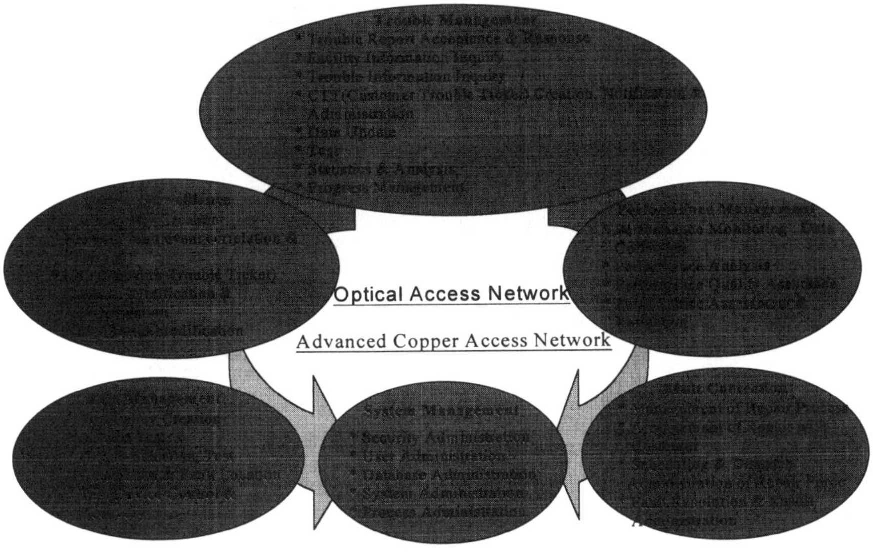

Figure 3. ANSWERS Maintenance Functions

Figure 3 shows 6 main maintenance functions such as trouble management, alarm surveillance, performance management, test management, fault correction and system management. For example, Trouble Management Functions can handle trouble report acceptance/response, facility information inquiry, trouble information inquiry, CTT(Customer Trouble Ticket) creation/notification/administration, test, statistics/analysis and progress management.

3. ANSWERS System Features

3.1 ANSWERS System Architecture

ANSWERS system has a 3-tier architecture which comprises Web-based PC Client, Application Server and Database Server. When operators connect to ANSWERS Web Server by using browser, ANSWERS Applet is downloaded, after that, operator's authorization is checked and ANSWERS services which consist of Java-based applications can be available using IIOP.

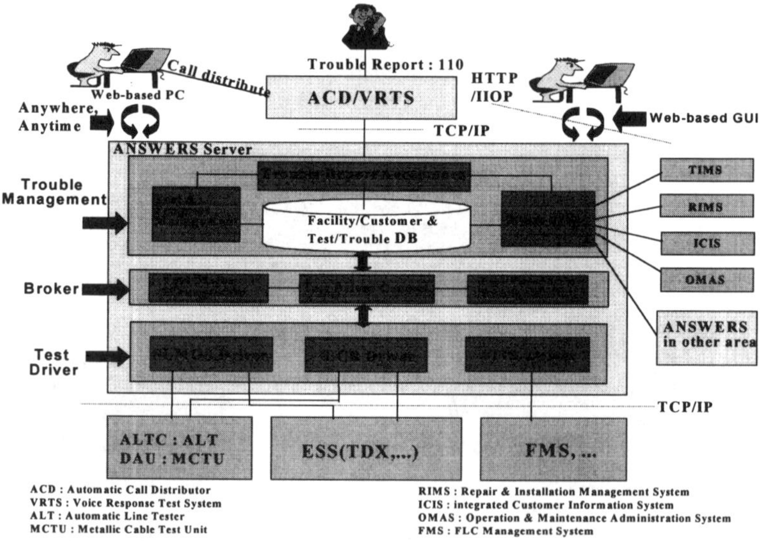

Figure 4. ANSWERS System Architecture

Figure 4 shows that ANSWERS system can be accessed by operators without any time and space limitations. If we need to modify or enhance ANSWERS system, all we have to do is just to upgrade ANSWERS System S/W in the server. We don't need to touch operator's PC. Maybe software maintenance cost and software upgrade time can be reduced much more than expected.

3.2 ANSWERS System Environment

As mentioned before, KT had already been using existing OSS like SLMOS for the maintenance of the conventional copper access network. So our operators were accustomed to the established business flows, in other words, maintenance flows, accordingly we tried to consider the existing maintenance activities in order to avoid a sudden change and take a step to move smoothly.

Figure 5 shows business flows relating to ANSWERS system and Table 1 shows the development & operation environment for the ANSWERS system. Figure 6 shows the screen of trouble report reception desk.

	Item	Contents
Client GUI Development	language	Java, HTML
	tool	Oracle Jdeveloper 2.0
Server Side Development	language	Java, C
	environment	OAS(Oracle Application Server)4.0.7, Oracle 8i
Operator PC Environment	H/W	Pentium II ± ₱350MHz), MM : 64MB À ì ó
	S/W	Netscape or Explorer Browser

Table 1. ANSWERS Development & Operational Environment

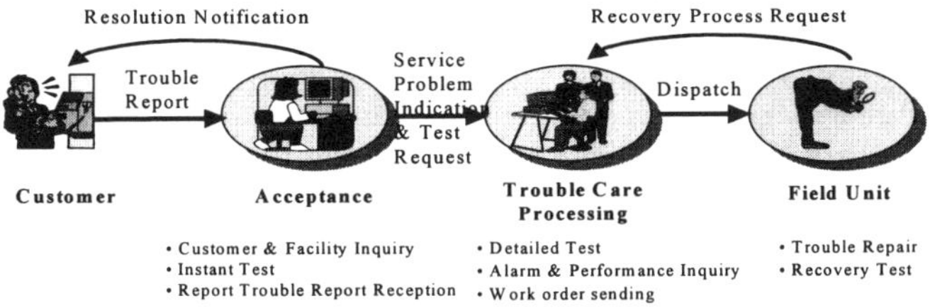

Figure 5. ANSWERS Business Flows

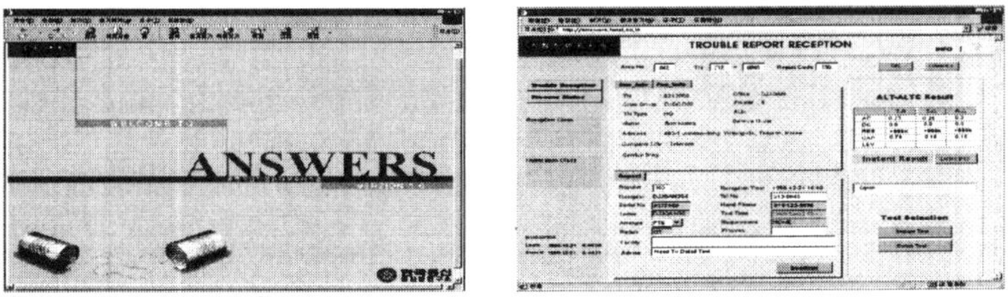

3.3 ANSWERS Server Configuration

ANSWERS System server can be expanded into a capacity to handle about 3 million customers. And ANSWERS System takes the HA(High Availability) dual configuration.

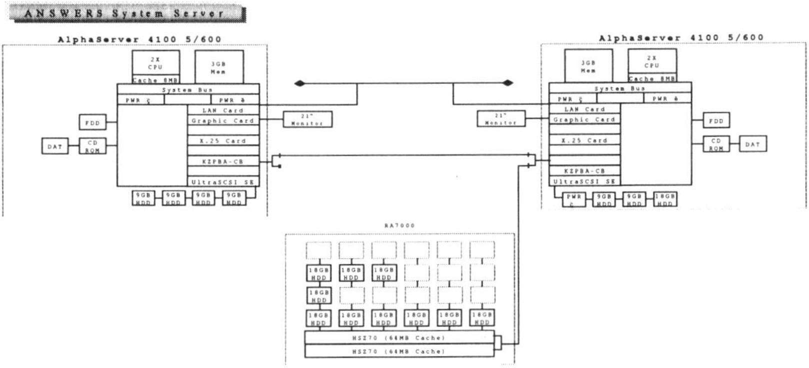

Figure 7. ANSWERS Server System Configuration

4. Maintenance Scenarios of ANSWERS System for the optical access network

ANSWERS System provides existing reactive maintenance process dealing with customer's complaints and also provides proactive maintenance process dealing with performance data & alarm from network.

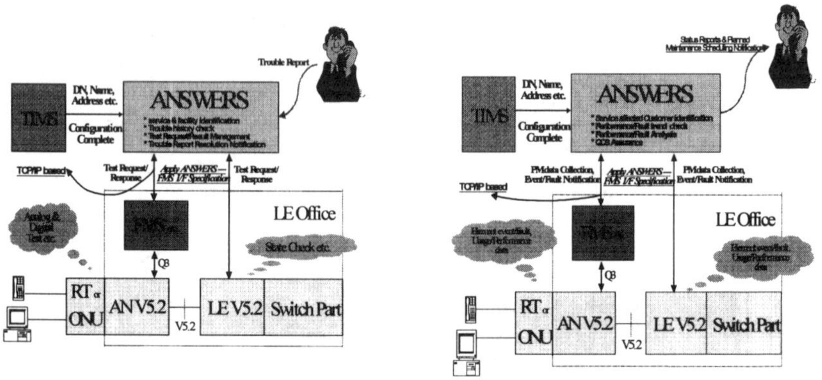

Figure 8. Maintenance Scenarios

5. Conclusion

This paper describes ANSWERS system which provides integrated maintenance services for the existing copper access network and broadband access network such as optical access network and high-speed copper access network.

KT expects ANSWERS system provides unified customer care services and enhanced business competitive power. As mentioned before, ANSWERS system has a 3-tier architecture which comprise Web-based PC Client, Application Server and Database Server, which will contribute to the cut-down of software maintenance cost and a speedy software upgrade. Therefore ANSWERS system will provide flexible maintenance framework to accommodate various access network transport technologies.

References

[1] Hyunmin Lim, "KT's Management System for the Optical Fiber Access Network", NFOEC'97, September, 1997

[2] Korea Telecom, "Management Framework for the Access Network", October, 1998

Author Index